AF389469

Multifunctional Nanomaterials:
Synthesis, Properties and Applications

Multifunctional Nanomaterials: Synthesis, Properties and Applications

Editor

Raghvendra Singh Yadav

MDPI • Basel • Beijing • Wuhan • Barcelona • Belgrade • Manchester • Tokyo • Cluj • Tianjin

Editor
Raghvendra Singh Yadav
Centre of Polymer Systems
Tomas Bata University in Zlin
Zlin
Czech Republic

Editorial Office
MDPI
St. Alban-Anlage 66
4052 Basel, Switzerland

This is a reprint of articles from the Special Issue published online in the open access journal *International Journal of Molecular Sciences* (ISSN 1422-0067) (available at: www.mdpi.com/journal/ijms/special_issues/multifunctional_nanomaterials).

For citation purposes, cite each article independently as indicated on the article page online and as indicated below:

LastName, A.A.; LastName, B.B.; LastName, C.C. Article Title. *Journal Name* **Year**, *Volume Number*, Page Range.

ISBN 978-3-0365-3139-7 (Hbk)
ISBN 978-3-0365-3138-0 (PDF)

Contents

About the Editor . . . . **vii**

Raghvendra Singh Yadav
Multifunctional Nanomaterials: Synthesis, Properties and Applications
Reprinted from: *Int. J. Mol. Sci.* **2021**, *22*, 12073, doi:10.3390/ijms222112073 . . . . **1**

Marika Musielak, Jakub Potoczny, Agnieszka Boś-Liedke and Maciej Kozak
The Combination of Liposomes and Metallic Nanoparticles as Multifunctional Nanostructures
in the Therapy and Medical Imaging—A Review
Reprinted from: *Int. J. Mol. Sci.* **2021**, *22*, 6229, doi:10.3390/ijms22126229 . . . . **3**

**Marjorie C. Zambonino, Ernesto Mateo Quizhpe, Francisco E. Jaramillo, Ashiqur Rahman,
Nelson Santiago Vispo and Clayton Jeffryes et al.**
Green Synthesis of Selenium and Tellurium Nanoparticles: Current Trends, Biological
Properties and Biomedical Applications
Reprinted from: *Int. J. Mol. Sci.* **2021**, *22*, 989, doi:10.3390/ijms22030989 . . . . **27**

Joan Estelrich and M. Antònia Busquets
Prussian Blue: A Nanozyme with Versatile Catalytic Properties
Reprinted from: *Int. J. Mol. Sci.* **2021**, *22*, 5993, doi:10.3390/ijms22115993 . . . . **61**

**Henrik Kahl, Theresa Staufer, Christian Körnig, Oliver Schmutzler, Kai Rothkamm and
Florian Grüner**
Feasibility of Monitoring Tumor Response by Tracking Nanoparticle-Labelled T Cells Using
X-ray Fluorescence Imaging—A Numerical Study
Reprinted from: *Int. J. Mol. Sci.* **2021**, *22*, 8736, doi:10.3390/ijms22168736 . . . . **77**

Jie Li, Shiyi Wang, Huayin Li, Yuan Tan and Yunjie Ding
Zn Promoted Mg-Al Mixed Oxides-Supported Gold Nanoclusters for Direct Oxidative
Esterification of Aldehyde to Ester
Reprinted from: *Int. J. Mol. Sci.* **2021**, *22*, 8668, doi:10.3390/ijms22168668 . . . . **93**

Chunxin Wang, Bo Cui, Yan Wang, Mengjie Wang, Zhanghua Zeng and Fei Gao et al.
Preparation and Size Control of Efficient and Safe Nanopesticides by Anodic Aluminum Oxide
Templates-Assisted Method
Reprinted from: *Int. J. Mol. Sci.* **2021**, *22*, 8348, doi:10.3390/ijms22158348 . . . . **107**

Tianming Song, Yawei Qu, Zhe Ren, Shuang Yu, Mingjian Sun and Xiaoyu Yu et al.
Synthesis and Characterization of Polyvinylpyrrolidone-Modified ZnO Quantum Dots and
Their In Vitro Photodynamic Tumor Suppressive Action
Reprinted from: *Int. J. Mol. Sci.* **2021**, *22*, 8106, doi:10.3390/ijms22158106 . . . . **119**

Haijie Sun, Yiru Fan, Xiangrong Sun, Zhihao Chen, Huiji Li and Zhikun Peng et al.
Effect of $ZnSO_4$, $MnSO_4$ and $FeSO_4$ on the Partial Hydrogenation of Benzene over Nano
Ru-Based Catalysts
Reprinted from: *Int. J. Mol. Sci.* **2021**, *22*, 7756, doi:10.3390/ijms22147756 . . . . **133**

Yuan-Chang Liang and Yu-Wei Hsu
Enhanced Sensing Ability of Brush-Like Fe_2O_3-ZnO Nanostructures towards NO_2 Gas via
Manipulating Material Synergistic Effect
Reprinted from: *Int. J. Mol. Sci.* **2021**, *22*, 6884, doi:10.3390/ijms22136884 . . . . **149**

Amjad Ali, Zainab Bukhari, Muhammad Umar, Muhammad Ali Ismail and Zaheer Abbas
Cu and Cu-SWCNT Nanoparticles' Suspension in Pulsatile Casson Fluid Flow via Darcy–Forchheimer Porous Channel with Compliant Walls: A Prospective Model for Blood Flow in Stenosed Arteries
Reprinted from: *Int. J. Mol. Sci.* **2021**, *22*, 6494, doi:10.3390/ijms22126494 . . . . **161**

Katarzyna Szurkowska, Paulina Kazimierczak and Joanna Kolmas
Mg,Si—Co-Substituted Hydroxyapatite/Alginate Composite Beads Loaded with Raloxifene for Potential Use in Bone Tissue Regeneration
Reprinted from: *Int. J. Mol. Sci.* **2021**, *22*, 2933, doi:10.3390/ijms22062933 . . . . **187**

M. Susana Gutiérrez, Alberto J. León, Paulino Duel, Rafael Bosch, M. Nieves Piña and Jeroni Morey
Effective Elimination and Biodegradation of Polycyclic Aromatic Hydrocarbons from Seawater through the Formation of Magnetic Microfibres
Reprinted from: *Int. J. Mol. Sci.* **2020**, *22*, 17, doi:10.3390/ijms22010017 . . . . **201**

About the Editor

Raghvendra Singh Yadav

Dr. Raghvendra Singh Yadav is working as a 'Senior Scientist/Researcher'at the Centre of Polymer Systems, Tomas Bata University in Zlin, Czech Republic.

He obtained his Ph.D. degree on the topic "Synthesis and Characterization of Nanophosphors (luminescent/optical nanoparticles)"from University of Allahabad, India. Dr. Yadav's current scientific activities are focused on "Lightweight, Flexible, Low-dimensional Electromagnetic Functional Nanocomposite Materials (MXene, MBene, Graphene, magnetic nanoparticles as nanofillers in a polymer matrix) and its Applications".

International Journal of
Molecular Sciences

Editorial

Multifunctional Nanomaterials: Synthesis, Properties and Applications

Raghvendra Singh Yadav

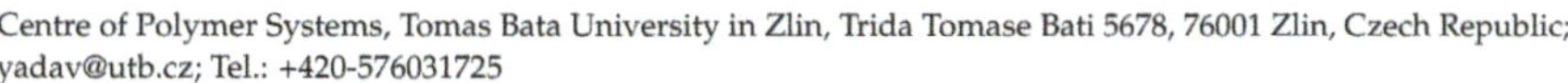

Centre of Polymer Systems, Tomas Bata University in Zlin, Trida Tomase Bati 5678, 76001 Zlin, Czech Republic; yadav@utb.cz; Tel.: +420-576031725

Citation: Yadav, R.S. Multifunctional Nanomaterials: Synthesis, Properties and Applications. *Int. J. Mol. Sci.* **2021**, *22*, 12073. https://doi.org/10.3390/ijms222112073

Received: 19 October 2021
Accepted: 2 November 2021
Published: 8 November 2021

Publisher's Note: MDPI stays neutral with regard to jurisdictional claims in published maps and institutional affiliations.

In this Special Issue *"Multifunctional Nanomaterials: Synthesis, Properties and Applications"*, we published three review papers and nine original research articles.

In the first article (review) [1], Si Amar Dahoumane et al. describe the recent advances in the emerging approaches applied for the biosynthesis of selenium and tellurium nanoparticles using different microorganisms, such as bacteria and fungi, and plant extracts. The green synthesis approach provides prospects to develop safer nanomaterials and also promote a better understanding of safety, health, and environment concerns. The second (review) article by Joan Estelrich et al. [2] reviews the enzyme-like properties of Prussian blue nanoparticles (PBNPs) and their applications in this field for biomedicine development. In the third (review) article [3], Marika Musielak et al. presented the most recent progress in the utilization of advanced nanostructures such as liposomes combined with metallic nanoparticles and other active substances such as drugs or contrast agents.

In the fourth article (research), Rafael Bosch et al. [4] reported the effective elimination and biodegradation of polycyclic aromatic hydrocarbons from seawater through the development of magnetic microfibers. Joanna Kolmas et al. [5] reported the Mg, Si-Co-substituted hydroxyapatite/alginate composite beads loaded with raloxifene for bone tissue regeneration application. The sixth article (research) by Amjad Ali et al. [6] discussed the Cu and Cu-SWCNT nanoparticle suspension in pulsatile casson fluid flow via a Darcy-Forchheimer porous channel with compliant walls. In the seventh article (research) [7], Yuan-Chang Liang et al. reported the development of a brush-like Fe_2O_3–ZnO nanostructures towards NO_2 gas sensing based on the control of appropriate material synergistic effects. Further, Zhikun Peng et al. [8] reported the influence of $ZnSO_4$, $MnSO_4$ and $FeSO_4$ on the partial hydrogenation of benzene over nano Ru-Based catalysts. The ninth published research article by Xiaoyu Yu et al. [9], discussed the development of an optimized synthesis approach for ZnO quantum dots (QDs) modified with polyvinylpyrrolidone (PVP40) which can potentially be applied in the fields of photoluminescence and photodynamic tumor suppression. In the tenth research article of this Special Issue, Xiang Zhao et al. [10] report a novel, versatile and controllable approach for efficient and safe nanopesticides by the anodic aluminum oxide templates-assisted method.

Furthermore, in the eleventh published research article, Yuan Tan et al. [11] reported that Zn promoted Mg-Al mixed oxides-supported gold nanoclusters for the direct oxidative esterification of aldehyde to ester. In the last, i.e., twelfth, published research article, Florian Grüner et al. [12] discuss a numerical study on the feasibility of monitoring tumor response by tracking nanoparticle-labelled T cells by utilizing the X-ray fluorescence imaging technique.

Funding: This research received no external funding.

Institutional Review Board Statement: Not applicable.

Informed Consent Statement: Not applicable.

Data Availability Statement: Not applicable.

Acknowledgments: As Guest Editor, I appreciate all the authors for their contributions in this Special Issue. I thank the reviewers for their support in evaluating the manuscripts. In addition, thanks also to the IJMS editorial staff members for their assistance.

Conflicts of Interest: The authors declare no conflict of interest.

References

1. Zambonino, M.C.; Quizhpe, E.M.; Jaramillo, F.E.; Rahman, A.; Santiago Vispo, N.; Jeffryes, C.; Dahoumane, S.A. Green Synthesis of Selenium and Tellurium Nanoparticles: Current Trends, Biological Properties and Biomedical Applications. *Int. J. Mol. Sci.* **2021**, *22*, 989. [CrossRef] [PubMed]
2. Estelrich, J.; Busquets, M.A. Prussian Blue: A Nanozyme with Versatile Catalytic Properties. *Int. J. Mol. Sci.* **2021**, *22*, 5993. [CrossRef] [PubMed]
3. Musielak, M.; Potoczny, J.; Boś-Liedke, A.; Kozak, M. The Combination of Liposomes and Metallic Nanoparticles as Multifunctional Nanostructures in the Therapy and Medical Imaging—A Review. *Int. J. Mol. Sci.* **2021**, *22*, 6229. [CrossRef] [PubMed]
4. Gutiérrez, M.S.; León, A.J.; Duel, P.; Bosch, R.; Piña, M.N.; Morey, J. Effective Elimination and Biodegradation of Polycyclic Aromatic Hydrocarbons from Seawater through the Formation of Magnetic Microfibres. *Int. J. Mol. Sci.* **2021**, *22*, 17. [CrossRef] [PubMed]
5. Szurkowska, K.; Kazimierczak, P.; Kolmas, J. Mg,Si—Co-Substituted Hydroxyapatite/Alginate Composite Beads Loaded with Raloxifene for Potential Use in Bone Tissue Regeneration. *Int. J. Mol. Sci.* **2021**, *22*, 2933. [CrossRef] [PubMed]
6. Ali, A.; Bukhari, Z.; Umar, M.; Ismail, M.A.; Abbas, Z. Cu and Cu-SWCNT Nanoparticles' Suspension in Pulsatile Casson Fluid Flow via Darcy–Forchheimer Porous Channel with Compliant Walls: A Prospective Model for Blood Flow in Stenosed Arteries. *Int. J. Mol. Sci.* **2021**, *22*, 6494. [CrossRef]
7. Liang, Y.-C.; Hsu, Y.-W. Enhanced Sensing Ability of Brush-Like Fe_2O_3-ZnO Nanostructures towards NO_2 Gas via Manipulating Material Synergistic Effect. *Int. J. Mol. Sci.* **2021**, *22*, 6884. [CrossRef]
8. Sun, H.; Fan, Y.; Sun, X.; Chen, Z.; Li, H.; Peng, Z.; Liu, Z. Effect of $ZnSO_4$, $MnSO_4$ and $FeSO_4$ on the Partial Hydrogenation of Benzene over Nano Ru-Based Catalysts. *Int. J. Mol. Sci.* **2021**, *22*, 7756. [CrossRef]
9. Song, T.; Qu, Y.; Ren, Z.; Yu, S.; Sun, M.; Yu, X.; Yu, X. Synthesis and Characterization of Polyvinylpyrrolidone-Modified ZnO Quantum Dots and Their In Vitro Photodynamic Tumor Suppressive Action. *Int. J. Mol. Sci.* **2021**, *22*, 8106. [CrossRef] [PubMed]
10. Wang, C.; Cui, B.; Wang, Y.; Wang, M.; Zeng, Z.; Gao, F.; Sun, C.; Guo, L.; Zhao, X.; Cui, H. Preparation and Size Control of Efficient and Safe Nanopesticides by Anodic Aluminum Oxide Templates-Assisted Method. *Int. J. Mol. Sci.* **2021**, *22*, 8348. [CrossRef] [PubMed]
11. Li, J.; Wang, S.; Li, H.; Tan, Y.; Ding, Y. Zn Promoted Mg-Al Mixed Oxides-Supported Gold Nanoclusters for Direct Oxidative Esterification of Aldehyde to Ester. *Int. J. Mol. Sci.* **2021**, *22*, 8668. [CrossRef] [PubMed]
12. Kahl, H.; Staufer, T.; Körnig, C.; Schmutzler, O.; Rothkamm, K.; Grüner, F. Feasibility of Monitoring Tumor Response by Tracking Nanoparticle-Labelled T Cells Using X-ray Fluorescence Imaging—A Numerical Study. *Int. J. Mol. Sci.* **2021**, *22*, 8736. [CrossRef] [PubMed]

International Journal of
Molecular Sciences

Review

The Combination of Liposomes and Metallic Nanoparticles as Multifunctional Nanostructures in the Therapy and Medical Imaging—A Review

Marika Musielak [1,2,3,*], Jakub Potoczny [4], Agnieszka Boś-Liedke [3] and Maciej Kozak [3]

1   Department of Electroradiology, Poznan University of Medical Sciences, 61-701 Poznań, Poland
2   Radiobiology Laboratory, Department of Medical Physics, Greater Poland Cancer Centre,
    61-866 Poznań, Poland
3   Department of Macromolecular Physics, Faculty of Physics, Adam Mickiewicz University,
    61-614 Poznań, Poland; agnieszkaboss@gmail.com (A.B.-L.); mkozak@amu.edu.pl (M.K.)
4   Heliodor Swiecicki Clinical Hospital in Poznan, 60-355 Poznań, Poland; potocznyjakub@op.pl
*   Correspondence: marikamusielak@gmail.com

Abstract: Nanotechnology has introduced a new quality and has definitely developed the possibilities of treating and diagnosing various diseases. One of the scientists' interests is liposomes and metallic nanoparticles (LipoMNPs)—the combination of which has introduced new properties and applications. However, the field of creating hybrid nanostructures consisting of liposomes and metallic nanoparticles is relatively little understood. The purpose of this review was to compile the latest reports in the field of treatment and medical imaging using of LipoMNPs. The authors focused on presenting this issue in the direction of improving the used conventional treatment and imaging methods. Most of all, the nature of bio-interactions between nanostructures and cells is not sufficiently taken into account. As a result, overcoming the existing limitations in the implementation of such solutions in the clinic is difficult. We concluded that hybrid nanostructures are used in a very wide range, especially in the treatment of cancer and magnetic resonance imaging. There were also solutions that combine treatments with simultaneous imaging, creating a theragnostic approach. In the future, researchers should focus on the description of the biological interactions and the long-term effects of the nanostructures to use LipoMNPs in the treatment of patients.

Keywords: liposomes; nanotechnology; nanotheragnostics; metallic nanoparticles

check for
updates

Citation: Musielak, M.; Potoczny, J.; Boś-Liedke, A.; Kozak, M. The Combination of Liposomes and Metallic Nanoparticles as Multifunctional Nanostructures in the Therapy and Medical Imaging—A Review. *Int. J. Mol. Sci.* **2021**, *22*, 6229. https://doi.org/10.3390/ijms22126229

Academic Editor: Raghvendra Singh Yadav

Received: 18 May 2021
Accepted: 6 June 2021
Published: 9 June 2021

Publisher's Note: MDPI stays neutral with regard to jurisdictional claims in published maps and institutional affiliations.

1. Introduction

At the core of the potential applications of lipid systems, and in particular liposomes, in medicine, is the ability of amphiphilic lipid molecules to self-organize and create a wide range of 3D structures. In aqueous environments, these molecules typically can form structures of a spherical shape (liposomes or micelles), various lamellar structures, or numerous phases characterized by cubic or hexagonal symmetry [1]. The structure of these systems and the conditions of their formation in the presence of a wide range of substances were characterized in detail using a wide range of physical methods [2,3]. Since 1970, liposomes (LPs) have been extensively studied in the category of a drug delivery system (DDS) due to their high compatibility and ability to deliver hydrophilic and hydrophobic substances [4,5].

Liposomes have been prepared as spherical nanostructures with an aqueous inner core surrounded by bilayers of phospholipids [6,7]. Moreover, LPs are at the center of interest related to nanotechnology because of the variety of multiple kinds of implementations [8,9] (Figure 1).

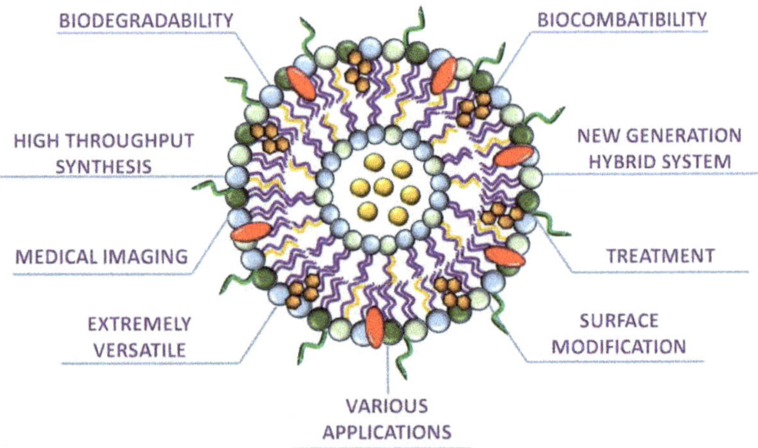

Figure 1. Benefits of liposomes with incorporated metallic nanoparticles in drug delivery systems. The picture presents the most interesting and crucial aspects of LipoMNPs according to usage in cancer treatment and medical imaging. The liposome of mixed lipid composition was selected as a simplified model. The Lipo-MNPs consist of phospholipids, cholesterol, DSPE-PEG, and MNPs.

An important application area of liposomes that have dominated their potential applications for a long time has been drug or genetic material transfer systems. Currently, the interest in such applications, especially as carriers of vaccines or therapeutic nucleic acids, is still serious [10–12].

Liposomes can be synthesized by different methods like the hydration film method [13], and then modified by incorporating various substances, e.g., metallic nanoparticles [13–16], or functionalized by different ligands, e.g., aptamers [17] (Figure 2). The combination of liposomes and metallic nanoparticles proved to be a potential technique that supports efficient medical imaging and treatment [18]. They have become the object of interest in the field of medical imaging, in particular for transporting contrast agents (CAs). There is a constant effort to improve the accumulation of contrast in the required area for better visualization. Research in this area has achieved a sophisticated level that is carefully suited to the diagnostic and treatment simultaneously. Particular attention is paid to the role of liposomes as adjuvants to the already known conventional therapies [19]. When designing liposomes, attention should be taken to avoid non-specific interactions, represent specific target binding, maintain stability and circulation time in the bloodstream as long as possible, escape the reticuloendothelial clearance system (RES), and avoid mononuclear phagocyte system (MPS) or penetration into the tumor interstitial fluid of TME with high pressure [20].

Appropriate functions and their properties should be considered depending on the liposomes' target application. Certain drug delivery systems are commonly used in the clinic. One example is Doxil®, a polyethylene glycol (PEG) liposome containing Doxorubicin the first nano-drug approved by U.S. Food and Drug Administration (FDA) [21–23]. It has been approved for the treatment of Kaposi's sarcoma and ovarian cancer [24]. Its admission as the first nano-drug opened the way to the use of other such systems [25–28].

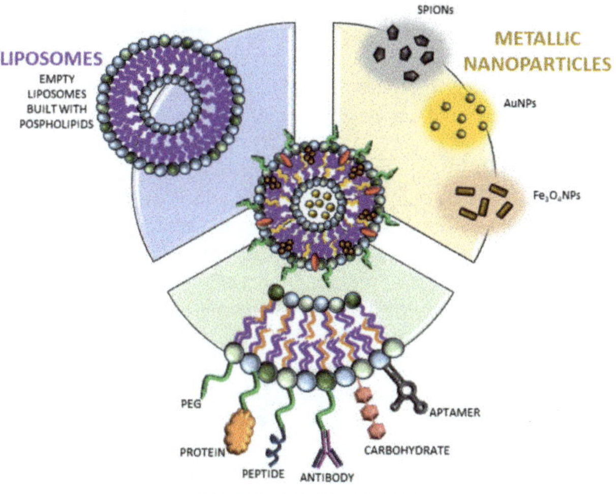

Figure 2. The scheme of liposomal modifications used in therapy support or medical imaging. Three of the most important components were highlighted as crucial parts of LipoMNPs. It presents the empty liposome, which is built with phospholipids, the core of nanostructures containing metallic nanoparticles, and different kinds of surface modifications. Each of these components play a key role in LipoMNPs and has various impacts depending on its target usage. The Lipo-MNPs consist of phospholipids, cholesterol, DSPE-PEG, and MNPs.

Conversely, structures designed for a drug delivery system will behave differently compared to those intended for imaging. Unfortunately, the combination of such nanostructures as liposomes and metallic nanoparticles is not widely used due to numerous limitations and the poorly understood nature of such compounds. There is a knowledge gap regarding the most efficient systems that could simultaneously image and treat cancer-related areas. There is no taxonomy that would allow the results to be compared in order to be able to design more efficient and improved systems.

Numerous reviews dedicated to liposomes have focused on a wide variety of biomedical applications [29]. However, the aim of this work is to present the most recent projects and published works focused on the use of advanced nanostructures such as liposomes in combination with metallic nanoparticles and other active substances such as drugs or contrast agents. We want to present their application in two areas: conventional therapy support and medical imaging. Especially, we want to focus on the advantages and disadvantages of this type of solution as well as indicate the further direction of work that should be undertaken to introduce such advanced nanostructures for clinical use.

2. Review Methodology

The literature search was conducted for papers related to metallic nanoparticles incorporated into liposomes as a new approach to the drug delivery system. Moreover, the search included papers dealing with using this kind of nanostructures in treatment and medical imaging. The publications were searched focusing on the PubMed database. The found articles were published in the years 1981–2021, which gives us a wide range of knowledge related to the presented topic of review. We used the following terms to find works cited in this review: "metallic nanoparticles in liposomes", "gold nanoparticles incorporated into liposomes", "metallic nanoparticles with liposomes in therapy", "metallic Nanoparticles with liposomes in medical imaging", "lipo-metallic nanoparticles", "metallic nanoparticles in combined therapy". We have established one main criterion for selecting a publication. The chosen articles should only be concerned with drug delivery systems consisting of metallic nanoparticles embedded in liposomes. The exclusion criterion was related to systems consisting of liposomes and attached drugs, or substances that were not metallic nanoparticles. Although there are many other complex nanostructures consisting of various components, our goal was to present only works focused on the medical therapy and imaging use of nanoparticles–liposome nanostructures.

3. Treatment

Here, we would like to focus on liposomes used in the transport of drugs and substances responsible for increasing the toxicity of conventional therapies. It is important to know all the crucial aspects, advantages, and limitations in order to properly design new cancer treatment strategies. One of the most frequently proposed methods involves the use of a combination of liposomes with metallic nanoparticles [30]. It is important to maintain appropriate conditions during synthesis and plan the effectiveness of such nanostructures, to obtain efficient transport to cells and the drug-releasing after reaching the target [31]. A challenge in this field of nanotechnology is the efficient incorporation of the drug into the liposome and the selection of the stimulating factor that does not destroy healthy tissues at the same time [32]. These types of advanced nanostructures have many biomedical applications. They are used for the controlled release of molecules or plasmids [33], cell activation [34], disease treatment [35], and imaging [36]. However, such strategies come with potential limitations. Very low drug incorporation is up to 5% [37]. The amount of drug administered is not effective enough to obtain a pharmacologically effective concentration in the body. Moreover, a significantly large amount of the drug may be released during transport before the liposome reaches its target site in the body, resulting in lower treatment activity and toxicity to healthy tissues [38].

The main interest of scientists is curing cancer due to the numerous difficulties that arise during the treatment process. Conventional treatment of cancer includes surgery, chemotherapy, and radiation therapy. Many different factors are used to sensitize cancer cells to ionizing radiation while protecting the percentage of normal cells in the field of the therapeutic beam [39]. Metallic nanoparticles are usually used as agents sensitizing cancer cells to ionizing radiation during radiotherapy [40]. In combination with liposomes as nanoparticle delivery carriers, it is possible to obtain a high-throughput treatment strategy. Success factors of this type of treatment system should also be taken into account [37]. Radiosensitizers incorporation must not have a destructive effect on the physical and chemical properties of the liposome. Usually, the process of realizing radiosensitizers is undertaken by using different factors. Under the influence of a stimulating factor such as temperature, pH or light energy, liposomes should decompose effectively. Moreover, the stimulus factor should not destroy healthy tissues (Figure 3).

The reason to look for improvements in conventional radiotherapy is the presence of hypoxia in the tumor area. Low oxygen concentration makes cancer cells significantly more resistant to the influence of drugs and ionizing radiation [41]. Moreover, hypoxia promotes the genetic transformation of cells into a more invasive phenotype, which further increases the predisposition to metastasis and disease recurrence [42]. In order to overcome those inconveniences, it has been proposed to use perfluorocarbons as O_2 carriers [43]. These types of compounds are characterized by high incorporation and easier penetration abilities through abnormal vascular structures in the tumor microenvironment. Cheng et al. [44] presented encapsulated perfluorohexane within liposomes to form 100 nm nanoparticles. Their aim was to enhance the effect of photodynamic therapy by sensitizing the tumor cells in the bearing mice model. Researchers observed tumor growth inhibition by injecting a designed sensitizing agent. In vitro and in vivo studies, exhibited significantly higher effectiveness of Oxy-PDT compared to the conventional PDT method. The results from in vivo studies showed complete inhibition of tumor growth in mice using Oxy-PDT at low dose photosensitization together with 20 s laser irradiation. Conventional PDT did not achieve comparable results as the inhibition of tumor growth was minimal. The group presented very promising results that may constitute the potential efficacy of the undertaken cancer treatment strategy in photodynamic therapy.

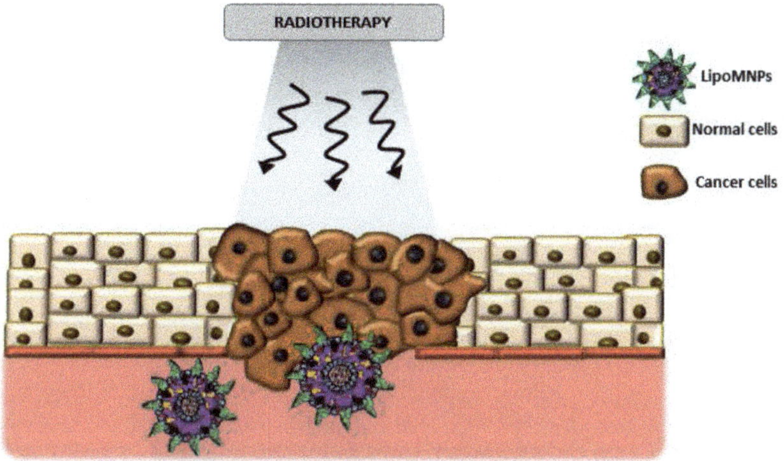

Figure 3. The schematic representation of LipoMNPs (Liposomes with metallic nanoparticles) usage as radiosensitizers in radiotherapy. Arrows are symbols of photons in ionizing radiation during radiotherapy.

In order to increase the biocompatibility and longer circulation time of liposomes in the bloodstream, they should be modified with poly(ethylene) glycol. Liu et al. [45] presented PEG-ylated liposomes to sensitize cells to radiation therapy. They proposed liposome-cholesterol-based nanoparticles conjugated with the radiosensitizer nitroimidazole by a hydrolyzable ester bond. Hybrid liposomes were introduced as a good carrier for promoting cargo release under hypoxic conditions. Doxorubicin-loaded liposomes were pharmacokinetically stable and scientists observed tumor growth inhibition using medical imaging. This experiment suggests that the hybrid liposomes are potential candidates as drug delivery systems for chemotherapeutics like doxorubicin, to improve the tumor response to treatment creating advanced synergistic chemo-radiotherapy. Zhang and coworkers [46] used cis-diamminedichloroplatinum(II) (cisplatin) attached to phospholipids as an oxygen generator in the tumor microenvironment. They highlighted the problems of implementation of the combined therapy in clinical usage. Catalase (CAT), an antioxidant enzyme, was incorporated in the liposome core constituted by cisplatin (IV)-prodrug-conjugated phospholipid, forming CAT@Pt(IV)-liposome. The designed structures caused a higher number of DNA double breaks after exposure of glioma cells to ionizing radiation. This approach integrated cisplatin-based chemotherapy and catalase-induced tumor hypoxia relief. Scientists have observed that their nanostructures are capable of accumulating sufficiently high in the tumor, while significantly reducing the hypoxia of hypoxic areas by triggering the decomposition of endogenous H_2O_2. It was found that CAT@Pt(IV) assessed the greatest results in combination with ionizing radiation. The designed "nano reactor" is a multifunctional structure with many benefits due to the simple method of synthesis and production, biocompatibility, good drug carrier, protecting the catalytic activity of the enzyme, and reduction of hypoxia. Their results constitute a good basis for further research in this direction, as the problem of resistance to radiotherapy and chemotherapy is still topical and difficult to overcome.

Gold nanoparticles (AuNPs) are the most frequently used to sensitize cancer cells. Authors [47] postulate, that radiosensitization with gold nanoparticles depends on the level of internalization. It takes place using two methods: micropinocytosis and passive diffusion membrane [48]. Chitchrani et al. [49] used liposomes as a "Trojan horse" to deliver 1.4 nm gold nanoparticles to cancer cells. Their aim was to overcome the energetically unfavorable endocytosis process for small nanoparticles. Researchers studied the efficiency of uptake and intracellular transport of fabricated nanostructures. The results demonstrated a 1000-fold enhancement of internalization using the liposome system. Moreover, the analysis showed that nanostructures made of liposomes and gold nanoparticles are delivered to the lysosomes after 40 min of incubation. This study provided much-needed information on the implantation of liposomes as a carrier of gold nanoparticles into the cell. Scientists [50] also investigated gold nanoparticle liposomes for their stability and the ability to release nanoparticles with light. Triggered drug release is a desirable form of substance administration due to its ability to control and regulate the release of the drug. The light energy is converted into thermal energy in gold nanoparticles, which is the factor causing the liposome breakdown. The aim of the research group was to analyze the drug delivery system and its release in the cytosol of the cell using visible light and near-infrared light signals. Moreover, scientists synthesized the liposome from components that were sensitive to changes in pH and temperature and incorporated gold nanoparticles in the form of rod and star shapes. Human retinal pigment epithelial cells (ARPE-19) and human umbilical vein endothelial cells (HUVECs) were used in this study. The designed nanostructures were non-toxic. The results demonstrated that the nanoparticles were not released without light interference. As a result, the light activated liposome formulations showed a controlled content release at the selected place and time.

Another research group [51] also used the combination of liposomes and gold nanoparticles, but to create a construction called the hybrid Cluster Bomb in the treatment of liver cancer. In addition, they used liposome-loaded paclitaxel in their study. The effectiveness of hybrid liposomes was proven by using xenograft Heps tumor-bearing mice. Researchers estimated the most efficient and effective ratio of individual components by presenting the accurate site and time-release mode for liver tumor treatment. They assessed the long-term effect of release on therapeutic efficacy by considering the results of a single dose for seven days by injection into the tail vein of mice. All paclitaxel (PTX) formulations were observed as effective in suppressing tumor growth after receiving a single dose. As assumed, the group treated with PTX/PTX-PEG400@GNPLips 1 (25:75) showed effective tumor inhibition. When comparing body weights between groups, it was found that the survival rate of the PTX/PTX-PEG400@GNPLips (25:75) treated group was significantly higher. At 7 days after injection, none of the mice died in the PTX/PTX-PEG400@GNPLips (25:75) group, while the survival rates in the saline and Taxol® groups were only 50% and 70%, respectively. The results indicate that the combined strategy of drug burst and sustained release from the nano-delivery system enables the precise control of drug concentration in tumor tissue and cells to achieve better anti-tumor activity and reduced systemic toxicity. Sharifabad et al. [52] tested an equally advanced nanostructure, however, using superparamagnetic iron oxide nanoparticles. They created liposome-capped core-shell mesoporous silica-coated superparamagnetic iron oxide nanoparticles called "magnetic protocells". Their aim was to prepare nanostructures and assess their toxicity on the MCF7 breast cancer cell line under the influence of an alternating magnetic field. Cells were treated with the synthesized compound and showed an approx. 20% decrease in proliferation compared to control cells. Researchers concluded that the use of their synthesis method would contribute to the treatment of tumors in animal models in the future in conjunction with magnetic hyperthermia. Bao et al. [53] designed the paclitaxel hybrid drug delivery system using liposomes and PTX-PEG400@GNPs gold nanoparticles. They used a covalent bond to conjugate thiol-terminated polyethylene glycol (PEG400)-PTX to gold nanoparticles and then incorporated the paclitaxel attached to the AuNPs into the liposomes. Due to this approach, the drug loading capacity increased. The encapsulated paclitaxel showed a longer circulation time and better targeting abilities than the commercial Taxol® (BRISTOL MYERS SQUIBB, BMS, Warsaw, Poland). They observed also that the combination of organic and inorganic structures resulted in better effects than commercially used drugs. However, this method is associated with some drug release limitations as no statistically significant differences between the designed hybrids and the conventional drug were noticed in other research groups. This study provides a lot of information on how to improve synthesis methods and what factors should be considered designing new strategies. Xing et al. [54] also used liposomes and gold nanoparticles to deliver doxorubicin but in order to support photothermal synergetic antitumor therapy, nanoparticles and doxorubicin were locked inside the liposome. The authors observed that under the influence of NIR (Near Infrared) light, complex liposome/metallic hybrids are more absorbed by the endocytosis process because the structure of the liposomes was broken down and the gold nanoparticles easily reached the inside of the cell. The designed nanostructures showed an excellent anti-cancer effect, inhibiting the development of cancer cells by approx. 80%. Moreover, in the mice bearing tumor model, the nanostructures also presented themselves as tumor suppressors. A research group led by Zheng et al. [55] focused on the study of tumor-specific, pH-responsive, peptide-modified, liposome-containing paclitaxel and superparamagnetic iron nanoparticles PTX/SPIO-SSL-$H_7K(R_2)_2$. Anticancer, imaging and targeting effect were assessed on the basis of an in vitro and in vivo model using the epithelial human breast cancer cell line MDA-MB-231. Their results confirmed the effectiveness of the produced nanostructures in selected issues, which constitute an advanced theragnostic approach in this field. Scientific publications about using LipoMNPs were collected in the Table 1.

Table 1. Selected applications of liposomes with metallic nanoparticles in cancer therapy.

Author	Year	Nanostructures	Synthesis Method	Nanostructures' Size	Cell/Tissue	Kind of Therapy	Results
Bromma [56]	2019	AuNPs entrapped in lipid nanoparticles	rapid-mixing method	53 nm	Breast cancer cells, MDA-MB-231	Radiotherapy	The addition of LNPs into tumor cells produced a 27% enhancement in tumor cell death
Bao [53]	2014	PTX-conjugated GNPs (PTX–PEG400@GNPs) in liposomes	thin film hydration	3.41 nm gold core	A murine liver cancer model	The drug delivery system	Maintains the superiority of both vehicles and improves the performance of hybrid systems
Chitchrani [49]	2010	AuNPs in liposome-based system	thin film hydration	105 nm	Cervical cancer cells, HeLa	The assessment of cellular uptake and transport	Au NP–liposomes demonstrated that they reside in lysosomes
Liu [45]	2020	Au nanoparticles and perfluorohexane nanoparticles encapsulated in lipid shell	film hydration method coupled with a double emulsion method	108 nm	Human anaplastic thyroid cancer cells, C643	Low-intensity focused ultrasound diagnosis ablation	An optional therapeutic platform for treating patients with drug-resistant cancer
Wang [57]	2017	Loading resveratrol (Res) in chitosan (CTS) modified liposome and coated by gold nanoshells (GNS@CTS@Res-lips).	mediation of CTS	115 nm	Cervical cancer cells, HeLa	Photothermal therapy	The nanocarriers displayed a synergistic antitumor effect of chemo photothermal therapy compared with PTT or chemotherapy alone
Zhu [16]	2018	Carboxyl-modified Au@Ag core-shell nanoparticles (Au@Ag@MMTAA) contained in the liposomes (DSPE-PEG2000-NH2)	thin film hydration	215 nm	Breast cancer cells, SKBR3	The assessment of cellular uptake and transport	The nanohybrids entered cells mainly through clathrin-mediated endocytosis and tended to attach on the cell, the highest mortality in vitro after laser treatment, surface before arriving in acidic lysosomes

Table 1. *Cont.*

Author	Year	Nanostructures	Synthesis Method	Nanostructures' Size	Cell/Tissue	Kind of Therapy	Results
Rengan [36]	2014	The Lipos Au particles	thin film hydration	100–150 nm	Breast cancer cells, MDA-MB-231	The drug delivery system and photothermal therapy	The efficient deployment for drug delivery application using NIR laser irradiation, enhanced parameters of drug delivery, and optical imaging, the Lipos Au NPs exhibited their true multifunctional ability by emitting good signals in CT X-ray analysis
Zhang [51]	2016	Gold conjugate-based liposomes with hybrid cluster bomb structure	thin film dispersion method,	115–150 nm	Xenograft Heps tumor-bearing mice	The multi-order drug delivery system	The time-release mode for tumor treatment using antitumor drugs
Sharifabad [52]	2016	Liposome-capped core-shell mesoporous silica-coated superparamagnetic iron oxide nanoparticles called 'magnetic protocells'	lipid hydration	53 nm	Breast cancer cells, MCF7 and likely glioblastoma cells, U87	The drug delivery system	Loaded nanoparticles under alternating magnetic field exhibited nearly 20% lower survival rate of cancer cells
Zheng [55]	2018	liposome-containing paclitaxel (PTX) and superparamagnetic iron oxide nanoparticles (SPIO NPs), PTX/SPIO-SSL-H$_7$K(R$_2$)$_2$,	thin film hydration	3.41 nm gold core	human breast cancer cell line, MDA-MB-231	The drug delivery system	Antitumor effect and enhancement of MRI parameters

AuNPs—Au Nanoparticles; LNPs—Lipos nanoparticles; PTX—paclitaxel; GNPs—Gold Nanoparticles; PEG—Polyethylene Glycol; GNS—gold nanoshells; PTT—Photothermal therapy; DSPE—1,2-Distearoyl-sn-glycero-3-phosphorylethanolamine; MMTAA—2-mercapto-4-methyl-5- thiazoleacetic acid; CT—computed tomography; SPIONPs—superparamagnetic iron oxide nanoparticles.

The combination of metallic nanoparticles and liposomes has also been used in vaccine design. Scientists [58] explored a biomaterial-based approach of converting tumor-derived antigenic microparticles (T-MPs) into a cancer vaccine presenting their potential in multiple murine tumor models. They used adjuvant CpG-loaded liposomes with Fe_3O_4/T-MP on the surface to get a vaccine (Fe_3O_4/T-MPs-CpG/Lipo). The group demonstrated that the engineered, complex vaccine targeted to antigen-presenting cells induced a strong tumor antigen-specific host immune response. Moreover, the vaccine in the tumor microenvironment could alter tumor-associated macrophages into a tumor-suppressive M1 phenotype by nano Fe_3O_4, resulting in the transformation of a "cold" tumor into a "hot" tumor. This study exemplifies the modulation of the tumor immunosuppressive network which potentially constitutes a personalized vaccine cellular strategy.

The future clinical success of nanoparticles depends on a more detailed understanding of the mechanisms of their physicochemical properties on the biological response. Researchers have investigated many studies to obtain nanostructures resulting in efficient cancer treatment, but further research is desired [56,57,59]. The nanoparticle-based radiosensitization method is used in proton and particle therapies [60,61]. However, these methods are based mainly on the wide use of metal nanoparticles, therefore the possible use of liposome-metal nanoparticle systems in the future is fully open. In our opinion, liposomes containing gold nanoparticles represent a potential factor that is a good complement to conventional therapies. Such nanostructures can be used as drug carriers in chemotherapy and at the same time make cancer cells more sensitive to radiation therapy. Using Lipo-MNPs, a new therapeutic approach in the form of radio-chemotherapy can be designed.

4. Medical Imaging

Lipid nanoparticles have proven to be excellent potential transporters of agents for imaging and therapy [13,62,63]. In order to create the best liposomal structure, it is mandatory to understand and know their behavior in vivo. Therefore, there is a constant effort to improve visualization, quantification, and monitoring of liposomes biodistribution using non-invasive imaging techniques that could further give information about drug release, for example. To image nanoliposomal structures in vivo, labeling ions are incorporated into the liposomal membrane, bound to the membrane, or are enclosed in the liposome interior. There are several contrast agents (CAs) widely used in radiological and nuclear medicine preclinical and clinical routine, improving the image quality of non-invasive imaging techniques [64]. In the group of radiological CAs, one can mention gadolinium-based, manganese-based, iron oxide, and iron platinum contrast agents for magnetic resonance imaging (MRI) [65], iodine- and lanthanide-based as well as gold, bismuth, and other metals used as X-ray-computed tomography (CT) contrast agent [66] and phospholipids with sulfur hexafluoride, octafluoropropane, or perfluorobutane for ultrasonography (USG) [67] (Figure 4).

Nuclear medicine (NM) uses radionuclides for the labeling of specific ligands (radiopharmaceutical) that are trapped and metabolized by cancer cells. That makes possible both functional imaging and therapy if certain conditions are met. Depends on the imaging technique, radionuclides decay with beta plus process like gallium-68 (^{68}Ga) or fluorine-18 (positron emission tomography, PET), or emit gamma rays at defined energy levels via isomeric transition or electron capture process like technetium-99m (^{99m}Tc) or iodine-123 (^{123}I) (single-photon emission computed tomography, SPECT; planar scintigraphy) emitting two or one gamma photon, respectively. The number of radiopharmaceuticals used for imaging is seemingly large [68]. However, going deeper into the subject one can recognize that this group is not so tremendous. The combination of liposomes and metallic nanoparticles used as contrast agents for mentioned technique proved to have the potential to supports efficient medical imaging. Research in this area [31,32] has achieved a sophisticated level that is carefully suited to the diagnosis and treatment. Although, most of the radiological modalities are freely available and used CAs are much cheaper than the production of radionuclides and their use for PET and SPECT scanning, their combination with liposomes suffers mostly from low sensitivity and signal to background. Contrary, nuclear techniques benefit from whole body capabilities, quantification, high sensitivity, and absence of tissue penetration issues, but presents limited spatial resolution [68]. For nuclear techniques, radiolabeling of liposomes was widely described in the literature [13,69–71] with the use of several different labeling methods [72]. According to Man et al. [68], in 2019, Tc-99m was the most commonly used radionuclide for liposome labeling. The second was indium-111 (In-111) and then slowly PET radionuclides started to play an important role with—copper-64 (Cu-64), manganese-52 (Mn-52), and zircon-89 (Zr-89).

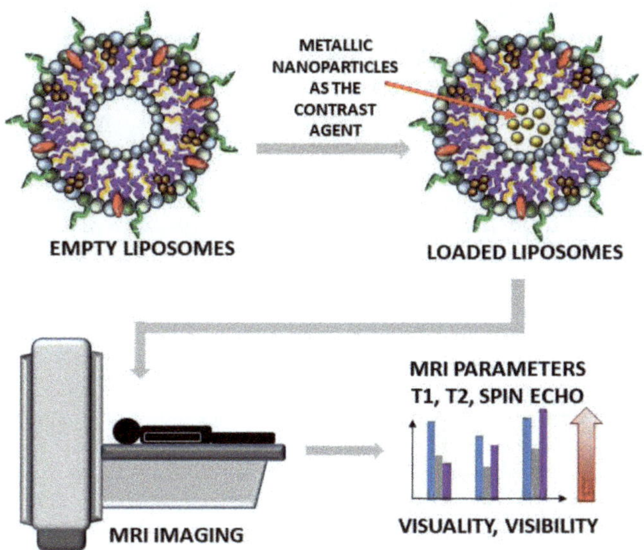

Figure 4. The scheme of future usage of liposomes loaded with metallic nanoparticles as the state-of-the-art contrast agents in medical imaging of patients will enhance the parameters of visuality and visibility. The red arrow is a symbol of increasing visuality and visibility in MRI imaging after nanostructures implementation.

The main approach which presented advantages of using LipoMNPs was introduced by German et al. [73]. The group assessed the efficiency of hybrid nanosystems (magnetite nanoparticles and liposomes, MFLs) as contrast agents. The magnetite nanoparticles (MNPs) were synthesized using the chemical precipitation form of Fe (II) and Fe(III) salts solution in a basic environment [74]. They showed their ability to perform MRI contrast enhancement in vitro and in vivo using renal carcinoma cells transplanted into rats during the administration of MFLs into the tumor. The scientists subcutaneously administered the suspension of renal cell carcinoma (Blokhin Russian Cancer Research Center) cells into twenty male albino Wistar rats. After the MRI study tissue samples of internal organs of the rats were taken for histological examination to assess the toxicity of the new contrast agent, the results revealed that MFLs increase the T1 (transverse relaxation time) parameter and decrease the T2 (longitudinal relaxation time). According to the authors, MFLs are able to effectively increase the contrast on T2-weighted images, allowing visualization of the tumor under both T1 and T2 sequences. The MFLs did not induce any significant changes in the internal organs due to MFLs' biocompatibility.

The newest reports [75,76] suggest using liposomes in the area of imaging diagnostics and treatment in neurology. However, it is worth mentioning that the interest in this topic started much earlier when Vieira et al. presented a wide review summarizing accomplishments in the field of liposome-based drug delivery across the blood–brain barrier [77].

Authors have concentrated on liposomal strategies for imaging drug accumulation for the treatment of ischemic zone and glioma diagnosis. Moreover, the advantage of the presence of targeting ligands over the liposomal surface was emphasized as an improvement of the agent retention time in the tumor and its uptake by the cancer cells. Recently, various neurological studies focus on Alzheimer's disease and amyloid plaque visualization [78], lesions in the brain due to HIV infection [79], Parkinson's disease [80], and stroke [76].

Comparing various medical topics using liposomal delivery systems for imaging and therapy, oncology is the area with the biggest collection of data referring to its broad application. Liposome-nanoparticle hybrids that include metallic components such as: gold nanoparticles [75,81] or gadolinium nanoparticles (GdNPs) [63,82,83], and ultrasuper- and superparamagnetic iron oxide nanoparticles (SPIONs)–called magnetoliposomes (MLs) [60,84,85]—can serve in the visualization of the cancer tissues using CT and MRI. Liposome-based CAs incorporating gadolinium were the first introduced nanostructure that has been widely used in various systems [86–88]. Scientists constantly broaden their research in terms of the administration of newly synthesized contrast agents containing liposomes and metallic nanoparticles into different types of cancer tissues.

Breast cancer was chosen by several authors as the target for possible imaging with the use of liposome-nanoparticle hybrids. Experts agree that mammography is the best available diagnostic tool for the early detection of breast cancer in patients of risk. Mammography is followed then by biopsy and MRI increasing costs of diagnosis. For that reason, scientists started to look for other alternatives to mammography showing better sensitivity and specificity [89]. One of them was computed tomography [90].

One year later, the same author [91] published an in vivo analysis of biodistribution and pharmacokinetics performed using Swiss albino mice proving at the same time multifunctional capabilities of Lipos Au NPs and their hepatic-biliary and renal clearance. Lipos were synthesized following the identical protocol [19] with small modifications to achieve a smaller size of liposomes. The estimated mean diameter of the whole structure was 100–120 nm, where the coating was created by gold nanoparticles of the 5–8 nm diameter ensuring the renal clearance after biodegradation of core liposome [92]. Lipos Au NPs were injected i.v. ~110 µg/400 µL through the tail vein and the analysis of various tissues, plasma, and urine was performed on days 1, 7, and 14 after injection. The authors observed that the injected particles were accumulated at the highest rate in the liver and spleen. A smaller percentage of 2–8 nm particles was accumulated in the kidney and further removed with urine. An in vivo study was carried out using the HT1080-f luc2-turboFP tumor xenograft model in BALB/c NUDE mice. When the tumor was around 70 mm^3 large, mice were segregated into three groups of five animals and treated as followed: the first group by injection of normal saline, the second was treated with laser only, and the third with laser and by injected Lipos Au NPs (0.5 µg/µL in 30 µL). Results of the experiment showed that the bioluminescence signal was significantly lower on day 30 in group II and III compared to group I. Additionally, in four out of five animals in group III, tumor regressed completely comparing to animals from group I or II which all died naturally.

The interest of breast cancer diagnostics was also focused on MRI and its CAs, namely superparamagnetic iron-oxide nanoparticles creating negative contrast in T2-weighted images (spin-spin relaxation time) and gadolinium providing positive contrast in T1-weighted images (spin-lattice relaxation time). SPIONs have shown great potential in theragnostics applications like therapy (hyperthermia cancer treatment), targeted drug delivery, and separation of biomolecules. They respond nicely to external magnetic fields that allow for directing them specifically to the region of interest. It was shown that SPIONs smaller in diameter than 10 nm, act as T1 agents whereas bigger than this acts as T2 [85]. Therefore, SPIONs size control is a critical challenge for their application in MR imaging. Regarding the agglomeration problem, it can be controlled by coating SPIONs, e.g., with polymers or lipids [92].

He et al. [70] conducted an experiment using magnetic iron-oxide nanoparticles (MIONs) bound to liposomes including chemotherapeutic anticancer drug mitoxantrone

(Mit) creating gonadorelin-functionalized Mit-loaded MLs (Mit-GML). They synthesized such hybrids in order to assess their effect in both therapy and MRI imaging of human MCF-7 breast cancer cells implanted into living animal models. A lipid film hydration method was used to prepare MION-loaded liposomes (MLs). They used the female athymic nude BALB/c mice with MCF-7 breast cancer cells implanted into their mammary fat pads. Mice underwent MRI examination after the tumor diameter reached 7–10 mm, with an aim of demonstrating the diagnostic ability of Mit-GML for targeted detection of cancer lesions. Images were obtained before and 2 h after contrast injection by a clinical MR scanner (GE Signa excite 1.5 T). Mit-GML significantly induced the targeted delivery of Mit to tumor cells with overexpression of LHRH receptors which resulted in inhibition of tumor cell growth. Results demonstrated an enhanced tumor accumulation of Mit-GML compared to other cells while analyzing biodistribution of the liposomes in various organs over 24 h. In a concentration-dependent manner in vitro the T2-weighted MR image of Mit- GML presented a dark MR contrast signal. The in vivo MRI T2-weighted images performed 2 h post-injection showed a decrease of signal intensity in the tumors. Cancer theranostics with Mit-GML presents a new strategy in breast cancer treatment with a strong need for further investigation.

Zhang et al. [93] created another combination of liposomes and metallic nanoparticles in the imaging diagnosis of breast cancer cells. They synthesized polyethylene glycol-coated liposomes. After that, SPIONs were incorporated into the core of liposomes and near-infrared dye (DiR) into the lipophilic bilayer creating hybrid nanostructures (PGN-L-IO/DiR). The authors used PGN635, a human monoclonal antibody that specifically targets phosphatidylserine (PS). They conjugated the antibody to the distant terminus of the polyethylene glycol chain. In vitro targeting specificity and toxicity was characterized with the use of adult bovine aortic endothelial cells (ABAE). Breast cancer cells MDA-MB-231 were injected subcutaneously on a thigh or both thighs of anesthetized mice. In vivo T2-weighted FSEMS MRI images were acquired at 9.4 T, acquired before and at different time points after i.v. injection of PGN-L-IO/DiR. After MRI at 24 h, near-infrared fluorescence imaging was performed and repeated at 48 h. The results confirmed the potential of the targeted liposomal–SPION hybrid for sensitive in vivo imaging of breast cancer cells in mice in a direct and accurate manner. The tumor visualization after i.v. infusion of PGN-L-IO/DiR was confirmed by both MRI examination and optical imaging. Irradiation resulted in an increase of PS exposure in MDA-MB-231 cells which confirms the use of this technique to facilitate the targeting of PGN-L-IO/DiR. The results provide perspectives for implementing both imaging and therapeutic agents into the PS- targeted liposomal platform. Recently, Patil-Sen et al. [94] reported on the fabrication of a three-component composite-magnetoliposome consisting of a SPION-silica core-shell coated with mesoporous silica and/or outer lipid layer. The aim of the construction of those structures was to show their potential theragnostic applications in drug delivery and magnetic hyperthermia. The capability of particles for hyperthermal therapy was also confirmed. Lipid coating enhanced the dispersion stability and lipid silica-coated SPIONs displayed significantly shorter T2 time as compared to the bare-SPIONs. The doxorubicin (DOX) release studies showed an efficiency of 35% for lipid-coated and 58% for silica-lipid-coated SPIONS. For the in vitro cytotoxicity study, the human fetal glial had a normal cell line (SVG) p12 and human breast cancer had a commercial cell line (MCF-7). Lipid-coated nanoparticles (without DOX) have shown excellent biocompatibility in vitro, against both cell lines. DOX-loaded nanocomposites showed high efficiency and selectivity as drug carriers against MCF-7 cell lines as compared to SVG p12 (indicative result). All those results show that a combination of lipid–silica dual coating for SPIONs creates a potential theragnostics system that can be used for both, diagnostic and targeted drug delivery, in hyperthermia-mediated cancer therapy.

A recent example of metallic liposomes use for breast cancer studies in nuclear medicine can be Lee et al. [95]. They reported a clinical trial of HER2-targeted PEGylated liposomal doxorubicin, labeled with Cu-64 radioisotope (Cu-64-MM-302) showing the enhanced permeability and retention effect in relation to treatment response of HER2-positive metastatic breast cancer. The research was driven in a cohort of 19 patients who had administrated Cu-64-MM-302 structures and afterward were imaged using PET/CT. Results showed that particles circulated for over 24 h and accumulated mostly in the liver and spleen. Retrospective analysis of outcomes related to the delivery of the drug into the tumor lesions and its high deposition were associated with more favorable treatment outcomes (HR = 0.42).

The imaging diagnostics of other cancers common in women that drew the attention of the scientists was visualization and monitoring of the therapy effects on ovarian cancer. According to Bray et al. [96], ovarian cancer is one of the most common cancers risking women's health and has the worst prognosis and mortality rate. Ravoori et al. [97] investigated a new contrast agent that could perform in both staging in the pre-surgery planning under MRI and optical aid in the surgical operation. They synthesized dual gadolinium liposomal contrast agent (DM-Dual-Gd-ICG) and used human ovarian cancer HeyA8 cells and OVCAR-3 cells injected into nude female mice. Increased T1-weighted MR signal and NIR signal were visualized in mice tumors injected with DM-Dual-Gd-ICG after two days. Results suggest potential clinical use in multimodal imaging that can be beneficial in the diagnosis and treatment of advanced ovarian cancer stages.

Chen et al. [98] synthesized a novel molecular imaging nanoprobe with the ability to perform as a dual modality in MRI and NIRF. The most important integrin for angiogenesis in many solid tumors, e.g., liver cancer, indocyanine green (ICG), was used as a NIRF dye. The synthesized probe consisted of SPIO@ Liposome bound to ICG and RGD (peptide: arg-gly-asp). They used HepG2 liver cancer cell lines for cytotoxicity assessment as well as integrin binding. In vivo imaging was performed in Balb/c male nude mice with a subcutaneous tumor acquired from injecting HepG2 cells. Scientists distributed intravenously the SPIO@Liposome-ICG-RGD in mice and observed their effect in both fluorescence imaging and MRI with a dual-modality technique. The study showed clear tumor delineation after SPIO@Liposome-ICG-RGD probe injection. The contrast-to-noise ratio obtained from MRI was helpful for detecting smaller tumors (0.9 ± 0.5 mm). The small tumors appeared to be much brighter than the surrounding normal liver tissue while using the fluorescence surgical navigating system. SPIOs represent the first nanoparticle MRI contrast agents in clinical use and show excellent biocompatibility and performance in the visualization of the tissues. Moreover, ICG is the only NIR dye approved by the FDA for diagnosis in clinical applications. The authors suggest a profound value in clinical practice by stating an example of patients with liver cancer exhibiting similar degrees of metastasis.

Very interesting results of gadolinium–lipid complexes containing PE-DTPA (chelating Gd+3) are presented by Šimečková et al. [99]. Structures were prepared using the lipid film hydration technique. Their cytotoxicity was tested in human liver cancer HepG2 cells, liver non-differentiated progenitor HepaRG cells, and HepaRG cells differentiated into both, hepatocyte-like and biliary epithelial cell populations. No toxicity and side effects were detected indicating the potential safety of the synthesized liposomes. Lorente et al. [100] chose colon cancer as a target of theragnostic therapy. Their MLPs (γ-Mag-NP-LPs) were synthesized using maghemite nanoparticles (γ-Fe2O3) incorporated in phosphatidylcholine (PC) liposomes. These magneto liposomes were tested in the human colon fibroblast CCD-18 cell line, human colon carcinoma T-84 cell line, and the murine macrophages RAW cell line. The cytotoxicity of NPs in macrophages, lymphocytes, and erythrocytes was also tested. The authors observed γ-Mag-NP-LPs internalization into tumor cells with the use of transmission electron microscopy (TEM) and Prussian Blue staining. In order to determine qualitatively the magnetic-induced mobility of cells treated with γ-Mag-NP-LPs, authors used T-84 cells. After 24 h, cells were treated with 10 and 100 μg/mL of γ-Mag-NP-LPs. The results presented satisfactory magnetic and biological properties of the MLPs in vitro, including a high degree of biocompatibility. Biosecurity in clinical use was confirmed by analyzing the ability to produce lysis of human blood cells and no increase in lysis was observed. The results suggest that the properties of γ-Mag-NP-LPs, make them acceptable for their loading with chemotherapeutic drugs or for hyperthermia treatment. In 2017, Blocker [101] published results of MM-DX-929 (Merrimack Pharmaceuticals, Inc., Cambridge, MA, USA) liposomes labeled with Cu-64 (64Cu-MM-DX-929) for PET imaging. The aim of the study was to capture the image of the effect of short-term Bevacizumab treatment on liposome delivery to the colon tumor. In vivo studies were carried out in HT-29 human colorectal adenocarcinoma tumors grown in severe combined immunodeficient mice (SCID). Animals received around 200–300 μCi of 64Cu-MM-DX-929 (20 μmol/kg lipid) i.v. via the tail vein. PET imaging was followed by CT scanning for the same period of time. As a consequence, authors showed that positron emission tomography can detect significant differences in liposome delivery to treated colon tumors when compared to untreated controls using MM-DX-929 labeled with Cu-64.

Thebault et al. [102,103] presented a novel approach with the implementation of theragnostic MRI liposomes for cancer imaging. In the work [102], the authors aimed to evaluate the magnetic targeting (MT) efficiency of synthesized ultra-magnetic liposomes (UML) for MRI imaging. They used CT26 murine colon tumor in Balb/C female mice. The MT method allowed an efficient semi-quantitative evaluation of targeting efficiency in tumors with a possibility of implementing this technique to different T2 contrast agents. In their next study [103], authors proposed a very compelling therapeutic strategy that consisted of the magnetic accumulation of thermosensitive ultra-magnetic liposomes encapsulating iron oxide nanoparticles (UML) and an antivascular disrupting agent combretastatin A4 phosphate (CA4P). They used the High-Intensity Focused Ultrasound (HIFU) technique to trigger the release of CA4P in a colon tumor. The whole process was monitored by MRI examination. Firstly, scientists studied the effect of CA4P-UML and their ability to damage the cytoskeleton of EAhy-926 endothelial cells. In order to evaluate the perfusion in the tumor 24 h after the treatment, the authors performed a Dynamic Contrast-Enhanced DCE perfusion protocol. CA4P-UML showed a dual effect of magnetic targeting and MRI monitoring. The full combination consisting of UML containing CA4P with magnetic targeting and local ultrasound to trigger the CA4P release provided a statistically significant decrease of the tumor volume. MRI bioimaging proved the benefits of this therapy in vivo with the functional decrease of the vasculature in terms of permeability as well as the reduction of growth of the tumor. Scientific publications about using LipoMNPs in medical imaging were collected in the Table 2.

In our point of view, Lipo-MNPs are much more often used in medical imaging than in therapy. Such nanostructures can be observed in almost all commonly used imaging due to the metallic core of the nanostructures. This design allows for better results and imaging parameters. However, we have noticed that the conducted research is also associated with many limitations, which should be taken into account in the future.

Table 2. Selected applications of liposomes with metallic nanoparticles in medical imaging.

Author	Year	Nanostructures	Synthesis Method	Nanostructures' Size	Cell/Tissue	Kind of Therapy	Results
German [72]	2015	Magnetite nanoparticles (MNPs) in magnetic fluid loaded liposomes (MFLs)	chemical precipitation form of Fe(II) and Fe(III) salts solution in basic environment; extrusion technique	NPs: 13 nm MFL size: 147 nm	Renal cell carcinoma administered subcutaneously into twenty male Wistar albino rats	MRI	Increase of T1, decrease of T2 time; visualization of the tumor under both T1 and T2 sequences
He [104]	2014	magnetic iron-oxide nanoparticles (MIONs); gonadorelin-functionalized Mit-loaded MLs (Mit-GML)	lipid film hydration	Mit-GML size-136 nm	MCF-7 breast cancer cells implanted into female athymic nude BALB/c mice	MRI	Enhanced tumor accumulation of Mit-GML; 2 h post injection decrease in T2 signal intensity in tumors
Zhang [92]	2014	super-paramegnetic iron-oxide nanoparticles (SPIONs), hybrid nanostructures (PGN-L-IO/DiR)	lipid film hydration	PGN-L-IO/DiR size: 111 nm	MDA-MB-231 breast cancer cells injected subcutaneously into nude BALB/c mice	MRI	tumor visualization: 24 h post injection hypointense intratumoral regions appeared
Patil-Sen [93]	2020	composite-magnetoliposome hybrid: SPION-silica	co-precipitation, thin film hydration, surfactant templating approach	hybrid size: 150 nm	MCF-7 breast cancer cells, fetal glial normal cell line SVG -12	MRI	proven use as a negative contrast in MRI imaging;
Lee [94]	2017	PEGylated liposomal doxorubicin, labeled with Cu-64 radioisotope (Cu-64-MM-302)	no information	no information	HER2-positive metastatic breast cancer cells—19 patients	PET/CT	Particle tumor accumulation and visualization in imaging techniques
Ravoori [96]	2016	dual gadolinium liposomal contrast agent (DM-Dual-Gd-ICG)	lipid film hydration	<150 nm	HeyA8 or OVCAR-3 ovarian cancer cells, intraperitone-al injection	MRI	Increased T1-weighted MR signal and NIR signal in tumors
Chen [97]	2017	SPIO@ Liposome bound to ICG and RGD	film method followed by extrusion	<150 nm	HepG2 liver cancer cells subcutaneously injected into ten Balb/c nude mice	MRI	clear tumor delineation after probe injection, contrast-to-noise ratio helpful for detecting smaller tumors

Table 2. *Cont.*

Author	Year	Nanostructures	Synthesis Method	Nanostructures' Size	Cell/Tissue	Kind of Therapy	Results
Blocker [100]	2017	MM-DX-929 liposomes labeled with Cu-64	empty MM-DX-929 liposomes were provided by Merrimack Pharmaceuticals	104 nm	HT-29 human colorectal adenocarcinoma cells grown in SCID mice	PET, CT	MM-DX-929 labeled with Cu-64 detect significant differences in liposomes delivery to treated colon tumors when compared to untreated controls.
Thebault [102]	2020	CA4P-loaded thermosensitive Ultra Magnetic Liposomes (CA4P-UML)	co-precipitation method, reverse-phase evaporation method	209 nm	CT-26 murine colon tumor in Balb/C female mice	MRI	decrease of the tumor volume and vasculature observed in MRI

Abbreviations: ICD—indocyanine green, RGD-Arginine-Glycine-Aspartic peptide, CA4P—combretastatin A4, PEG-polyethylene glycol, MRI—magnetic resonance imaging, PET—positron emission tomography, CT—computed tomography PGN—human antibody PGN635, L-liposome, IO—iron oxide, DiR—near-infrared dye, SPION- super-paramegnetic iron-oxide nanoparticle, MION—magnetic iron-oxide nanoparticle, UML—ultra magnetic liposomes, MM-DX-929—64Cu-liposomal doxorubicin PET Agent (Merrimack Pharmaceuticals, Inc. Cambridge, MA, USA), MM-DX-302 -HER2-targeted antibody–liposomal doxorubicin conjugate (Merrimack Pharmaceuticals, Inc. Cambridge, MA, USA), SCID—Severe combined immunodeficient mice.

5. Future Perspectives

The combination of liposomes and metallic nanoparticles has allowed the practical design of nanoscale devices combined with numerous functional molecules, including tumor-specific ligands, antibodies, anti-cancer drugs, and imaging probes. Moosavian et al. [30] presented the challenges and limitations in the development of liposomal drug delivery systems in anti-cancer therapies. They emphasized that researchers, in the future, should focus on better preparation of liposomes that would be more efficient in clinical trials. Moreover, it was suggested that results obtained on selected animal models are overestimated and show inadequate effects of liposomes, which will not occur in clinical implementation. A similar tendency can be observed in the above-presented works related to the use of systems consisting of liposomes and metallic nanoparticles. None of the teams considered several models or the optimization process of the method of obtaining liposomes. It would be recommended to describe the differences between individual variants of liposomes that differ in size, composition, or other biofunctionalizations. Moreover, researchers focus only on describing the effect of therapy of liposomes on cancer cells without taking into account the description of their biointeractions with biological matter [105]. Considering the radiosensitization effect of metallic nanoparticles on cells, it is important to clear the spatial relationship between the distribution of oxygen molecules and hypoxic tumor regions. It is necessary to optimize the conditions and time of reoxygenation in the area of the tumor microenvironment after the application of active nanostructures and radiotherapy. The relationship between liposome internalization and hypoxia is still unknown.

Furthermore, scientists should focus on targeting the transport of nanostructures towards specific cell organelles. Most studies only consider liposome targeting due to ligands and receptors on the surface of cancer cells. This is a logical approach to target engineered nanostructures directly to enter the cell, but liposome performance inside the cell is poorly understood. Hybrid liposomes action should show selectivity for given organelles depending on the desired effect. Researchers observe that nanoparticles released from the liposomes should reach the mitochondria to impair the functioning of the cell, or directly the nucleus, especially in gene therapy or in the regulation of protein metabolism.

We assume that in the future, scientists should use more models in experiments, using different cell lines, normal and cancer ones. However, when doing in vivo experiments, they should also focus on the influence of nanostructures on healthy tissues, highlighting the given key parameters that have a selective effect between normal and cancer cells. In addition, more attention should be paid to the more careful planning of fabricated nanostructures in order to be able to present the reasons why one component was used for synthesis. Our conclusions should help scientists learn from previous studies and plan better, more refined treatment or imaging strategies.

6. The Various Development Methods of Lipo-MNPs

The nanotechnology era has introduced various novel synthetic strategies in the area of controlled drug delivery. Liposomes are attractive biomimetic nanocarriers characterized by their biocompatibility and high loading capacity [106]. The constantly growing literature database in the field of liposomology, covering complementary studies from different fields, is an indication of increasing interest in this research area. Many techniques and methodologies have evolved for the preparation of liposomes, on small and large scales, since their introduction to the scientific community around 40 years ago [107]. Incorporating drug molecules into nanocarriers offers exciting opportunities to redefine the pharmacokinetic behavior of the drug, improving its therapeutic efficiency and reducing side effects [108–110]. Several types of drug delivery nanocarriers based on organic platforms such as liposomes, polymers, and dendrimers have been used as "smart" systems that can release therapeutic agents under physiological conditions [109]. Synthetic liposomes are small artificial vesicles of spherical shape that can be created from cholesterol and natural nontoxic phospholipids. Liposome properties differ considerably with lipid composition, surface charge, size, and the method of preparation [M1]. These nanostructures increased the efficacy and therapeutic index of the drug, stability via encapsulation. They are non-toxic, flexible, biocompatible, completely biodegradable, and non-immunogenic for systemic and non-systemic administrations [111].

We have highlighted here various methods of liposome preparation. General methods of preparation include four basic stages: (1) drying down lipids dissolved before in the organic solvent, (2) dispersing the lipid in aqueous media, (3) purifying the resultant liposome, (4) analyzing the final product (polydispersity, stability, etc.) [112]. It should be noted that only a few of the conventional liposome preparation procedures are capable of entrapping large quantities of water-soluble agents [113]. Bioactive agents can be entrapped in lipid vesicles by the conventional methods of reverse-phase evaporation technique [114], ether injection/vaporization technique [115,116], and freeze-thaw method [117], just to name a few.

While there are already existing drug delivery systems—e.g., liposomes, available for treatment—the effective loading and retention of the expected drug ratio can be challenging. Liposome preparation protocols vary depending on their intended application. Moreover, the continuous development of synthesis methods and the advancement of the resulting nanostructures is noticeable. In addition to conventional liposome synthesis methods, we would like to present examples of other systems consisting of metals and liposomes. Illes et al. [118] proposed a new type of drug carrier: liposome-coated metal-organic framework (MOF) nanoparticles. These systems combine the advantages of liposomes with an easy and efficient loading process. They also presented the successful synthesis of liposome-coated MOF nanoparticles via the fusion method. The resulting nanostructure, once loaded, show no premature leakage and an efficient release. The successful loading of these nanovehicles with both single and multiple drugs at the same time makes them a promising candidate for use in combination therapy. Wuttke et al. [119] presented novel metal–organic framework nanoparticles encapsulated by a lipid membrane. They demonstrated that the MOF@lipid system can effectively keep dye molecules inside the porous scaffold of the MOF while the lipid bilayer precludes their premature release. The group employed fluorescence microscopy in order to demonstrate the high uptake of lipid-coated nanoparticles by cancer cells. Considering the various ways to synthesize different functionalized MOF nanoparticles as well as the richness of lipids with diverse functions, MOF@lipid nanoparticles have great potential as a novel hybrid nanocarrier system. It is worth noting that, the MOF core is capable of storing different active ligands such as drugs or molecules used for imaging and diagnosis. Moreover, the lipid shell could be used for the incorporation of targeting or shielding molecules (e.g., PEG) as well as for the creation of controlled release mechanisms.

Metal nanocarriers are emerging structures that can enhance the therapeutic activity of many drugs by improved release and targeted potential, however numerous barriers, such as colloidal instability, cellular toxicity, and poor cellular uptake, restrain their applicability in vivo. The nanohybrid systems offer to overcome these limitations and to combine the properties of liposomes and metal nanocarriers for effective theragnostic drug delivery systems [106].

7. Conclusions

The term "theragnostic", introduced in 1998 by J. Funkhouser, means the ability to combine therapy and diagnostics of disease at the same time. Determining this was an important step forward towards personalized medicine. The possibility of combining nanoparticles bound to liposomes with drugs offers a chance of implementing this idea of the approach to the disease [71,120]. LipoMNPs are a promising step forward in a more accurate visualization of the tissues and structures in imaging diagnostics [31]. It is possible because of the biological properties of liposomes and the ability of metallic nanoparticles to facilitate the imaging process [5]. Researchers in recent years analyzed such combination and their effect on several cancer cells of different types both in vitro and in vivo assessing the cytotoxicity and safety of this novel type of contrast agents [70].

The results [6] show the true potential of LipoMNPs in the novel approach to the idea of theragnostic treatment. Before an implementation to the clinic, there is a strong need for further investigation. Researchers should focus on multiple implementations of once synthesized liposome-metallic particles contrast agents in different types of cancer cells. There is a need for finding an optimal synthesis technique, resulting in a more universal hybrid that can be applied to different cancerous tissues with the ability to target drugs suited to the cancer characteristics. Future progress requires possible clinical trials preceded by numerous in vivo studies as well as more research towards the use of clinical diagnostic imaging machines in visualizing the liposome-nanoparticle hybrids [62]. Moreover, there is still a lack of knowledge about biointeractions between nano hybrids and cancer cells. The level of conducted experiments is advanced because investigators have designed new strategies on the basis of previous results.

Author Contributions: Conceptualization, M.M. and A.B.-L.; methodology, M.M.; investigation, M.M., A.B.-L., J.P., M.K.; writing—original draft preparation, M.M., J.P., A.B.-L., M.K.; writing—review and editing, M.M., M.K.; visualization, M.M.; funding acquisition, M.K. All authors have read and agreed to the published version of the manuscript.

Funding: This research and the APC were funded by Adam Mickiewicz University.

Data Availability Statement: MDPI Research Data Policies.

Acknowledgments: This research was supported by the Greater Poland Cancer Centre and the Department of Macromolecular Physics, Faculty of Physics, Adam Mickiewicz University.

Conflicts of Interest: The authors declare no conflict of interest.

References

1. Fahy, E.; Cotter, D.; Sud, M.; Subramaniam, S. Lipid classification, structures and tools. *Biochim. Biophys. Acta BBA Mol. Cell Biol. Lipids* **2011**, *1811*, 637–647. [CrossRef]
2. Flores, J.; White, B.M.; Brea, R.J.; Baskin, J.M.; Devaraj, N.K. Lipids: Chemical tools for their synthesis, modification, and analysis. *Chem. Soc. Rev.* **2020**, *49*, 4602–4614. [CrossRef]
3. Marsh, D. *Handbook of Lipid Bilayers [Internet]*; CRC Press: Boca Raton, FL, USA, 2013; Available online: https://www.taylorfrancis.com/books/9781420088335 (accessed on 13 May 2021).
4. Wang, A.; Lin, W.; Liu, D.; He, C. Application of liposomal technologies for delivery of platinum analogs in oncology. *Int. J. Nanomed.* **2013**, *8*, 3309. [CrossRef]
5. Mi, P. Stimuli-responsive nanocarriers for drug delivery, tumor imaging, therapy and theranostics. *Theranostics* **2020**, *10*, 4557–4588. [CrossRef] [PubMed]
6. Kotouček, J.; Hubatka, F.; Mašek, J.; Kulich, P.; Velínská, K.; Bezděková, J.; Fojtíková, M.; Bartheldyová, E.; Tomečková, A.; Stráská, J.; et al. Preparation of nanoliposomes by microfluidic mixing in herring-bone channel and the role of membrane fluidity in liposomes formation. *Sci. Rep.* **2020**, *10*, 1–11. [CrossRef] [PubMed]
7. El-Shafie, S.; Fahmy, S.A.; Ziko, L.; Elzahed, N.; Shoeib, T.; Kakarougkas, A. Encapsulation of Nedaplatin in Novel PEGylated Liposomes Increases Its Cytotoxicity and Genotoxicity against A549 and U2OS Human Cancer Cells. *Pharmaceutics* **2020**, *12*, 863. [CrossRef] [PubMed]
8. Song, M.; Liu, C.; Chen, S.; Zhang, W. Nanocarrier-Based Drug Delivery for Melanoma Therapeutics. *Int. J. Mol. Sci.* **2021**, *22*, 1873. [CrossRef] [PubMed]
9. Chen, L.; Zhang, J.; Zhou, X.; Yang, S.; Zhang, Q.; Wang, W.; You, Z.; Peng, C.; He, C. Merging metal organic framework with hollow organosilica nanoparticles as a versatile nanoplatform for cancer theranostics. *Acta Biomater.* **2019**, *86*, 406–415. [CrossRef] [PubMed]
10. Ponti, F.; Campolungo, M.; Melchiori, C.; Bono, N.; Candiani, G. Cationic lipids for gene delivery: Many players, one goal. *Chem. Phys. Lipids* **2021**, *235*, 105032. [CrossRef]
11. De Leo, V.; Milano, F.; Agostiano, A.; Catucci, L. Recent Advancements in Polymer/Liposome Assembly for Drug Delivery: From Surface Modifications to Hybrid Vesicles. *Polymers* **2021**, *13*, 1027. [CrossRef]
12. Pietralik, Z.; Krzysztoń, R.; Kida, W.; Andrzejewska, W.; Kozak, M. Structure and Conformational Dynamics of DMPC/Dicationic Surfactant and DMPC/Dicationic Surfactant/DNA Systems. *Int. J. Mol. Sci.* **2013**, *14*, 7642–7659. [CrossRef]
13. Capriotti, G.; Varani, M.; Lauri, C.; Franchi, G.; Pizzichini, P.; Signore, A. Copper-64 labeled nanoparticles for positron emission tomography imaging: A review of the recent literature. *Q. J. Nucl. Med. Mol. Imaging* **2020**, *64*, 346–355. [CrossRef] [PubMed]
14. Lee, J.-H.; Shin, Y.; Lee, W.; Whang, K.; Kim, D.; Lee, L.P.; Choi, J.-W.; Kang, T. General and programmable synthesis of hybrid liposome/metal nanoparticles. *Sci. Adv.* **2016**, *2*, e1601838. [CrossRef] [PubMed]

15. Alwattar, J.K.; Mneimneh, A.T.; Abla, K.K.; Mehanna, M.M.; Allam, A.N. Smart Stimuli-Responsive Liposomal Nanohybrid Systems: A Critical Review of Theranostic Behavior in Cancer. *Pharmaceutics* **2021**, *13*, 355. [CrossRef] [PubMed]
16. Zhu, D.; Wang, Z.; Zong, S.; Zhang, Y.; Chen, C.; Zhang, R.; Yun, B.; Cui, Y. Investigating the Intracellular Behaviors of Liposomal Nanohybrids via SERS: Insights into the Influence of Metal Nanoparticles. *Theranostics* **2018**, *8*, 941–954. [CrossRef]
17. Jahangirian, H.; Kalantari, K.; Izadiyan, Z.; Rafiee-Moghaddam, R.; Shameli, K.; Webster, T.J. A review of small molecules and drug delivery applications using gold and iron nanoparticles. *Int. J. Nanomed.* **2019**, *14*, 1633–1657. [CrossRef]
18. Cadinoiu, A.N.; Rata, D.M.; Atanase, L.I.; Daraba, O.M.; Gherghel, D.; Vochita, G.; Popa, M. Aptamer-Functionalized Liposomes as a Potential Treatment for Basal Cell Carcinoma. *Polymers* **2019**, *11*, 1515. [CrossRef]
19. Thanekar, A.M.; Sankaranarayanan, S.A.; Rengan, A.K. Role of nano-sensitizers in radiation therapy of metastatic tumors. *Cancer Treat Res. Commun.* **2021**, *26*, 100303. [CrossRef]
20. Narang, A.S.; Varia, S. Role of tumor vascular architecture in drug delivery. *Adv. Drug Deliv. Rev.* **2011**, *63*, 640–658. [CrossRef]
21. Barenholz, Y.C. Doxil®—The first FDA-approved nano-drug: Lessons learned. *J. Control. Release* **2012**, *160*, 117–134. [CrossRef]
22. Duggan, S.T.; Keating, G.M. Pegylated Liposomal Doxorubicin: A Review of its use in Metastatic Breast Cancer, Ovarian Cancer, Multiple Myeloma and AIDS-Related Kaposi's Sarcoma. *Drugs* **2011**, *71*, 2531–2558. [CrossRef]
23. Wei, X.; Cohen, R.; Barenholz, Y. Insights into composition/structure/function relationships of Doxil® gained from "high-sensitivity" differential scanning calorimetry. *Eur. J. Pharm. Biopharm.* **2016**, *104*, 260–270. [CrossRef] [PubMed]
24. Lytton-Jean, A.K.R.; Kauffman, K.J.; Kaczmarek, J.C.; Langer, R. Cancer Nanotherapeutics in Clinical Trials. In *Nanotechnology-Based Precision Tools for the Detection and Treatment of Cancer*; Mirkin, C.A., Meade, T.J., Petrosko, S.H., Stegh, A.H., Eds.; Cancer Treatment and Research; Springer: Cham, Switzerland, 2015; Volume 166, pp. 293–322. Available online: http://link.springer.com/10.1007/978-3-319-16555-4_13 (accessed on 9 March 2021).
25. Anselmo, A.C.; Mitragotri, S. Nanoparticles in the clinic: An update. *Bioeng. Transl. Med.* **2019**, *4*, e10143. [CrossRef] [PubMed]
26. Beltrán-Gracia, E.; López-Camacho, A.; Higuera-Ciapara, I.; Velázquez-Fernández, J.B.; Vallejo-Cardona, A.A. Nanomedicine review: Clinical developments in liposomal applications. *Cancer Nanotechnol.* **2019**, *10*, 1–40. [CrossRef]
27. Dri, D.A.; Marianecci, C.; Carafa, M.; Gaucci, E.; Gramaglia, D. Surfactants, Nanomedicines and Nanocarriers: A Critical Evaluation on Clinical Trials. *Pharmaceutics* **2021**, *13*, 381. [CrossRef]
28. Mohamed, M.; Lila, A.S.A.; Shimizu, T.; Alaaeldin, E.; Hussein, A.; Sarhan, H.A.; Szebeni, J.; Ishida, T. PEGylated liposomes: Immunological responses. *Sci. Technol. Adv. Mater.* **2019**, *20*, 710–724. [CrossRef] [PubMed]
29. Patra, J.K.; Das, G.; Fraceto, L.F.; Campos, E.V.R.; Rodriguez-Torres, M.d.P.; Acosta-Torres, L.S.; Diaz-Torres, L.A.; Grillo, R.; Swamy, M.K.; Sharma, S.; et al. Nano based drug delivery systems: Recent developments and future prospects. *J. Nanobiotechnol.* **2018**, *16*, 1–33. [CrossRef]
30. Moosavian, S.A.; Bianconi, V.; Pirro, M.; Sahebkar, A. Challenges and pitfalls in the development of liposomal delivery systems for cancer therapy. *Semin. Cancer Biol.* **2021**, *69*, 337–348. [CrossRef]
31. Singh, R.; Lillard, J.W. Nanoparticle-based targeted drug delivery. *Exp. Mol. Pathol.* **2009**, *86*, 215–223. [CrossRef]
32. Attia, M.F.; Anton, N.; Wallyn, J.; Omran, Z.; Vandamme, T.F. An overview of active and passive targeting strategies to improve the nanocarriers efficiency to tumour sites. *J. Pharm. Pharmacol.* **2019**, *71*, 1185–1198. [CrossRef]
33. Mahor, S.; Collin, E.; Dash, B.C.; Pandit, A. Controlled Release of Plasmid DNA from Hyaluronan Nanoparticles. *Curr. Drug Deliv.* **2011**, *8*, 354–362. [CrossRef]
34. Mi, Y.; Smith, C.C.; Yang, F.; Qi, Y.; Roche, K.C.; Serody, J.S.; Vincent, B.G.; Wang, A.Z. A Dual Immunotherapy Nanoparticle Improves T-Cell Activation and Cancer Immunotherapy. *Adv. Mater.* **2018**, *30*, 1706098. [CrossRef] [PubMed]
35. Bera, K.; Maiti, S.; Maity, M.; Mandal, C.; Maiti, N.C. Porphyrin–Gold Nanomaterial for Efficient Drug Delivery to Cancerous Cells. *ACS Omega* **2018**, *3*, 4602–4619. [CrossRef] [PubMed]
36. Rengan, A.K.; Jagtap, M.; De, A.; Banerjee, R.; Srivastava, R. Multifunctional gold coated thermo-sensitive liposomes for multimodal imaging and photo-thermal therapy of breast cancer cells. *Nanoscale* **2014**, *6*, 916–923. [CrossRef] [PubMed]
37. Mathiyazhakan, M.; Wiraja, C.; Xu, C. A Concise Review of Gold Nanoparticles-Based Photo-Responsive Liposomes for Controlled Drug Delivery. *Nano Micro Lett.* **2018**, *10*, 1–10. [CrossRef]
38. Lungu, I.I.; Grumezescu, A.M.; Volceanov, A.; Andronescu, E. Nanobiomaterials Used in Cancer Therapy: An Up-To-Date Overview. *Molecules* **2019**, *24*, 3547. [CrossRef] [PubMed]
39. Musielak, M. Ocena wpływu dawki i mocy promieniowania jonizującego na komórki raka piersi. The assessment of the effect of ionizing radiation dose and dose rate for breast cancer cells. *Lett. Oncol. Sci.* **2019**, *15*, 117–125. [CrossRef]
40. Musielak, M.; Boś-Liedke, A.; Piotrowski, I.; Kozak, M.; Suchorska, W. The Role of Gold Nanorods in the Response of Prostate Cancer and Normal Prostate Cells to Ionizing Radiation—In vitro Model. *Int. J. Mol. Sci.* **2020**, *22*, 16. [CrossRef]
41. Rey, S.; Schito, L.; Koritzinsky, M.; Wouters, B.G. Molecular targeting of hypoxia in radiotherapy. *Adv. Drug Deliv. Rev.* **2017**, *109*, 45–62. [CrossRef] [PubMed]
42. Brown, J.M.; Wilson, W.R. Exploiting tumour hypoxia in cancer treatment. *Nat. Rev. Cancer* **2004**, *4*, 437–447. [CrossRef] [PubMed]
43. Li, J.; Shang, W.; Li, Y.; Fu, S.; Tian, J.; Lu, L. Advanced nanomaterials targeting hypoxia to enhance radiotherapy. *Int. J. Nanomed.* **2018**, *13*, 5925–5936. [CrossRef]
44. Cheng, Y.; Jiang, C.; Qiu, X.; Wang, K.; Huan, W.; Yuan, A.; Wu, J.; Hu, Y. Perfluorocarbon nanoparticles enhance reactive oxygen levels and tumour growth inhibition in photodynamic therapy. *Nat. Commun.* **2015**, *6*, 8785. [CrossRef]

45. Liu, H.; Xie, Y.; Zhang, Y.; Cai, Y.; Li, B.; Mao, H.; Liu, Y.; Lu, J.; Zhang, L.; Yu, R. Development of a hypoxia-triggered and hypoxic radiosensitized liposome as a doxorubicin carrier to promote synergetic chemo-/radio-therapy for glioma. *Biomaterials* **2017**, *121*, 130–143. [CrossRef] [PubMed]

46. Zhang, R.; Song, X.; Liang, C.; Yi, X.; Song, G.; Chao, Y.; Yang, Y.; Yang, K.; Feng, L.; Liu, Z. Catalase-loaded cisplatin-prodrug-constructed liposomes to overcome tumor hypoxia for enhanced chemo-radiotherapy of cancer. *Biomaterials* **2017**, *138*, 13–21. [CrossRef] [PubMed]

47. Cui, L.; Tse, K.; Zahedi, P.; Harding, S.M.; Zafarana, G.; Jaffray, D.A.; Bristow, R.G.; Allen, C. Hypoxia and Cellular Localization Influence the Radiosensitizing Effect of Gold Nanoparticles (AuNPs) in Breast Cancer Cells. *Radiat. Res.* **2014**, *182*, 475–488. [CrossRef]

48. Yen, H.; Hsu, S.; Tsai, C. Cytotoxicity and Immunological Response of Gold and Silver Nanoparticles of Different Sizes. *Small* **2009**, *5*, 1553–1561. [CrossRef]

49. Chithrani, D.B.; Dunne, M.; Stewart, J.; Allen, C.; Jaffray, D.A. Cellular uptake and transport of gold nanoparticles incorporated in a liposomal carrier. *Nanomed. Nanotechnol. Biol. Med.* **2010**, *6*, 161–169. [CrossRef]

50. Lajunen, T.; Viitala, L.; Kontturi, L.-S.; Laaksonen, T.; Liang, H.; Vuorimaa-Laukkanen, E.; Viitala, T.; Le Guével, X.; Yliperttula, M.; Murtomäki, L.; et al. Light induced cytosolic drug delivery from liposomes with gold nanoparticles. *J. Control. Release* **2015**, *203*, 85–98. [CrossRef]

51. Zhang, N.; Chen, H.; Liu, A.-Y.; Shen, J.-J.; Shah, V.; Zhang, C.; Hong, J.; Ding, Y. Gold conjugate-based liposomes with hybrid cluster bomb structure for liver cancer therapy. *Biomaterials* **2016**, *74*, 280–291. [CrossRef] [PubMed]

52. Sharifabad, M.E.; Mercer, T.; Sen, T. Drug-loaded liposome-capped mesoporous core–shell magnetic nanoparticles for cellular toxicity study. *Nanomedicine* **2016**, *11*, 2757–2767. [CrossRef] [PubMed]

53. Bao, Q.-Y.; Zhang, N.; Geng, D.-D.; Xue, J.-W.; Merritt, M.; Zhang, C.; Ding, Y. The enhanced longevity and liver targetability of Paclitaxel by hybrid liposomes encapsulating Paclitaxel-conjugated gold nanoparticles. *Int. J. Pharm.* **2014**, *477*, 408–415. [CrossRef]

54. Xing, S.; Zhang, X.; Luo, L.; Cao, W.; Li, L.; He, Y.; An, J.; Gao, D. Doxorubicin/gold nanoparticles coated with liposomes for chemo-photothermal synergetic antitumor therapy. *Nanotechnology* **2018**, *29*, 405101. [CrossRef]

55. Zheng, X.-C.; Ren, W.; Zhang, S.; Zhong, T.; Duan, X.-C.; Yin, Y.-F.; Xu, M.-Q.; Hao, Y.-L.; Li, Z.-T.; Li, H.; et al. The theranostic efficiency of tumor-specific, pH-responsive, peptide-modified, liposome-containing paclitaxel and superparamagnetic iron oxide nanoparticles. *Int. J. Nanomed.* **2018**, *13*, 1495–1504. [CrossRef] [PubMed]

56. Bromma, K.; Rieck, K.; Kulkarni, J.; O'Sullivan, C.; Sung, W.; Cullis, P.; Schuemann, J.; Chithrani, D.B. Use of a lipid nanoparticle system as a Trojan horse in delivery of gold nanoparticles to human breast cancer cells for improved outcomes in radiation therapy. *Cancer Nanotechnol.* **2019**, *10*, 1. [CrossRef]

57. Wang, M.; Liu, Y.; Zhang, X.; Luo, L.; Li, L.; Xing, S.; He, Y.; Cao, W.; Zhu, R.; Gao, D. Gold nanoshell coated thermo-pH dual responsive liposomes for resveratrol delivery and chemo-photothermal synergistic cancer therapy. *J. Mater. Chem. B* **2017**, *5*, 2161–2171. [CrossRef]

58. Zhao, H.; Zhao, B.; Wu, L.; Xiao, H.; Ding, K.; Zheng, C.; Song, Q.; Sun, L.; Wang, L.; Zhang, Z. Amplified Cancer Immunotherapy of a Surface-Engineered Antigenic Microparticle Vaccine by Synergistically Modulating Tumor Microenvironment. *ACS Nano* **2019**, *13*, 12553–12566. [CrossRef] [PubMed]

59. Ma, Y.; Peng, X.; Wang, L.; Li, H.; Cheng, W.; Zheng, X.; Liu, Y. Cetuximab-conjugated perfluorohexane/gold nanoparticles for low intensity focused ultrasound diagnosis ablation of thyroid cancer treatment. *Sci. Technol. Adv. Mater.* **2020**, *21*, 856–866.

60. Martínez-González, R.; Estelrich, J.; Busquets, M. Liposomes Loaded with Hydrophobic Iron Oxide Nanoparticles: Suitable T2 Contrast Agents for MRI. *Int. J. Mol. Sci.* **2016**, *17*, 1209. [CrossRef]

61. Lacombe, S.; Porcel, E.; Scifoni, E. Particle therapy and nanomedicine: State of art and research perspectives. *Cancer Nanotechnol.* **2017**, *8*, 1–17. [CrossRef] [PubMed]

62. Aryasomayajula, B.; Salzano, G.; Torchilin, V.P. Multifunctional Liposomes. In *Cancer Nanotechnology*; Methods in Molecular Biology; Zeineldin, R., Ed.; Springer: New York, NY, USA, 2017; Volume 1530, pp. 41–61. Available online: http://link.springer.com/10.1007/978-1-4939-6646-2_3 (accessed on 10 March 2021).

63. Erdogan, S.; Torchilin, V.P. Gadolinium-Loaded Polychelating Polymer-Containing Tumor-Targeted Liposomes. In *Liposomes*; Weissig, V., Ed.; Methods in Molecular Biology; Humana Press: Totowa, NJ, USA, 2010; Volume 605, pp. 321–334. Available online: http://link.springer.com/10.1007/978-1-60327-360-2_22 (accessed on 10 March 2021).

64. Carvalho, A.; Gonçalves, M.C.; Corvo, M.L.; Martins, M.B.F. Development of New Contrast Agents for Imaging Function and Metabolism by Magnetic Resonance Imaging. *Magn. Reson. Insights* **2017**, *10*. [CrossRef]

65. Xiao, Y.-D.; Paudel, R.; Liu, J.; Ma, C.; Zhang, Z.-S.; Zhou, S.-K. MRI contrast agents: Classification and application (Review). *Int. J. Mol. Med.* **2016**, *38*, 1319–1326. [CrossRef] [PubMed]

66. Lusic, H.; Grinstaff, M.W. X-ray-Computed Tomography Contrast Agents. *Chem. Rev.* **2013**, *113*, 1641–1666. [CrossRef] [PubMed]

67. Ignee, A.; Atkinson, N.S.; Schuessler, G.; Dietrich, C. Ultrasound contrast agents. *Endosc. Ultrasound* **2016**, *5*, 355. [CrossRef] [PubMed]

68. Man, F.; Gawne, P.; de Rosales, R.T. Nuclear imaging of liposomal drug delivery systems: A critical review of radiolabelling methods and applications in nanomedicine. *Adv. Drug Deliv. Rev.* **2019**, *143*, 134–160. [CrossRef]

69. Pellico, J.; Gawne, P.J.; de Rosales, R.T.M. Radiolabelling of nanomaterials for medical imaging and therapy. *Chem. Soc. Rev.* **2021**, *50*, 3355–3423. [CrossRef]

70. He, Y.; Zhang, L.; Song, C.; Zhu, D. Design of multifunctional magnetic iron oxide nanoparticles/mitoxantrone-loaded liposomes for both magnetic resonance imaging and targeted cancer therapy. *Int. J. Nanomed.* **2014**, *9*, 4055. [CrossRef]

71. Prasad, R.; Jain, N.K.; Yadav, A.S.; Chauhan, D.S.; Devrukhkar, J.; Kumawat, M.K.; Shinde, S.; Gorain, M.; Thakor, A.S.; Kundu, G.C.; et al. Liposomal nanotheranostics for multimode targeted in vivo bioimaging and near-infrared light mediated cancer therapy. *Commun. Biol.* **2020**, *3*, 1–14. [CrossRef] [PubMed]

72. Aranda-Lara, L.; Morales-Avila, E.; Luna-Gutiérrez, M.A.; Olivé-Alvarez, E.; Isaac-Olivé, K. Radiolabeled liposomes and lipoproteins as lipidic nanoparticles for imaging and therapy. *Chem. Phys. Lipids* **2020**, *230*, 104934. [CrossRef] [PubMed]

73. German, S.V.; Navolokin, N.A.; Kuznetsova, N.R.; Zuev, V.V.; Inozemtseva, O.A.; Anis'kov, A.A.; Volkova, E.K.; Bucharskay, A.B.; Maslyakova, G.N.; Fakhrullin, R.F. Liposomes loaded with hydrophilic magnetite nanoparticles: Preparation and application as contrast agents for magnetic resonance imaging. *Colloids Surf. B Biointerfaces* **2015**, *135*, 109–115. [CrossRef]

74. Massart, R. Preparation of aqueous magnetic liquids in alkaline and acidic media. *IEEE Trans. Magn.* **1981**, *17*, 1247–1248. [CrossRef]

75. Sonkar, R.; Jha, A.; Viswanadh, M.K.; Burande, A.S.; Pawde, D.M.; Patel, K.K.; Singh, M.; Koch, B.; Muthu, M.S. Gold liposomes for brain-targeted drug delivery: Formulation and brain distribution kinetics. *Mater. Sci. Eng. C* **2021**, *120*, 111652. [CrossRef] [PubMed]

76. Landowski, L.M.; Niego, B.; Sutherland, B.A.; Hagemeyer, C.E.; Howells, D.W. Applications of Nanotechnology in the Diagnosis and Therapy of Stroke. *Semin. Thromb. Hemost.* **2020**, *46*, 592–605. [CrossRef] [PubMed]

77. Vieira, D.; Gamarra, L. Getting into the brain: Liposome-based strategies for effective drug delivery across the blood–brain barrier. *Int. J. Nanomed.* **2016**, *11*, 5381–5414. [CrossRef]

78. Tanifum, E.A.; Ghaghada, K.; Vollert, C.; Head, E.; Eriksen, J.L.; Annapragada, A. A Novel Liposomal Nanoparticle for the Imaging of Amyloid Plaque by Magnetic Resonance Imaging. *J. Alzheimers Dis.* **2016**, *52*, 731–745. [CrossRef] [PubMed]

79. Tomitaka, A.; Arami, H.; Huang, Z.; Raymond, A.; Rodriguez, E.; Cai, Y.; Febo, M.; Takemura, Y.; Nair, M. Hybrid magneto-plasmonic liposomes for multimodal image-guided and brain-targeted HIV treatment. *Nanoscale* **2018**, *10*, 184–194. [CrossRef] [PubMed]

80. Ji, B.; Wang, M.; Gao, D.; Xing, S.; Li, L.; Liu, L.; Zhao, M.; Qi, X.; Dai, K. Combining nanoscale magnetic nimodipine liposomes with magnetic resonance image for Parkinson's disease targeting therapy. *Nanomedicine* **2017**, *12*, 237–253. [CrossRef] [PubMed]

81. Mieszawska, A.J.; Mulder, W.J.M.; Fayad, Z.A.; Cormode, D.P. Multifunctional Gold Nanoparticles for Diagnosis and Therapy of Disease. *Mol. Pharm.* **2013**, *10*, 831–847. [CrossRef]

82. Zhang, K.; Du, X.; Yu, K.; Zhang, K.; Zhou, Y. Application of novel targeting nanoparticles contrast agent combined with contrast-enhanced computed tomography during screening for early-phase gastric carcinoma. *Exp. Ther. Med.* **2018**, *15*, 47–54. [CrossRef]

83. Cao, Y.; Xu, L.; Kuang, Y.; Xiong, D.; Pei, R. Gadolinium-based nanoscale MRI contrast agents for tumor imaging. *J. Mater. Chem. B* **2017**, *5*, 3431–3461. [CrossRef]

84. Kostevšek, N.; Cheung, C.C.L.; Serša, I.; Kreft, M.E.; Monaco, I.; Franchini, M.C.; Vidmar, J.; Al-Jamal, W.T. Magneto-Liposomes as MRI Contrast Agents: A Systematic Study of Different Liposomal Formulations. *Nanomaterials* **2020**, *10*, 889. [CrossRef]

85. Frascione, D.; Diwoky, C.; Almer, G.; Opriessnig, P.; Vonach, C.; Gradauer, K.; Leitinger, G.; Mangge, H.; Stollberger, R.; Prassl, R. Ultrasmall superparamagnetic iron oxide (USPIO)-based liposomes as magnetic resonance imaging probes. *Int. J. Nanomed.* **2012**, *7*, 2349.

86. Yang, C.-T.; Hattiholi, A.; Selvan, S.T.; Yan, S.X.; Fang, W.-W.; Chandrasekharan, P.; Koteswaraiah, P.; Herold, C.J.; Gulyás, B.; Aw, S.E.; et al. Gadolinium-based bimodal probes to enhance T1-Weighted magnetic resonance/optical imaging. *Acta Biomater.* **2020**, *110*, 15–36. [CrossRef]

87. Marasini, R.; Thanh Nguyen, T.D.; Aryal, S. Integration of gadolinium in nanostructure for contrast enhanced-magnetic resonance imaging. *Wiley Interdiscip. Rev. Nanomed. Nanobiotechnol.* **2020**, *12*, e1580. [CrossRef] [PubMed]

88. Cheung, C.; Al-Jamal, W.T. Liposomes-Based Nanoparticles for Cancer Therapy and Bioimaging. In *Nanooncology*; Gonçalves, G., Tobias, G., Eds.; Nanomedicine and Nanotoxicology; Springer: Cham, Switzerland, 2018; pp. 51–87. Available online: http://link.springer.com/10.1007/978-3-319-89878-0_2 (accessed on 10 March 2021).

89. Wang, L. Early Diagnosis of Breast Cancer. *Sensors* **2017**, *17*, 1572. [CrossRef] [PubMed]

90. Virani, N.; Hendrick, A.; Wu, D.; Southard, B.; Babb, J.; Liu, H.; Awasthi, V.; Harrison, R.G. Enhanced computed tomography imaging of breast cancer via phosphatidylserine targeted gold nanoparticles. *Biomed. Phys. Eng. Express* **2019**, *5*, 065019. [CrossRef]

91. Rengan, A.K.; Bukhari, A.; Pradhan, A.; Malhotra, R.; Banerjee, R.; Srivastava, R.; De, A. In vivo Analysis of Biodegradable Liposome Gold Nanoparticles as Efficient Agents for Photothermal Therapy of Cancer. *Nano Lett.* **2015**, *15*, 842–848. [CrossRef]

92. Musielak, M.; Piotrowski, I.; Suchorska, W.M. Superparamagnetic iron oxide nanoparticles (SPIONs) as a multifunctional tool in various cancer therapies. *Rep. Pract. Oncol. Radiother.* **2019**, *24*, 307–314. [CrossRef] [PubMed]

93. Zhang, L.; Zhou, H.; Belzile, O.; Thorpe, P.; Zhao, D. Phosphatidylserine-targeted bimodal liposomal nanoparticles for in vivo imaging of breast cancer in mice. *J. Control. Release* **2014**, *183*, 114–123. [CrossRef]

94. Patil-Sen, Y.; Torino, E.; De Sarno, F.; Ponsiglione, A.; Chhabria, V.N.; Ahmed, W.; Mercer, T. Biocompatible superparamagnetic core-shell nanoparticles for potential use in hyperthermia-enabled drug release and as an enhanced contrast agent. *Nanotechnology* **2020**, *31*, 375102. [CrossRef]
95. Lee, H.; Shields, A.F.; Siegel, B.A.; Miller, K.D.; Krop, I.; Ma, C.X.; Lorusso, P.M.; Munster, P.N.; Campbell, K.; Gaddy, D.F.; et al. 64 Cu-MM-302 Positron Emission Tomography Quantifies Variability of Enhanced Permeability and Retention of Nanoparticles in Relation to Treatment Response in Patients with Metastatic Breast Cancer. *Clin. Cancer Res.* **2017**, *23*, 4190–4202. [CrossRef]
96. Bray, F.; Ferlay, J.; Soerjomataram, I.; Siegel, R.L.; Torre, L.A.; Jemal, A. Global cancer statistics 2018: GLOBOCAN estimates of incidence and mortality worldwide for 36 cancers in 185 countries. *CA Cancer J. Clin.* **2018**, *68*, 394–424. [CrossRef]
97. Ravoori, M.K.; Singh, S.; Bhavane, R.; Sood, A.K.; Anvari, B.; Bankson, J.; Annapragada, A.; Kundra, V. Multimodal Magnetic Resonance and Near-Infrared-Fluorescent Imaging of Intraperitoneal Ovarian Cancer Using a Dual-Mode-Dual-Gadolinium Liposomal Contrast Agent. *Sci. Rep.* **2016**, *6*, 38991. [CrossRef] [PubMed]
98. Chen, Q.; Shang, W.; Zeng, C.; Wang, K.; Liang, X.; Chi, C.; Liang, X.; Yang, J.; Fang, C.; Tian, J. Theranostic imaging of liver cancer using targeted optical/MRI dual-modal probes. *Oncotarget* **2017**, *8*, 32741. [CrossRef] [PubMed]
99. Šimečková, P.; Hubatka, F.; Kotouček, J.; Turánek Knötigová, P.; Mašek, J.; Slavík, J.; Kováč, O.; Neča, J.; Kulich, P.; Hrebík, D.; et al. Gadolinium labelled nanoliposomes as the platform for MRI theranostics: In vitro safety study in liver cells and macrophages. *Sci. Rep.* **2020**, *10*, 1–13. [CrossRef]
100. Lorente, C.; Cabeza, L.; Clares, B.; Ortiz, R.; Halbaut, L.; Delgado, Á.V.; Perazzoli, G.; Prados, J.; Arias, J.L.; Melguizo, C. Formulation and in vitro evaluation of magnetoliposomes as a potential nanotool in colorectal cancer therapy. *Colloids Surf. B Biointerfaces* **2018**, *171*, 553–565. [CrossRef] [PubMed]
101. Blocker, S.J.; Douglas, K.A.; Polin, L.A.; Lee, H.; Hendriks, B.S.; Lalo, E.; Chen, W.; Shields, A.F. Liposomal 64 Cu-PET Imaging of Anti-VEGF Drug Effects on Liposomal Delivery to Colon Cancer Xenografts. *Theranostics* **2017**, *7*, 4229–4239. [CrossRef]
102. Thébault, C.J.; Ramniceanu, G.; Michel, A.; Beauvineau, C.; Girard, C.; Seguin, J.; Mignet, N.; Ménager, C.; Doan, B.-T. In vivo Evaluation of Magnetic Targeting in Mice Colon Tumors with Ultra-Magnetic Liposomes Monitored by MRI. *Mol. Imaging Biol.* **2019**, *21*, 269–278. [CrossRef]
103. Thébault, C.J.; Ramniceanu, G.; Boumati, S.; Michel, A.; Seguin, J.; Larrat, B.; Mignet, N.; Ménager, C.; Doan, B.-T. Theranostic MRI liposomes for magnetic targeting and ultrasound triggered release of the antivascular CA4P. *J. Control. Release Off. J. Control. Release Soc.* **2020**, *322*, 137–148. [CrossRef]
104. Mitchell, N.; Kalber, T.L.; Cooper, M.S.; Sunassee, K.; Chalker, S.L.; Shaw, K.P.; Ordidge, K.L.; Badar, A.; Janes, S.M.; Blower, P.J.; et al. Incorporation of paramagnetic, fluorescent and PET/SPECT contrast agents into liposomes for multimodal imaging. *Biomaterials* **2013**, *34*, 1179–1192. [CrossRef]
105. Verma, A.; Stellacci, F. Effect of Surface Properties on Nanoparticle–Cell Interactions. *Small* **2010**, *6*, 12–21. [CrossRef]
106. Skupin-Mrugalska, P.; Minko, T. Development of Liposomal Vesicles for Osimertinib Delivery to EGFR Mutation—Positive Lung Cancer Cells. *Pharmaceutics* **2020**, *12*, 939. [CrossRef]
107. Mozafari, M.R. Liposomes: An Overview of Manufacturing Techniques. *Cell. Mol. Biol. Lett.* **2005**, *10*, 711.
108. Peer, D.; Karp, J.M.; Hong, S.; Farokhazad, O.C.; Margalit, R.; Langer, R. Nanocarriers as an emerging platform for cancer therapy. *Nat. Nanotechnol.* **2007**, *2*, 751–760. [CrossRef] [PubMed]
109. Davis, M.E.; Chen, Z.; Shin, D.M. Nanoparticle therapeutics: An emerging treatment modality for cancer. *Nat. Rev. Drug Discov.* **2008**, *7*, 771–782. [CrossRef] [PubMed]
110. Cati, K.K.; Belowich, M.E.; Liong, M.; Ambrogio, M.W.; Lau, Y.A.; Khativ, H.A.; Zink, J.I.; Khashab, N.M.; Stoddart, J.F. Mechanised nanoparticles for drug delivery. *Nanoscale* **2009**, *1*, 16–39. [CrossRef]
111. Himanshu, A.; Sitasharan, P.; Singhai, A.K. Liposomes as drug carriers. *Life Sci.* **2011**, *2*, 945–951.
112. Riaz, M. Liposome preparation method. *Pak. J. Pharm. Sci.* **1996**, *9*, 65–77. [PubMed]
113. Vemuri, S.; Rhodes, C.T. Preparation and characterization of liposomes as therapeutic delivery systems: A review. *Pharm. Acta Helv.* **1995**, *70*, 95–111. [CrossRef]
114. Ann, N.Y. Acad Liposomes and their uses in biology and medicine. *Science* **1978**, *308*, 1–462.
115. Szoka, F., Jr.; Papahadjopoulos, D. Procedure for preparation of liposomes with large internal aqueous space and high capture by reverse-phase evaporation. *Proc. Natl. Acad. Sci. USA* **1978**, *75*, 4194–4198. [CrossRef]
116. Deamer, D.; Bangham, A.D. Large volume liposomes by an ether vaporization method. *Biochim. Biophys. Acta* **1976**, *443*, 629–634. [CrossRef]
117. Schieren, H.; Rudolph, S.; Finkelstein, M.; Coleman, P.; Weissmann, G. Comparison of large unilamellar vesicles prepared by a petroleum ether vaporization method with multilamellar vesicles: ESR, diffusion and entrapment analyses. *Biochim. Biophys. Acta* **1978**, *542*, 137–153. [CrossRef]
118. Illes, B.; Wuttke, S.; Engelke, H. Liposome-Coated Iron Fumarate Metal-Organic Framework Nanoparticles for Combination Therapy. *Nanomaterials* **2017**, *7*, 351. [CrossRef] [PubMed]
119. Wuttke, S.; Braig, S.; Preiß, T.; Zimpel, A.; Sicklinger, J.; Bellomo, C.; Rädler, J.O.; Aollmarb, A.M.; Bein, T. MOF nanoparticles coated by lipid bilayers and their uptake by cancer cells. *Chem. Commun.* **2015**, *51*, 15752–15755. [CrossRef]
120. Sangtani, A.; Nag, O.K.; Field, L.D.; Breger, J.C.; Delehanty, J.B. Multifunctional nanoparticle composites: Progress in the use of soft and hard nanoparticles for drug delivery and imaging. *Wiley Interdiscip. Rev. Nanomed. Nanobiotechnol.* **2017**, *9*, e1466. [CrossRef] [PubMed]

International Journal of
Molecular Sciences

Review

Green Synthesis of Selenium and Tellurium Nanoparticles: Current Trends, Biological Properties and Biomedical Applications

Marjorie C. Zambonino [1], Ernesto Mateo Quizhpe [1], Francisco E. Jaramillo [1], Ashiqur Rahman [2,3], Nelson Santiago Vispo [1], Clayton Jeffryes [3] and Si Amar Dahoumane [1,4,*]

[1] School of Biological Sciences and Engineering, Yachay Tech University, Hacienda San José s/n, San Miguel de Urcuquí 100119, Ecuador; marjorie.zambonino@yachaytech.edu.ec (M.C.Z.); ernesto.quizhpe@yachaytech.edu.ec (E.M.Q.); francisco.jaramillo@yachaytech.edu.ec (F.E.J.); nvispo@yachaytech.edu.ec (N.S.V.)

[2] Center for Midstream Management and Science, Lamar University, Beaumont, TX 77710, USA; arahman2@lamar.edu

[3] Center for Advances in Water and Air Quality & The Dan F. Smith Department of Chemical Engineering, Lamar University, Beaumont, TX 77710, USA; cjeffryes@lamar.edu

[4] Department of Chemical Engineering, Polytechnique Montréal, C.P. 6079, Succ. Centre-ville, Montréal, QC H3C 3A7, Canada

* Correspondence: sa.dahoumane@gmail.com or si-amar.dahoumane@polymtl.ca

Citation: Zambonino, M.C.; Quizhpe, E.M.; Jaramillo, F.E.; Rahman, A.; Santiago Vispo, N.; Jeffryes, C.; Dahoumane, S.A. Green Synthesis of Selenium and Tellurium Nanoparticles: Current Trends, Biological Properties and Biomedical Applications. *Int. J. Mol. Sci.* **2021**, *22*, 989. https://doi.org/10.3390/ijms22030989

Received: 28 December 2020
Accepted: 18 January 2021
Published: 20 January 2021

Publisher's Note: MDPI stays neutral with regard to jurisdictional claims in published maps and institutional affiliations.

Abstract: The synthesis and assembly of nanoparticles using green technology has been an excellent option in nanotechnology because they are easy to implement, cost-efficient, eco-friendly, risk-free, and amenable to scaling up. They also do not require sophisticated equipment nor well-trained professionals. Bionanotechnology involves various biological systems as suitable nanofactories, including biomolecules, bacteria, fungi, yeasts, and plants. Biologically inspired nanomaterial fabrication approaches have shown great potential to interconnect microbial or plant extract biotechnology and nanotechnology. The present article extensively reviews the eco-friendly production of metalloid nanoparticles, namely made of selenium (SeNPs) and tellurium (TeNPs), using various microorganisms, such as bacteria and fungi, and plants' extracts. It also discusses the methodologies followed by materials scientists and highlights the impact of the experimental sets on the outcomes and shed light on the underlying mechanisms. Moreover, it features the unique properties displayed by these biogenic nanoparticles for a large range of emerging applications in medicine, agriculture, bioengineering, and bioremediation.

Keywords: SeNPs; TeNPs; nanofactories; biosynthesis; biomass; mechanistic aspects; bioactivity; bioapplications; sustainability

1. Introduction

Nanotechnology has become one of the most promising interdisciplinary technologies, connecting physics, chemistry, biology, materials science, electronics, and medicine [1]. The quantity of engineered nanoparticles (NPs) is expected to increase significantly in the years to come as they receive growing global attention due to their attractive properties, multifunctionalities, unique characteristics, and innovative applications in different industrial and scientific domains [2–6]. Several physical and chemical methods have been extensively explored to fabricate NPs, such as laser ablation [7,8], coprecipitation [9,10], hydrothermal route [11,12], solvothermal route [13,14], sol-gel process [15,16], polyol process [17,18], electrochemical methods [19,20], sonochemistry [21,22], and microwave-assisted methods [23,24]. However, the use of toxic chemicals and/or the generation of harmful byproducts limit their application in clinical fields. Thus, materials scientists rely on a plethora of precursors and reducing/stabilizing agents from biological resources to produce environmentally friendly NPs to lower or eliminate the use and generation of hazardous chemicals [25–27]. Such biosystems include natural biomolecules [28–31], plants [32–34], algae [35–40], bacteria [41,42], yeast and fungi [43,44]; these biological entities exhibit high reductive capacities due to the presence of enzymes, proteins, lipids, sugars, and metabolites. Overall, the biological-mediated synthesis of metallic and metalloid nanoparticles is a single-step, bioreductive process that follows a bottom-up approach and involves the reduction of metal ions dissolved usually in aqueous solutions at room or mild temperature and atmospheric pressure [33,45,46].

Nanoparticles have remarkable advantages over bulk materials, such as a larger surface area, higher surface energy, spatial confinement and reduced imperfections [47]. Their features, such as the size, morphology, chemical composition, surface functionality, and crystallinity, play an important role in determining their potential applications in numerous fields, such as biomedicine, nanobiotechnology, agriculture, pharmacology, optoelectronics, etc. [48–50]. Over the past few years, selenium and tellurium have become chalcogenides of great interest owing to their unique photoconductive and thermoconductive properties [51]. They are known as "E-tech" elements with characteristics similar to that of sulfur and are fundamental constituents of photovoltaic solar panels, electronic devices, and alloys [51,52].

Selenium is an essential trace element for life [53,54]. It is an allotropic nonmetal usually red and grey present in nature under three forms: amorphous, crystalline trigonal with helical chains, and crystalline monoclinic (α, β, γ) with Se8 rings [6]. The synthesis of selenium nanoparticles (SeNPs) by microorganisms and plants induces variations in their crystallinity, morphology, and size due to the diversity of the followed biological methodologies, reducing enzymes and biosurfactants [55]. Although some investigations have reported the biosynthesis of SeNPs under aerobic and anaerobic conditions, aerobic microorganisms have generated the ideal outcomes [56]. The process typically reduces selenite (Se(IV)) or selenate (Se(VI)) species into elemental selenium (Se(0)). Se-based nanomaterials exhibit chemotherapeutic and chemopreventive features, antioxidant properties, low cytotoxicity, and anticancer efficacy, making them a useful tool in nanomedicine [57,58]. They also have a strong, dose-dependent antimicrobial effect on various microorganisms' growth and propagation [56].

Tellurium is a metalloid present in nature as a soluble oxyanion under four oxidation states: -2 (H_2Te), $+2$ (TeO_2^{2-}), $+4$ (TeO_3^{2-}), and $+6$ (TeO_4^{2-}). It can be toxic in very low concentrations (1 µg mL^{-1}) [59]. Recently, the conversion of tellurite to black elemental tellurium including extra/intracellular accumulation, volatilization, and methylation, has piqued the interest of researchers [60]. Tellurium nanoparticles (TeNPs) have become of interest in research and industry due to their excellent biocompatibility [61], antimicrobial, antioxidant and anticancer activity [62,63], and their ability to reduce cholesterol and triglyceride levels [64]. The high efficiency of microorganisms to transform metalloid oxyanions to less toxic elemental forms results in toxicity reduction and increased selenium and tellurium bioavailability [65]. Moreover, the same microorganisms provide exceptional bioremediation tools and technological applications due to their ability to biorecover the cations of these metalloids and promote the subsequent production of Se and Te nanomaterials [51,66–70]. The principal applications of biogenic SeNPs and TeNPs are summarized in Figure 1.

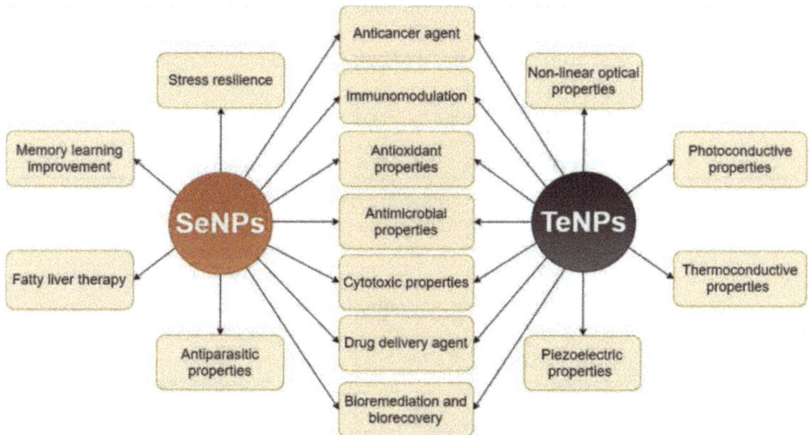

Figure 1. Applications of selenium nanoparticles (SeNPs) and tellurium nanoparticles (TeNPs).

The present review aims at providing a comprehensive insight upon the emerging routes implemented for the biosynthesis of SeNPs and TeNPs using various microorganisms and plants via different methodologies. It also elaborates on the underlying mechanisms that govern these bioprocesses, describes the unique biological properties of these metalloids' nanomaterials, and discusses their diverse applications in the biomedical field.

2. Green Synthesis of Inorganic Nanoparticles Using Microorganisms

The holy grail in nanotechnology consists in elaborating cost-effective and environmentally friendly approaches for the synthesis of nanomaterials that modulate their size, morphology, assembly, and colloidal stability [71]. The biosynthesis of inorganic nanoparticles is generally implemented in aqueous media at room temperature or mild heating and atmospheric pressure [26]. Those are simple conditions that engage the production of high-quality nanomaterials. In that sense, these NP biosynthetic methods that rely on microorganisms, such as bacteria, fungi, microalgae, yeast and viruses, and plants are fully eco-friendly approaches [42]. These microbial and plant-assisted methodologies provide easy, inexpensive, and nontoxic routes to yield NPs that exhibit a diversity of sizes, shapes, and composition along with unique physicochemical attributes and outstanding biological properties.

Nature has devised several reliable, cost-effective, nontoxic, clean, and ecofriendly biological techniques to produce SeNPs and TeNPs [72,73]. Green nanotechnology employs natural biological resources, such as bacteria, fungi, yeast, algae, plants, and viruses, and, most often, water as the solvent. To achieve the fabrication of monodispersed, highly stable NPs with a desired size and controlled morphology, the biomolecular machinery availability is needed [74]. The main benefit is that microorganisms are effective tools that act as nanofactories avoiding thus the use of and/or generation of harsh, toxic chemicals. They also have the ability to accumulate and detoxify heavy metals due to various reductase enzymes that reduce metal salts to metallic nanoparticles with a narrow size distribution and, therefore, less polydispersity [75,76]. Biological processes usually occur at mild conditions, i.e., ambient temperature and atmospheric pressure, and do not require skilled professionals nor sophisticated equipment making them amenable to controlled and scale-up procedures [74]. However, they also present some limitations related to NP composition, crystallinity, morphology, and size distribution.

Recently, the extra- and intra-cellular microbial production of metallic/metalloid NPs have been studied [27,33,41,43–45,77–79]. In extracellular formation, the added metal salts are transformed into NPs in the culture broth or attached to the cell membrane. Conversely, the intracellular process first transports the metal ions through the cell membrane, i.e., internalization, to the cell interior where the nanoparticles are formed. Then, these internally formed NPs are released to the supernatant using several procedures, such as the cell lysis, to be recovered and purified [72,74,80]. The following sections describe the outstanding role played by different microorganisms, namely bacteria, fungi and yeast, and plants in the biosynthesis of SeNPs and TeNPs.

3. Parameters Affecting the Green Synthesis of Metalloid Nanoparticles

Various factors, such as the precursor, biomass type, temperature, pH, and reaction time, govern the production and stabilization of SeNPs and TeNPs by microorganisms. The pH is an important factor that determines the shape, size, and composition of the NPs [80,81]. For instance, Wu et al. reported the formation, at pH 8, of effectively dispersed spherical SeNPs of 60 nm in diameter in epigallocatechin-3-gallate (EGCG). However, the protonation of the EGCG in acidic conditions (pH 1.0) rapidly induced the aggregations of these NPs as their dimensions reached 300 nm within the first 3 min resulting eventually in the loss of their nanoscale features [82]. According to Akçay and Avcı, the maximum yield occurred at pH 7 and 8 [83] while Kuroda et al. reported the optimum reduction rate at pH values of 6–9 for selenite and 7–9 for selenate [84]. Wadhwani et al. demonstrated the synthesis of SeNPs in a pH range of 4–10 [58]. No synthesis occurred at pH 2 and 1.5 mM of sodium selenite due to the presence of less functional groups that are required for the reduction process. The precursor concentration can also control the NP shape and size. For example, the same study by Wadhwani et al. proved that spherical and rod morphologies of the SeNPs appear at 3.0 mM Na_2SeO_3 while only spheres are observed at 1.5 mM of the same precursor [58].

Green approaches for the synthesis of SeNPs and TeNPs are cost- and energy-efficient, requiring lower temperatures compared to their chemical or physical counterparts [58]. The temperature is found to be a factor that leads to the formation and then aggregation of SeNPs [85]. For instance, the reduction process occurs at temperatures up to 40 °C using *Acinetobacter* sp. SW30 and higher temperatures (around 80 °C and 100 °C) may lead to the aggregation of the SeNPs into nanorods [58]. It is relevant to indicate that, in the case of bacteria, elevated temperatures (>45 °C) may block the normal biosynthesis of SeNPs [86]. Likewise, high temperatures (over 60 °C) and low temperatures (below 25 °C) reduce the efficiency of inorganic NP production using fungi [87,88]. Moreover, the incubation time plays a significant role in the quality and morphology of the NPs. In the case of most bacteria, the average incubation time ranges from 24 to 72 h, but long incubation periods may cause NPs to aggregate, grow, or shrink [89]. The properties of NPs may have a lifetime, but extended exposure times can induce metastable changes to the surface morphology, crystallinity, and optical absorption of nanostructures [90].

The concentration of precursors and reducing/surfactant agents are also critical to control the growth and morphology of the nanoparticles [26,91–93]. The precursor concentration can have a strong influence on the color intensity and rate of change during the NP formation process [94,95]. Se (Na_2SeO_4, Na_2SeO_3, SeO_2) and Te (Na_2TeO_3, K_2TeO_3) precursors along with the pH and reaction time are tuned to produce metalloid nanostructures of different sizes [49,96] and shapes (e.g., SeNPs, Te nanorods (TeNRs), Te nanowires (TeNWs), and Te nanotubes (TeNTs)) [97,98]. Additionally, the size of SeNPs is determined by the initial precursor concentration [99]. The tolerance towards selenium oxyanions can be evaluated by exposing the microorganisms to different precursor concentrations. For example, Presentato et al. evaluated the bioconversion yield and rate of 0.5 and 2 mM of SeO_3^{2-} into thermodynamically stable Se(0) nanostructures considering unconditioned and conditioned physiological states of the actinomycete *Rhodococcus aetherivorans* BCP1 [99]. The results showed that the initial precursor concentration had a strong effect on the size and size evolution of the obtained SeNPs. For instance, the smallest Se NPs that are obtained at the lowest concentration evolve to form Se nanorods (SeNRs). On the other hand, the longest SeNPs obtained at the highest concentration eventually form the shortest SeNRs. The strain *Phomopsis viticola* has the same degree of inhibition, in terms of biomass production, when incubated in the presence of SeO_3^{2-} or TeO_3^{2-} [100] whereas two strains of *Aspergillus*, *A. flavus* DSMZ 1959 and *A. parasiticus* DSMZ 1300, were less inhibited by SeO_3^2 compared to TeO_3^{2-} [100]. However, Wang et al. found that different sodium selenite concentrations did not affect the size and morphology of the produced SeNPs using *Bacillus subtilis* [101].

To optimize SeNP bioproduction, the selenium precursor concentration (sodium selenite) varied from 10 to 30 mM and the impact of the pH and reaction time was assessed [102]. Besides, statistical optimization techniques might be used for the design of the experiment, such as the response surface methodology (RSM) [102,103]. Overall, the yield of NP synthesis has a direct correlation with the precursor concentration: the higher the concentration, the greater the production. Moreover, it can be suggested that the lower the precursor concentration and temperature, the smaller the size of produced NPs (*vide infra*).

4. Techniques of Characterization

The characterization of metalloid NPs is needed to correlate their physicochemical properties to their biological effects and toxicity [49,104–107]. The initial physicochemical characterization of these NPs is carried out by using a myriad of routine lab techniques to analyze their shape, size and size distribution, porosity, surface chemistry, crystallinity, and dispersion pattern [108]. The most widely used techniques include UV-visible (UV-Vis) spectroscopy, luminescence spectroscopy (LS), scanning electron microscopy–energy dispersive X-ray spectroscopy (SEM-EDX), transmission electron microscopy (TEM), Fourier transform infra-red spectroscopy (FT-IR), X-ray diffraction (XRD). XRD confirms the presence of NPs and determines their lattice structure, crystallinity, and crystallite size using the Debye–Scherrer equation [21]. Electron microscopy techniques, such as TEM and SEM, enable the study of NP shape and size to deduce their size distribution along with elemental composition (EDX) [21,109]. According to Kapur et al., magnified field emission scanning electron microscopy (FESEM) images provide information about the nature and composition of the NPs [108]. The FTIR is an efficient technique that provides reproducible analyses used to reveal the presence of functional groups at the NP surface. These groups may be involved in the reduction of the metal ions and/or the NP capping that ensures the colloidal stability [58,95]. In addition to determining the surface charge (z-potential) of the NPs, the dynamic light scattering (DLS) provides the NP hydrodynamic diameter and good insight into their stability/aggregation by measuring their Brownian motion [108]. The atomic force microscopy (AFM) provides quantitative information about length, width, height, morphology, and surface texture of NPs through a tridimensional visualization [56].

5. Microbial Biosynthesis of Selenium Nanoparticles
5.1. Using Bacteria

In recent years, the biosynthesis of Se-containing NPs using bacteria has been reported as a new environmentally friendly route that offers tremendous advantages, such as easy handling, short synthesis times, and simple genetic manipulation [101]. Various bacteria reduce inorganic selenite (SeO_3^{2-}) or selenate (SeO_4^{2-}) to elemental red selenium Se(0) nanoparticles of various morphologies including spherical, hexagonal, polygonal, and triangular ones [109]. The academic community has extensively explored the aerobic and anaerobic bacteria involved in the production of SeNPs (Table 1) through various reduction pathways under both aerobic and anaerobic conditions [56,73,110–113]. However, further investigations are required to fully determine the underlying biochemical pathways and the biochemicals that govern these processes.

Table 1. Biosynthesis of SeNPs using bacteria.

Species	Localization	Precursor	Concentration (mM)	Incubation Temperature and Time	Size (nm) *	Color and Shape	Z-Potential (mV)	Sample Quantification	Activity/Application	Ref.
Staphylococcus carnosus	Intracellular	Na_2SeO_3	1–5	37 °C for 72 h	439–525	Red Spherical	−26.13 and −20.40	Cocktail of proteins derived from *S. carnosus*	Agriculture Future medicine	[109]
Bacillus mycoides Stenotrophomonas maltophilia	Cell free extract	Na_2SeO_3	2	27 °C for 6 h or 24 h	160–171	Spherical	−70 and −80	C: 73–75% O: 10–11% Se: 9–11% P: 3–5% S: 1%	Antibacterial Antibiofilm	[114]
Acinetobacter schindleri Staphylococcus sci-uriExiguobacterium acetylicum Enterobacter cloacae	Near the cell membrane	Na_2SeO_3	10–50	25 or 37 °C for 24 h	~100	Spherical Transformation to nanowires	N/A	Se: 83.9%	Antibacterial	[115]
Stenotrophomonas bentonitica	Intracellular Extracellular	Na_2SeO_3	2	28 °C for 48 h	30–400 (~34)	Orange-red Spherical Hexagonal Polygonal Nanowires	N/A	Extracellular flagella-like proteins	Bioremediation, Safety of deep geological repository systems	[74]
Shewanella sp.	N/A	Na_2SeO_3	0.01–1.0	30 °C for 24 h	1–20	Spherical	N/A	N/A	N/A	[116]
Bacillus sp.	Intracellular. Associated to cell debris	SeO_2	1.26	30 °C for 24 h	80–220	Red Spherical	−16.3	Se: 100%	Anticancer Antibiofilm Antiparasitic Antioxidant	[117–120]
Azoarcus sp.	Extracellular Associated to cell debris	Na_2SeO_3	1–8	30 °C for 24 h	123	Orange Spherical	N/A	N/A	Agriculture Bioremediation	[121]
Acinetobacter sp.	Intracellular	Na_2SeO_3	0.1–4	30 °C for 24 h	~100	Red Spherical Rod shaped polygonal	+10	Proteins Amines Amides	Anticancer	[58]
Duganella sp. *Agrobacterium* sp.	Cell surface Extracellular polymeric substances (EPS)Culture medium	Na_2SeO_3 Na_2SeO_4	4 g L^{-1} 2 g L^{-1}	28 ± 2 °C	100–220	Red Spherical	N/A	Proteins	Agriculture	[110]

Table 1. *Cont.*

Species	Localization	Precursor	Concentration (mM)	Incubation Temperature and Time	Size (nm) *	Color and Shape	Z-Potential (mV)	Sample Quantification	Activity/Application	Ref.
Burkholderia fungorum	Mostly extracellular	Na_2SeO_3	0.5–2	27 °C for 96 h	170–200	Red-orange Spherical	From −25 to +20	Proteins	Bioremediation	[122]
Comamonas testosteroni	Intracellular: cytoplasm or periplasm	Se(IV) and Se(VI)	5	28 °C for 48 h	100–200	Red fine-grained	N/A	Selenium content 100%	Bioremediation	[123]
Bacillus subtilis	Extracellular	Selenite	4	48 °C for 48 h	50–400	Red Spherical monoclinic that can transform to anisotropic 1D trigonal structure (nanowires)	N/A	Proteins Biopolymers	Biosensing	[101]
Alishewanella sp.	Intracellular	Na_2SeO_3	1	37 °C for 4 h	100–220	Spherical	−28.7	Proteins Lipids Organic substances Inorganic ions	Bioremediation	[75]
Azospirllum brasilense	Intracellular Extracellular	Na_2SeO_3	10	31 °C for 24 h	50–100	Spherical	−21 to −24	Proteins Polysaccharides Lipids	N/A	[124]
Azospirllum brasilense	Extracellular	Na_2SeO_3 Na_2SeO_4	1–5	30 °C	400	Red Spherical	−18	Proteins Carbohydrates EPS	Bioremediation Biotechnological applications	[125]
Pseudomonas aeruginosa	Cell surface	Selenite	0.25–1.0	37 °C for 24–72 h	47–165 (~96)	Red Spherical	251.8	Proteins	Bioremediation	[126]
Stenotrophomonas maltophilia	Intracellular Released to the medium	Na_2SeO_3	0.5–5.0	27 °C for 24 and 48 h	160–250	Spherical	140	Proteins Carbohydrates Lipids	Bioremediation	[113]
Bacillus cereus	Intracellular	Na_2SeO_3	0.5–1200	30 °C for 24 h	170	Red Spherical	N/A	N/A	Medicine Veterinary medicine	[127]
Zooglea ramigera	Extracellular	Na_2SeO_3	3	30 °C for 48 h	30–150	Red Spherical Nanorods (trigonal)	N/A	Enzymes Proteins Bacterial material	N/A	[128]

Table 1. *Cont.*

Species	Localization	Precursor	Concentration (mM)	Incubation Temperature and Time	Size (nm) *	Color and Shape	Z-Potential (mV)	Sample Quantification	Activity/Application	Ref.
Pseudomonas sp. *Lysinibacillus* *Thauera selenatis*	N/A	Na$_2$SeO$_3$	200	30 °C for 40 days	N/A	Red Spherical	N/A	Reduced in the presence of nitrate	Denitrification of mine wastewater	[129]
Escherichia coli	Intracellular Extracellular	Na$_2$SeO$_3$	1	N/A	50–100	Spherical	N/A	Quinone-mediated	N/A	[97]
Acinetobacter sp.	Intracellular	Na$_2$SeO$_3$	1	37 °C for 24 h	100 ± 10	Orange Spherical amorphous	N/A	Lignin peroxidase	N/A	[130]
Enterococcus faecalis	Extracellular	Na$_2$SeO$_3$	0.19–2.97	37 and 42 °C for 24 and 48 h	29–195	Red/light red Spherical	N/A	N/A	Antibacterial	[55]
Streptomyces minutiscleroticus	Extracellular	Na$_2$SeO$_3$	1	48–72 h	100–250	Red Spherical	N/A	Proteins	Wound ointment Anticancer drug Coating for medical instruments	[131]
Streptomyces griseobrunneus	N/A	N/A	N/A	30 °C	48–136	Red Trigonal	N/A	Proteins Enzymes	Photocatalytic	[132]
Vibrio natriegens	Intracellular Associated to cell debris	Na$_2$SeO$_4$ Na$_2$SeO$_3$	1	30 °C for 24 h	136 ± 31	Red Spherical	N/A	Proteins	Bioremediation	[133]
Staphylococcus aureus Methicillin-resistant *Staphylococcus aureus* (MRSA) *Escherichia coli* *Pseudomonas aeruginosa*	Intracellular Associated to cell debris	Na$_2$SeO$_3$	2	37 °C for 72 h	90–150	Orange-red	N/A	Lipids Proteins	Antimicrobial	[61]
Rhodococcus aetherivorans	Extracellular	Na$_2$SeO$_3$	0.5–2	40 °C for 40 min then cooled to RT	53–97	Spherical Nanorods	−13 to −32	Organic material	N/A	[99]
Pseudomonas stutzeri	Intracellular	Na$_2$SeO$_3$	2.5	28 °C	100–250	Reddish Spherical	−19.5	Proteins Lipids Other organic substances	N/A	[46]
Lactobacillus casei	Intracellular	Na$_2$SeO$_3$	1.2	37 °C for 24 h	50–80	Red Spherical	N/A	Polysaccharides Proteins	Antioxidant Anticancer	[134]

Table 1. *Cont.*

Species	Localization	Precursor	Concentration (mM)	Incubation Temperature and Time	Size (nm) *	Color and Shape	Z-Potential (mV)	Sample Quantification	Activity/Application	Ref.
Streptomyces enissocaesilis	Extracellular	SeO_2	5	30 °C for 72 h	20–211	Brown, orange and deep yellow Spherical	−220	Proteins	Antimicrobial	[135]
Pseudomonas stutzeri	N/A	Na_2SeO_3	1–3	37 °C for 48 h	75–200	Bright red Spherical	−46.2	Proteins Organic molecules	Antiangiogenic Antiproliferative	[103]
Streptomyces sp.	Extracellular	Na_2SeO_3	1	28 °C for 72–96 h	20–150	Red Spherical	N/A	Free amines Aromatic rings Cysteine residues Amides	Antibacterial Larvicidal Anthelminthic	[136]
Lysinibacillus sp.	Extracellular	Na_2SeO_3	1	37 °C for 3 days	130	Red Spherical	−19.1 to −28.8	Proteins Polysaccharides Fatty acids	Antibiofilm Antimicrobial	[137]
Lactobacillus acidophilus L. plantarum L. rhamnosus	Extracellular	Na_2SeO_3	4	35° for 48 h	20–80	Red	N/A	Proteins	N/A	[76]
Idiomarina sp.	Intracellular	Na_2SeO_3	4 and 8	37 °C for 48 h	35 and 150–350	Brick red Spherical/Hexagonal	N/A	N/A	Antineoplastic Anticancer	[138]
Ralstonia eutropha	Extracellular	Na_2SeO_4	1.5	30 °C for 48 h	40–120	Red Spherical/Nanorods	−7.7	N/A	Antibacterial	[139]
Pseudomonas stutzeri	Extracellular Cell surface	Na_2SeO_4 Na_2SeO_3	5 and 11 mM 4 and 9 mM	34 °C for 7 days	≤200	Red Spherical	N/A	N/A	Bioremediation	[84]
Enterobacter cloacae	Intracellular Extracellular	Na_2SeO_3	0.5–15	37 °C for 8 h	100–300	Red Rod-shaped	N/A	Organic material	N/A	[140]
Bacillus cereus	Intracellular Extracellular	Na_2SeO_3	0.5–10	37 °C for 48 h	150–200	Spherical	−46.86	Proteins	N/A	[56]
Stenotrophomonas maltophilia Ochrobactrum sp.	N/A	Na_2SeO_3	0.5	27 °C for 24 and 48 h	357	Spherical	N/A	Organic compounds	Antimicrobial Antibiofilm	[71]
Shewanella oneidensis	Cell surface Extracellular	Selenite	0.5	30 °C for 6–48 h	20	Red Spherical	N/A	EPS	N/A	[141]

Table 1. *Cont.*

Species	Localization	Precursor	Concentration (mM)	Incubation Temperature and Time	Size (nm) *	Color and Shape	Z-Potential (mV)	Sample Quantification	Activity/Application	Ref.
Synechococcus leopoliensis	Intracellular Extracellular	Na_2SeO_3	5	35 °C	254 ± 52 200 ± 37	Red-brown Fused spheres Elongated rods	N/A	N/A	N/A	[142]
Comamonas testosteroni	Extracellular	Na_2SeO_3	0.2–50	28 °C for 24 h	100–200	Red Round Rod-shaped	N/A	Proteins	Bioremediation	[143]
Azospirillum brasilense	Extracellular	Na_2SeO_3	10–50	31–32 °C for 24 h	25–80	Red-orange Spherical	−21 to −24	N/A	N/A	[144]
Bacillus cereus	Cell surface	Na_2SeO_3	0.25–1.0	37 °C for 24–72 h	50–150 (~93)	Red Rod-shaped	−31.1 ± 4.9	N/A	Bioremediation	[145]
Bacillus sp.	Extracellular	SeO_2	6.4	33 °C for 72 h	31–335 (~126)	Red-orange Spherical	N/A	Alcohols Phenols Amides Amines Amino acids	Antioxidant	[83]

* An inorganic particle is considered as a nanomaterial if one of its dimensions ranges between 1 and 100 nm.

The following species have been screened under aerobic conditions: *Streptomyces minutiscleroticus* M10A62 [131], *Comamonas testosteroni* S44 [143], *Lactobacillus* sp., *Bifidobacterium* sp. and *Streptococcus thermophilus* [146], *Enterobacter cloacae* Z0206 [140], *Azospirillum brasilense* [125] and the gram + bacteria *Bacillus* strains: *Bacillus* sp. MSh-1 [117,147], *B. subtilis* [101], and *B. cereus* [127]. On the other hand, several species of anaerobic bacteria have been screened for their ability to promote the production of SeNPs, such as *Shewanella* sp. HN-41 [116], *S. oneidensis* MR-1 [141], *Stenotrophomonas bentonitica* [71], *Alishewanella* sp. WH16-1 [75], *Vibrio natriegens* [133], and the facultative anaerobic bacteria *L. casei* 393 [134,148]. Moreover, anaerobic upflow sludge blanket reactors are used to fabricate SeNPs [53,149–151]. Besides, some species are able to biosynthesize SeNPs under aerobic and/or anaerobic conditions, such as *Azoarcus* sp. CIB [121].

The aerobic Se-reducing bacteria are simpler, faster, and more effective synthesizers of SeNPs as they grow rapidly and produce more cells [123]. They also possess greater advantages in agriculture and bioremediation over anaerobic bacteria since the soil and water treatment occurs aerobically [152–154]. Other benefits lie in their ability to identify the functional microbiota and the molecular homeostatic mechanisms responsible for Se oxyanion reduction. For example, in the case of the aerobic strain *C. testosteroni* S44, which can resist the toxicity of some heavy metal cations, such as Cu^{2+}, Zn^{2+}, As^{4+}, and Se^{4+}, the reduction of Se(VI) to SeNPs is carried out by the sulfite reductase (CysIJ) enzyme in the sulfate assimilation pathway [123]. This pathway has been suggested to be the general mechanism of selenate (Se(VI)) reduction in aerobic organisms related to the selenium biocycle. Moreover, the Cr(VI) reductase (known as CsrF) in the genome of *Alishewanella* sp. WH16-1 has been reported as a novel bacterial aerobic selenite reductase [75]. Due to its similarities with the structure and reduction activity of the flavoenzymes ChR, FerB and ArsH, CsrF may also act as a Se(IV) reductase.

In anaerobic bacteria, Se(VI)/Se(IV) reduction can occur on the cell surface via a two-step process; first, Se(VI) is reduced to Se(IV), then Se(IV) is reduced to subsequently give rise to SeNPs [155]. Conversely, in aerobic bacteria, it is more challenging to reduce Se-oxyanions on the surface of cells due to the tendency of oxygen to accept the electrons prior to Se(IV) [123,156]. Therefore, the reduction occurs intracellularly and then Se(0)/SeNPs are exported extracellularly by cell lysis [53,157], rapid expulsion pathway [158], efflux via a vesicular secretion system [155], vesicular transport [159], and hyphal lysis or fragmentation [160]. Nevertheless, the specific efflux system is still unknown.

Estevam et al. produced SeNPs using *Staphylococcus carnosus* TM300 that were harvested by first sonicating the pellet and then separating the NPs by ulterior centrifugations [109]. Cocktails of proteins were attached to the SeNP surface to act as potential natural stabilizers that prevent the formation of precipitates at the flask's bottom. Moreover, these SeNPs exhibited nematicidal activity against the nonpathogenic nematode *Steinernema feltiae* and biological activity against *E. coli* and *S. cerevisiae*, for bacterial and yeast infections, respectively. Wadhwani et al. detailed the SeNP synthesis by challenging the cell suspension and total cell proteins (TCP) of *Acinetobacter* sp. SW30 with sodium selenate [58]. This cell suspension formed spherical SeNPs of 78 nm in diameter after 6 h incubation and transformed into rod-like structures after 48 h. These selenium structures were observed at different pH values ranging from 6 to 10 and two precursor concentrations (1.5 and 3.0 mM) (Figure 2). On the other hand, polygonal-shaped SeNPs of 79 nm in size were obtained in the supernatant at 4 mg mL^{-1} of TCP.

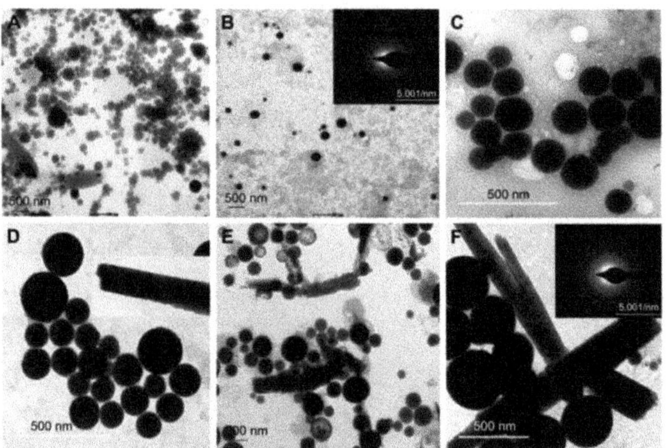

Figure 2. Transmission Electron Microscopy images of biogenic SeNPs synthesized by incubating the cell suspension of *Acinetobacter* sp. at 37 °C with 1.5 mM Na_2SeO_3 at: (**A**) pH 6, (**B**) pH 7 and (**C**) pH 9. TEM micrographs of the same experiment when the Na_2SeO_3 concentration is brought to 3.0 mM at (**D**) pH 6, (**E**) pH 7, and (**F**) pH 9. Reproduced from [58] with permission from Dove Medical Press.

Moreover, Fernández-Llamosas et al. reported that the anaerobic beta-proteobacteria *Azoarcus* sp. CIB is tolerant to selenite oxyanions and acts as a good biocatalyst synthesizing electron-dense SeNPs in its stationary growth phase [121]. This study proposed the existence of an energy-dependent selenite exporter to minimize the intracellular accumulation of the as-produced SeNPs by transporting them out of the cell. Tugarova et al. suggested a general mechanism of SeNP biosynthesis by *Aspergillus brasilense* [144]. The process involves the transport of Se ions to the cell interior where they are reduced into elemental Se(0) nuclei; these nuclei are then released to the supernatant where the extracellular biosynthesis of SeNPs occurs. The synergistic inhibition effect of these SeNPs in combination with six antibiotics was tested against pathogenic bacteria. Furthermore, the rhizobacterium *A. brasilense* appears to biotransform selenite to mixed selenium-sulfur NPs with a sulfate concentration of 800 mg L^{-1}; this mechanism is suitable for bioremediation, agriculture, nanobiotechnology, and medical applications [125].

Figueroa et al. reported the in vivo and in vitro synthesis of Se and Te nanostructures using *Acinetobacter schindleri* and *Staphylococcus sciuri* from a total of 47 bacterial strains [115]. Triangular, spherical, and rod-like Se nanostructures were also efficiently fabricated in vitro using *E. cloacae* glutathione reductase (GorA) in both crude extracts and purified protein. Similar studies investigated biomolecules involved in mediating the reduction of selenium oxyanions to elemental selenium or SeNPs, such as glutathione (GSH) [140,156,161], glutathione reductase [162], proteins [75,163], thioredoxin reductase [162,164], SerABC reductase [165], fumarate reductase [140,141], NADH-dependent enzymes [166], NADH flavin oxidoreductase [84,166], membrane-bound SrdBCA amino acid sequence [167], DMSO reductase family of molybdoproteins [168], sulfite reductase [169], hydrogenase I [170], nitrite reductase [171], chromate selenite reductase flavoenzyme (CsrF) [75], and other enzymes and biosurfactants [172,173].

In addition, some biomolecules have been found to act as reducing, capping, and/or stabilizing agents and play a fundamental role in altering the features of SeNPs and controlling their size distribution [56,145]. For instance, Ruiz Fresneda et al. indicated that extracellular flagella-like proteins can biotransform the amorphous Se(0) nanospheres to crystalline and polycrystalline one dimensional (1D) trigonal Se(0) nanostructures with distinct shapes, such as nanowires and polygons [74]. Moreover, Wang et al. used *Bacillus subtilis* to obtain semiconducting spherical monoclinic SeNPs that could be transformed into 1-D trigonal nanowires with an actinomorphic nature [101]. This process might involve an oriented attachment mechanism based on the Ostwald ripening mechanism. Moreover, proteins present in the solution are thought to provide long-term stability to the SeNPs and prevent their agglomeration. Another study using *Burkholderia fungorum* strain DBT1 determined aerobic selenite reduction can be attributed to cytoplasmic enzymatic activation mediated by electron donors [122]. The same study suggested that an organic layer surrounding the SeNPs, composed of extracellular matrix (ECM) that includes carbohydrates, proteins, and humic-like substances, stabilizes the particles by modifying their zeta potential.

Previous studies also highlighted the importance of the protein fraction released by microorganisms to externally coat nanoparticles to increase electrostatic repulsions and, consequently, increase their colloidal stability [174–176]. This characteristic is essential to maintain the long-term stability, avoid the aggregation and prevent the transformation of colloidal SeNPs into the black amorphous Se form [56]. This is evidenced in high negative z-potential values that are indicative of particle repulsion. For example, carbonyl groups of amino acid residues [142] and SH groups of L-cysteine [177] can strongly bind to metal NPs and form a biomolecular, stabilizing, and protecting cap.

5.2. Using Fungi

The mycogenic biosynthesis of inorganic NPs has been extensively investigated due to the advantages of fungi over bacteria and actinomycetes [178,179]. Fungi are easy to culture and manipulate, and can grow in highly concentrated media with heavy metal cations. They can also survive and reproduce in high selenium concentrations. The main advantages of NP mycosynthesis are easy scaling-up, low-cost downstream processing and easy manipulation, low-cost and viability of the fungal biomass [180]. Furthermore, fungi release reductive proteins and enzymes into the extracellular medium; these biomolecules reduce Se ions into harmless, precipitating SeNPs [181]. The general process of microbially assisted synthesis of SeNPs and TeNPs is shown in Figure 3.

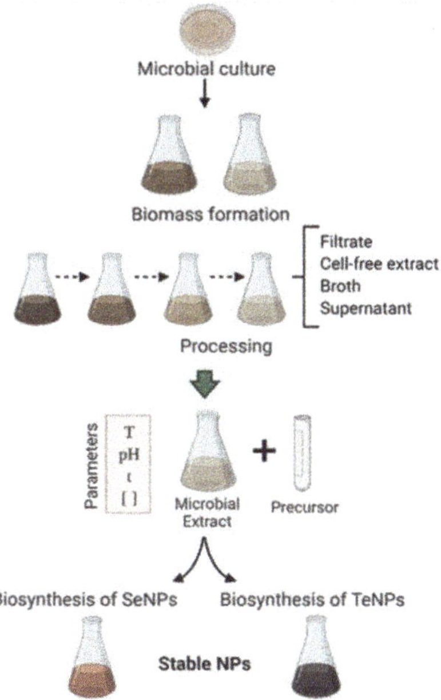

Figure 3. Schematic diagram detailing the microbially assisted procedure of metalloid nanoparticles.

Numerous fungal species reduce selenite/selenate to intra- or extracellular SeNPs (Table 2). Under extracellular conditions, Diko et al. reported the synthesis of spherical and pseudospherical SeNPs using the supernatant of *Trichoderma* sp. WL-Go in culture broth [182]. Liang et al. used four fungal species: *Aureobasidium pullulans*, *Mortierella humilis*, *Trichoderma harzianum*, and *Phoma glomerata*, to produce SeNPs and TeNPs and provide nucleation sites with extracellular protein and polymeric substances [183]. Mosallam et al. combined γ-rays and the supernatant of *A. oryzae* to produce SeNPs and found a strong correlation between the antioxidant capacity and both the phenolic content and SeNP yield [184]. Moreover, the biomimetic mycosynthesis of SeNPs with simple preparation protocols from, for instance, *Alternaria alternata* yields uniform and stable SeNPs [180].

Table 2. Biosynthesis of SeNPs by fungi.

Species	Location	Size (nm)	Shape	Activity/Application	Ref.
Trichoderma sp.	Extracellular	20–220	Spherical Pseudospherical	N/A	[182]
Pleurotus ostreatus	Aqueous extract	7–28	Spherical	Antioxidant Antimicrobial Anticancer	[185]
Penicillium chrysogenum	Cell-free supernatant	48–50	Spherical	Antimicrobial Antibiofilm	[186]
Phanerochaete chrysosporium	Intracellular Extracellular	50–600	Spherical	Bioremediation	[65]
Polyporus umbellatus	N/A	212 ± 23 82 ± 1	Spherical	Anticancer Antiproliferative	[187]
Auricularia auricula-judae	Embedded in triple helix β-(1,3)-D-glucan	60	Hollow nanotubes	Acute myeloid leukemia (AML) therapy	[188]
Trichoderma atroviride	Culture filtrate (CF) Cell lysate (CL) Cell wall debris (CW)	60–123	Spherical	Production of crop plants (tomatoes) Management of plant diseases	[181]
Aureobasidium pullulans *Mortierella humilis* *Trichoderma harzianum* *Phoma glomerata*	Extracellular	48–78	Spindle-shaped	Bioremediation	[51]
Dictyophora indusiata	Intracellular	89	Spherical	Anticancer	[189]
Catathelasma ventricosum	N/A	50	Spherical	Antidiabetic	[190]
Aspergillus oryzae	N/A	55	Spherical	Antimicrobial	[184]
Pyrenochaeta sp. *Acremonium strictum* *Plectosphaerella cucumerina* *Stagonospora* sp. *Alternaria alternata* *Paraconiothyrium sporulosum*	Fungal hyphae Intracellular Extracellular	50–300	Spherical	N/A	[191]
Alternaria alternata	Extracellular	30–150	Spherical	N/A	[180]

Table 2. *Cont.*

Species	Location	Size (nm)	Shape	Activity/Application	Ref.
Pleurotus ostreatus *Lentinus edodes* *Ganoderma lucidum* *Grifola frondosa*	Intracellular Extracellular	50–150	Spherical	N/A	[192]
Lentinula edodes	Intracellular (fungal hyphae)	180 ± 17	N/A	N/A	[193]
Pleurotus ostreatus *Ganoderma lucidum* *Grifola frondosa*	Intracellular Cell-free filtrate	20–550	N/A	N/A	[194]
Cordyceps sinensis	N/A	80–125	Spherical	Antioxidant	[195]
Mariannaea sp.	Intracellular Extracellular	45 213	Spherical	N/A	[196]
Gliocladium roseum	Cell-free filtrate	20–80	Spherical	N/A	[197]

The medicinal basidiomycete *Lentinus edodes* F-249 can transform selenium within organic and inorganic compounds into spherical SeNPs of ~180 nm [193]. *Dictyophora indusiata* is a saprophytic fungus able to form a hybrid Se nanostructure by exploiting its novel polysaccharide (DP1) [189]. The DP1-functionalized SeNPs proved to have an antiproliferative effect against HepG2 cancer cells via death receptor- and mitochondria-mediated apoptotic mechanisms.

Some studies have also depicted both the intracellular and extracellular synthesis of SeNPs using fungi [191,196]. For example, three fractions of the fungus *Trichoderma atroviride*, namely the culture filtrate (CF), cell lysate (CL), and cell wall debris (CW), produced bioactive SeNPs that were able to form aggregate fungal spores, thus avoiding the adhesion of the pathogen *Phytophthora infestans* to the host cell and blocking its infection of tomato plants [181]. A similar mechanism has been reported for *Lentinula edodes* [193], *Mariannaea* sp. [196], *Fusarium* sp., and *T. reesei* [198]. Other researchers exploited intra- and extracellular extracts of the xylotrophic basidiomycetes *Pleurotus ostreatus*, *L. edodes*, *Ganoderma lucidum*, and *Grifola frondosa* to produce SeNPs of various sizes and shapes [192]. Along with basidiomycetes, other fungal groups, such as *Ascomycota* and *Zygomycota*, can also produce nanoparticles, but these mushrooms are known to be allergenic and/or pathogenic to animals and plants [199–201]. Therefore, nontoxic, edible, and cultivated basidiomycetes are a better alternative for biotechnological applications including nanotechnology as the NP synthesis can occur in their mycelia and culture media [192]. Under both extra- and intracellular conditions, the toxicity effects and the removal mechanisms vary according to the fungal species and Se precursors. Rosenfeld et al. demonstrated that six fungal species (*P. sporulosum*, *A. strictum*, *A. alternata*, *P. cucumerina*, *Pyrenochaeta* sp., and *Stagonospora* sp.) constitute an excellent detoxification biosystem that tolerates high Se concentrations and reduces selenite/selenate to Se(0) [191].

El-Sayyad et al. fabricated SeNPs by employing two different eco-friendly green synthetic methodologies: either using *Penicillium chrysogenum* filtrate or combining *P. chrysogenum* filtrate with gentamicin drug (CN) as the stabilizing agent after application of γ-irradiation [186]. The second process resulted in the highest synthesis yield and enhanced antipathogenic and antibiofilm potential. It is also easy to produce Se-based nanocomposites. For instance, Jin et al. prepared SeNPs embedded and homogeneously dispersed in black fungus-extracted BFP nanotubes (triple helix β-(1,3)-D-glucan) that possess hydrophilic hydroxyl groups. These nanocomposites showed interesting cytotoxic and antitumor properties [188].

5.3. Using Yeast

Yeast is a relevant model system to investigate the metabolic detoxification pathways of selenite/selenate and their conversion to selenomethionine [202,203]. Thus, Se-rich yeasts are used as a food supplement because they accumulate up to 3000 ppm of selenium [203] and can be used as a cancer treatment at elevated doses (>200 µg Se per day) [204]. However, further analyses are needed to identify and quantify the chemical forms of selenium should these Se-rich yeasts be commercialized. For example, Jiménez-Lamana et al. used single particle inductively coupled plasma mass spectrometry (SP-ICPMS) to detect, characterize, and quantify putative nanoparticles in Se-rich yeasts [205]. Bartosiak et al. calculated the accurate yield of SeNP synthesis mediated by *Saccharomyces boulardii* using continuous photochemical vapor generation (PCVG) coupled with microwave-induced plasma optical emission spectrometry (MIP-OES) and UV-Vis spectrophotometry (PCVG-MIP-OES) [206]. This efficient method enabled the selective identification and quantification of both the unreacted Se(IV) and the final water-soluble SeNPs without the need to separate them. Lian et al. synthesized spherical and quasi-spherical SeNPs of 70–90 nm in size utilizing the yeast cell-free extract of *Magnusiomyces ingens* LH-F1; some surface proteins played a significant role during the synthesis, acting as reducing or capping agents [207]. Nevertheless, the mechanisms of SeNP formation are not fully understood.

S. cerevisiae primarily reduces selenium ions through metabolism [208,209]. Owing to its high selenium tolerance, *S. cerevisiae* constitutes a promising and cost-effective alternative for the removal of selenium ions from aqueous solutions [210]. Additionally, it is postulated that SeNPs are expelled from *S. cerevisiae* cells by vesicle-like structures under microaerophilic conditions followed by the ulterior capping of these NPs with residual organic components from the vesicle-like structures [211]. As the SeNPs are stabilized by the natural organic molecules of yeast cultures, there is no need for additional stabilizing agents [206].

The reduction of selenite/selenate to elemental selenium in yeasts forms SeNPs either extra- or intracellularly. In intracellular routes, a genetically engineered, metal-resistant *Pichia pastoris* clone carrying Cyb5R gene has been found to be a safe bioreactor to produce homogeneous and stable selenium and silver NPs. This yeast used a versatile and simple mechanism of biosorption and biotransformation of metals with less toxic waste than physicochemical synthesis [50]. On the other hand, the extracellular processes have the advantage of easy biogenic NP recovery over their intracellular counterparts [211]. According to Rassouli, the general procedure for the extraction and purification of yeast-produced SeNPs consists of (i) applying some enzymatic, chemical, or mechanic method to destroy the cell wall; (ii) collecting the biomass by centrifugation at 8000 rpm for 10 min; (iii) crushing the cells using liquid nitrogen and ultrasounds; (iv) incubating the broken cells with added buffer at 60 °C for 10 min; (v) mixing the pellet containing the cell fragments and NPs with octanol and distilled water to give rise to two phases of which (vi) the SeNP-containing top phase is recovered and further washed with ethanol and chloroform [48].

6. Microbial Synthesis of Tellurium Nanoparticles

Tellurium is highly toxic to living beings and is not essential in biological metabolism. This may explain why TeNP biosynthesis using microbes is more limited when compared to SeNP [212]. Few articles have been published that detail the biosynthesis of TeNPs using microorganisms (Figure 4) [51,97,213–219]. Generally, K_2TeO_3 or Na_2TeO_3 precursors are used to produce TeNPs since they are least toxic when compared to other precursors [97,212,220–222]. Tellurium has different oxidation states: telluride (Te^{2-}), tellurite (TeO_3^{2-}), tellurate (TeO_4^{2-}). In general, the agglomeration of Te(0) is associated with the respiration of the microorganisms, such as yeast (*S. cerevisiae*), where the fermentation increases the production [223]. On the other hand, a decrease in NP production is observed in bacteria when the oxygen is limited [213].

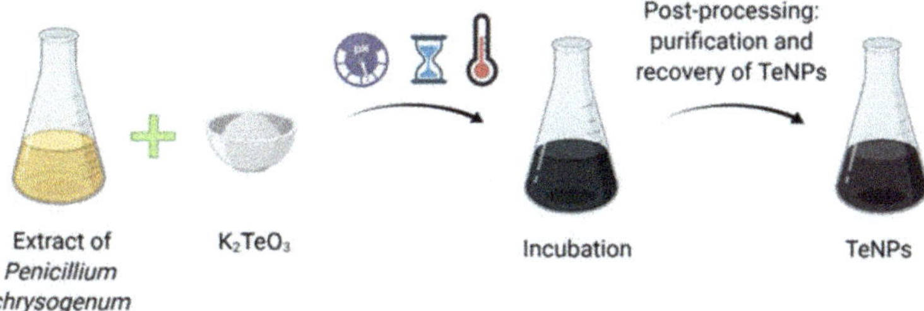

Figure 4. Tellurium nanoparticles (TeNPs) synthesis using microorganisms.

Considering that tellurium is in the same group as selenium, Yang et al. studied the antioxidant activity of TeNPs recovered from tellurium-enriched *Spirulina platensis* cultures where tellurium interacts with two phycobiliproteins, the phycocyanin (Te-PC) and allophycocyanin (Te-APC) (Figure 5) [224].

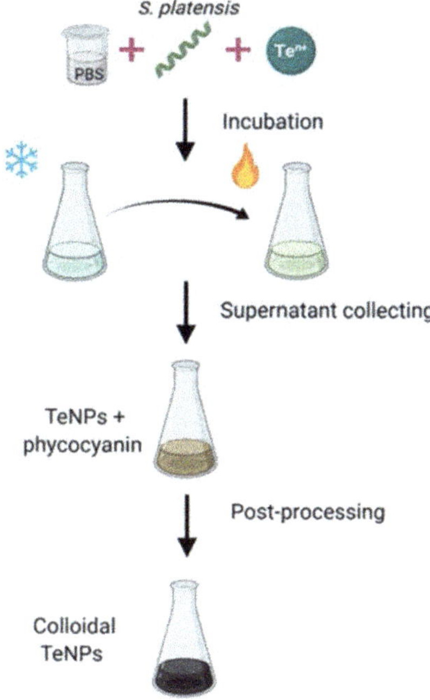

Figure 5. Purification of tellurium-containing phycocyanin (Te-PC) and allophycocyanin (Te-APC) from Te-enriched *S. platensis* using a chromatographic method.

From a mechanistic point-of-view, a correlation has been established between the growth, size, and shape of TeNPs and the proteins and enzymes present in the media, in addition to other small molecules, such as pyruvate, lactate, and NADH [51,213,225]. Furthermore, the formation of elemental tellurium can be inhibited by other molecules, such as nitrate, nitrite, and fumarate [226]. Since the conditions affecting the TeNP formation can vary as a function of the used organism, there are also great variations in microbe growth time (1–9 days), precursor concentrations (12–600 mg L^{-1}), and reaction time (1–8 days).

7. Plant-Mediated Synthesis of Metalloid Nanoparticles

Phytonanotechnology is of special interest for synthesizing SeNPs since it is a simple, eco-friendly, high-throughput, and inexpensive route [227–229]. The biofabrication of NPs via plants involves proteins, amino acids, organic acids, vitamins, as well as secondary metabolites that act as reducers and stabilizers, such as polysaccharides, alkaloids, flavonoids, phenols, saponins, quinine, steroids, and glycosides [230,231]. Plant-mediated NP synthesis may be carried out through two ways. Via the in vivo route, the NP morphology and size depend strongly on the biosynthesis location, e.g., roots, leaves, fruits, peels, buds, etc., and the implicated metabolites [27]. A chelation-mediated detoxification faculty may explain the mechanism of NP synthesis [232]. The enzymatic antioxidant system is also activated to provide a reactive oxygen species (ROS) balance [233]. Generally, inorganic Se salts (selenite and selenate) taken up by plants are biotransformed into organic Se forms, such as SeCys2, SeMet, and MeSeCys bounded with proteins [234,235]. Hu et al. demonstrated the bioavailability of SeNPs in roots and shoots where they could be biotransformed into organic Se compounds, selenite and selenate to generate Se-biofortified plants [236]. However, the in vitro synthesis using plant extracts is better since it eliminates the lengthy process of cultivation, but still allows for screening the experimental parameters, such as the biomass choice, extraction process and amount, the pH, and temperature [237].

7.1. Plant-Based Synthesis of Selenium Nanoparticles

Several papers have reported the plant-derived biosynthesis of SeNPs with varying sizes and morphologies (Table 3). For instance, *Hibiscus sabdariffa* fabricated spherical, triangular, and hexagonal SeNPs with a size of 20–50 nm [238] whereas *Azadirachta indica* has been used as a rapid and efficient biosystem to produce crystalline and spherical SeNPs with a smooth surface [239]. *Withania somnifera* was the best adaptogen herb with active withanolide and flavonoids, used as a bioreductant system to fabricate SeNPs of 40–90 nm [240]. Although plants offer the most suitable green synthesis protocols, the mode of action of plant-produced SeNPs against bacteria remains unknown; it is suggested that the nanoparticles interact with the peptidoglycan layer and break up the bacterial cell wall [227]. Besides, SeNPs are able to induce apoptosis or programmed cell death [174]. Anu et al. reported spherical SeNPs produced by a cheap aqueous extract of garlic cloves, *Allium sativum*, that acted as both the reducing and capping agent [241]. These biogenic SeNPs showed lower cytotoxicity against the Vero cell line than those chemically synthesized. The same group took advantage of the medicinal properties of *Cassia auriculata* to synthesize functional SeNPs that displayed interesting anticancer and antiproliferative characteristics [241]. Similar studies have reported the use of *Vitis vinifera* [32], broccoli extract [108], and *Capsicum annum* [242] to fabricate Se nanorods and nanoballs. Importantly, Ramamurthy et al. presented a combination of SeNPs, made using fenugreek seed extract, and doxorubicin to form a chemoprotective agent against cancer [243]; Vennila et al. studied the antibacterial, anticancer, and anti-inflammatory activity of SeNPs biofabricated by *Spermacoce hispida* and functionalized with apigenin, quinoline, quinazoline, and synaptogenin B [244]; Kokila et al. reported on Se-NPs using the leaves of *Diospyros montana* as a biocidal agent against both Gram+ *S. aureus* and Gram– *E. coli* and the fungus *A. niger* [245].

Table 3. Different species of plants used for the biosynthesis of SeNPs.

Plant Species	Part	Metabolites	Shape	Size (nm)	Activity/Application	Ref.
Withania somnifera	Leaves	Flavonoids Phenolics Tannins	Spherical	40–90	Antibacterial Antioxidant Anticancer	[240]
Psidium guajava	Leaves	N/A	Spherical	8–20	Antibacterial	[227]
Allium sativum	Cloves	N/A	Spherical	40–100	Cytotoxicity	[241]

Table 3. *Cont.*

Plant Species	Part	Metabolites	Shape	Size (nm)	Activity/Application	Ref.
Cassia auriculata	Leaves	N/A	Amorphous	10–20	Anti-leukemia	[237]
Momordica charantia	Roots and shoots	Terpenoids Phenolics	Spherical	10–30	Toxicological studies	[246]
Hawthorn fruit	Fruit	N/A	Spherical	113	Antitumor	[247]
Hibiscus sabdariffa	Leaves	Phenols Alcohols	Spherical Triangular Hexagonal	20–50	Antioxidant	[238]
Pelargonium zonale	Leaves	N/A	Spherical	40–60	Antibacterial Antifungal	[248]
Aloe vera	Leaves	Hydroxyls Amides	Spherical	121–3243	Antibacterial Antifungal	[249]
Emblica officinali	Fruit	Phenolics Flavonoids Tannins	Spherical	20–60	Antimicrobial	[228]
Moringa oleifera	Leaves	Phenolics Flavones	Spherical	23–35	Anticancer	[250]
Triticum aestivum	Roots	N/A	Spherical	140 ± 40	Biofertilizer	[236]
Broccoli	N/A	Carotenes Glucosinolates Polyphenols	Spherical	50–150	Antioxidant Anticancer	[108]
Diospyros montana	Leaves	Phenolics Flavonoids	Spherical	4–16	Antibacterial Anticancer	[245]
Ocimum tenuiflorum	Leaves	Polyphenols	Spherical	15–20	Inhibition of nephrolithiasis	[183]
Theobroma cacao	Shell	Polysaccharides Proteins Phenolics	Spherical Trigonal	1–3	N/A	[251]
Zingiber officinale	Roots	Flavonoids Terpenoids	Spherical	100–150	Antimicrobial Antioxidant	[252]
Mucuna pruriens	Seed	Phytochemicals	Spherical Nanorods	100–120	Antioxidant Anticancer	[102]
Azadirachta indica	Leaves	Polyphenols Flavonoids Proteins	Spherical	142–168 221–328	Antibacterial	[239]
Vitis vinifera	N/A	Lignin	Spherical	3–18	N/A	[32]
Clausena dentata	Leaves	Flavonoids Triterpenoids Polyphenols	Spherical	46–79	Larvicidal	[229]
Spermacoce hispida	Leaves	Polyols Saponins	Rod-shaped	120 ± 15	Anti-inflammatory Antibacterial Anticancer	[244]
Rosa roxburghii	N/A	Polysaccharide (RTFP-3)	Spherical	105	Antioxidant	[253]
Lycium barbarum	Berries	Flavonols (catechins)	Spherical Triangular	83–160	Antioxidant	[254]
Fenugreek	Seeds	Phenol Flavonol	Oval	50–150	Anticancer	[243]
Allium sativum	Bulbs	Alcohols Phenols	Spherical	205	Antioxidant Anticancer	[255]

The application of SeNPs in toxicological studies is relevant due to their association with DNA cytosine methylation, chromatin structure, and transcription processes. It is advantageous for the manipulation and study of cellular division, tissue differentiation, metabolism, and transcription programs [246]. Cui et al. (2018) reported on the production of monodispersed and stable SeNPs from hawthorn fruit extract (HE-SeNPs) whose antitumor activity was evidenced by the apoptosis induced in HepG2 cells through the overproduction of intracellular ROS and mitochondrial membrane potential (MMP) loss or disruption [247]. Additionally, HE-SeNPs induced the upregulation of caspase-9 and downregulation of Bcl-2. Fardsadegh et al. detailed the hydrothermal synthesis of SeNPs using *Aloe vera* leaf extract and determined a prediction model and optimal conditions using response surface methodology (RSM) [249].

7.2. Plant-Based Synthesis of Tellurium Nanoparticles

Tellurium is not essential for plant metabolism besides being toxic in most cases [256]. Despite this, it has been documented that some plants have the ability to metabolize Te and transform it into telluroamino acids [257] and organotellurium [258]. *A. sativum*, commonly known as garlic, can assimilate chalcogens to give rise to Te-methyltellurocysteine (MeTe-Cys) and S-methyltellurosulfide metabolites [256]. The TeNP size is found to be 40–55 nm. The majority of these metabolites were found highly concentrated at the tips of their gloves and in the initial part of the roots. In some cases, TeNPs produced by plants may appear as spheres, rod-shaped, and plates [259].

8. Biosynthesis of Bimetallic Se-Te Alloy Nanoparticles

Bimetallic Se-Te alloy NPs possess unique and enhanced properties including optical, semiconductive electroresistance, and magnetoresistance [90,260,261]. A few studies have reported the bacterial synthesis of Se-Te nanostructures by *B. beveridgei* [262] and soil isolates of heterotrophic aerobic bacteria [263]. The simultaneous formation of trigonal-hexagonal Se(0)–Te(0) nanostructures from the bioreduction of Se and Te oxyanions in a lab-scale upflow anaerobic sludge blanket reactor (UASB) was also described [149]. A layer of extracellular polymeric substances (EPS) capped the nanoparticles to immobilize them in the granular sludge. Besides crystalline hexagonal TeNPs, the fungus *Phanerochaete chrysosporium* biofabricated unique Se-Te nanospheres and needle-like nanoparticles of 500–600 nm (Figure 6) [65].

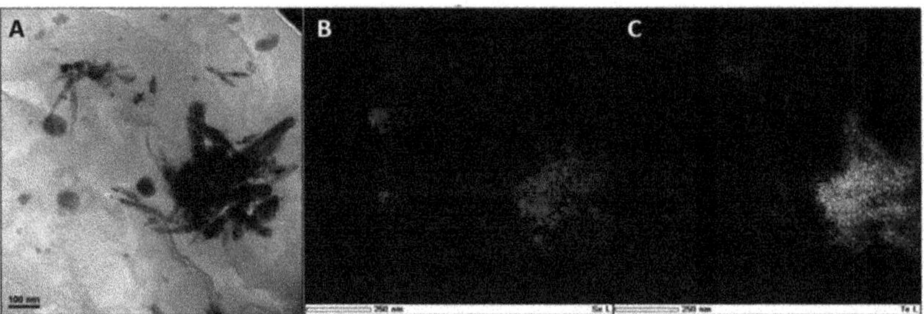

Figure 6. (**A**) TEM image of the hyphae of *Phanerochaete chrysosporium* that depicts Se-Te alloy NPs. STEM-EDS elemental mapping for Se (**B**) and Te (**C**) that confirms the alloy character of these Se-Te NPs. Adapted from [65] with permission from Elsevier.

Additionally, Asghari-Paskiabi et al. reported the formation of stable Se-S NPs inside *S. cerevisiae* [209]; Vogel et al. investigated the extracellular synthesis of Se-S NPs by *Azospirillum brasilense* mainly attributable to the high negative surface charge due to the covering organic layer made of proteins and carbohydrates [125].

9. Bioapplications of SeNPs and TeNPs

In the field of nanobiotechnology, nanoparticles represent the core of a nano-biomaterial; they can be functionalized with different moieties to reduce the toxicity and improve the effects of the drugs [264–266]. Moreover, nanoparticles can be used for various medical, industrial, or biological applications. For instance, in nanomedicine, a wide number of surface structures to functionalize the NP surface have been developed for imaging, sensing, and drug delivery applications [267]; the as-obtained NPs can be used for the detection of pathogens and biomolecules or the hyperthermia treatment of cancer [268].

Nanoscale selenium has attracted the attention of scientists due to its bioavailability and lower toxicity compared to the other forms of selenium [269]. Gao et al. studied the antioxidant properties of SeNPs and demonstrated the reduced risk of selenium toxicity [187]. Moreover, SeNPs can be used as an antioxidant in food additives due to their lower risk of toxicity. Besides their antioxidant activity, SeNPs are also an excellent chemopreventive agent against cancer as well as a potential anticancer drug [270]. Specifically, the efficacy and specificity of using nanoselenium at a concentration as low as 2 µg mL^{-1} against prostate cancer has been reported [174]. Other studies highlighted the antimicrobial properties [114] and antifungal activity [271] of SeNPs.

The antimicrobial, antioxidant, antifungal, and anticancer properties of TeNPs have been well documented. For instance, Shakibaie et al. described the antioxidant and antimicrobial properties of biologically synthesized tellurium nanorods (TeNRs) [272]. Moreover, another study reveals the antimicrobial and anticancer properties of citrus juice-mediated synthesized TeNPs [62] while the *S. baltica*-synthesized TeNRs exhibit an excellent photocatalytic and anti-biofilm activity to counter potential human pathogens [59]. The next graphic summarizes the main applications of SeNPs and TeNPs (Figure 7).

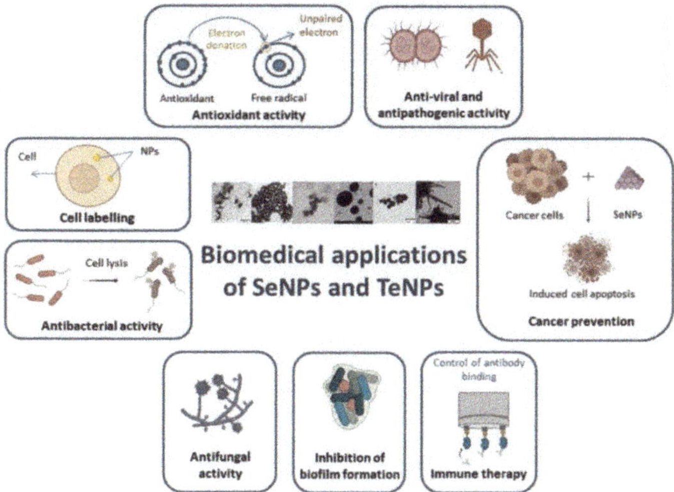

Figure 7. Schematic representation of the various bioapplications of biogenic SeNPs and TeNPs.

10. Human Cell-Cytotoxicity and Immune Response Induced by SeNPs and TeNPs

According to several studies, various nanoparticles may be cytotoxic and cause harmful effects or even irreversible damage to human cells [264,265]. Therefore, it is necessary to determine how synthesized nanoparticles affect the immune cells [273–275]. Selenium nanomaterials have attracted considerable attention as a novel anticancer and chemopreventive agent due to their exceptional biocompatibility and low toxicity [276]. For instance, Cremonini et al. studied the effect of biogenic SeNPs synthesized using *Stenotrophomonas maltophilia* (−) and *B. mycoides* (+) on the viability and function of the antigen-presenting cells, DCs, and cultured fibroblasts (nonimmune cells) [114]. As a result, the as-produced SeNPs did not cause any damage to human cells since there was no stimulation or increase in the release of proinflammatory and immunostimulatory cytokines including IL-12, IL-6, IL-8, and TNF-α. Other studies indicate the SeNPs synthesized by bacteria can induce apoptosis or inhibit both growth and proliferation of cancer cells in culture [276–279]. SeNPs synthesized by *Acinetobacter* sp. SW30 seem to display a greater anticancer activity when compared to their chemically synthesized counterparts; in fact, they reveal a strong antiproliferative activity against 4T1 cells, MCF-7, NIH/3T3, and HEK293 cell lines [58]. SeNPs synthesized by *B. oryziterrae* also showed potential anticancer activity against H157 lung cancer cell lines [280].

An assay carried out using the SeNPs produced by *Bacillus* sp. MSh-1 against the human fibrosarcoma cell line (HT-1080) demonstrated that the higher the concentration, the higher the cytotoxicity [117]. Moreover, the same study showed the anti-invasive property of HT-1080 cells and the moderate inhibition of MMP-2 expression, a good insight for the treatment and prevention of tumor metastasis. The MTT assay has been used to assess the cell viability, proliferation, and cytotoxicity of breast cancer cells.

One possible explanation for the anticancer activity of SeNPs was reported by Ahmed et al. which encompasses the mobilization of endogenous copper, possibly chromatin-bound copper, and the subsequent prooxidant action [276]. The authors suggested that cancer cells are more subject to electron shuttling between copper ions and selenium nanostructures which release reactive oxygen species (ROS) and thereby kill cancer cells such as Hep-G2 and MCF-7 cell lines. The precise mechanism of anticarcinogenic actions of SeNPs is not totally understood. Since it possesses a high bioactivity and represents the major component of selenoproteins, selenium may increase the carcinogen detoxification, inhibit tumor cell invasion and angiogenesis, enhance immune surveillance, and provide antioxidant protection [281–283].

The cytotoxic effects of biosynthesized TeNPs have also been investigated due to their ability to act as an anticancer and antiviral agent [283–285]. For instance, Forootanfar et al. demonstrated the lower cytotoxic effect of biogenic TeNRs compared to potassium tellurite on four cell lines of MCF-7, HT1080, HepG2, and A549 [286]. Overall, the toxicity of Te nanostructures depends on the employed synthesis method and their size/morphology [287].

11. Conclusions and Perspectives

The present review extensively describes different green methodologies used for the biofabrication of SeNPs and TeNPs. A variety of microorganisms, such as bacteria, fungi and yeast, and plant extracts have become novel, sustainable, risk-free, and cost-effective bionanofactories that reduce selenite/selenate and tellurite/tellurate into their nanosized zero-valent counterparts. To achieve simple, fast, and efficient biological syntheses, these eco-friendly procedures leverage the different organic molecules and metabolites that act as reducing, chelating, and stabilizing agents, such as proteins, EPS, lipids, flavonoids, phenols, and alcohols. The bioreduction and biotransformation of different Se and/or Te species into elemental Se/Te have emerged as an important pursuit in biomedicine, chemistry, nanotechnology, and engineering. Some experimental parameters including the pH, temperature, reaction time, and precursor concentration, along with biosurfactants, play an active role in determining the shape, size, crystallinity, dispersion, and properties of the as-obtained metalloid NPs. This review found that most of the biogenic SeNPs were spherical while their TeNP counterparts were rod-shaped; this constitutes a remarkable outcome in bionanotechnology. However, it is necessary to carry out deeper research on the specifically involved production and transformation mechanisms. Although the toxicity effect of bioresources (i.e., plants) or the nanoparticles synthesized have not been fully explored yet, green production opens up opportunities to manufacture safer nanomaterials and foster better understanding of safety, health, and environment issues.

A myriad of literature shows research at the laboratory scale using living or dead biomass. An important challenge lies in developing large-scale production processes, where larger amounts of templates, surfactants, and other auxiliary substances are required. Then, the use of continuous-flow microreactors and other sources such as waste materials and algae/microalgae may provide significant advantages for industrial level and nanotechnology applications. The development of greener methods that enhance the bioavailability, longevity, and composition-control of NPs could be carried out by computational, synthetic biology and genetic engineering techniques. The employment of natural "nanofactories" is still at an early stage; however, further research would enable the development of straightforward approaches to create potential solutions in nanomedicine, biomedical devices, energy crises, water pollution, and optoelectronics.

Funding: This research received no external funding.

Conflicts of Interest: The authors declare no conflict of interest.

References

1. Hosnedlova, B.; Kepinska, M.; Skalickova, S.; Fernandez, C.; Ruttkay-Nedecky, B.; Peng, Q.; Baron, M.; Melcova, M.; Opatrilova, R.; Zidkova, J.; et al. Nano-selenium and its nanomedicine applications: A critical review. *Int. J. Nanomed.* **2018**, *13*, 2107–2128. [CrossRef]
2. Li, X.; Xu, H.; Chen, Z.-S.; Chen, G. Biosynthesis of Nanoparticles by Microorganisms and Their Applications. *J. Nanomater.* **2011**, *2011*, 270974. [CrossRef]
3. Rashidi, L.; Khosravi-Darani, K. The applications of nanotechnology in food industry. *Crit. Rev. Food Sci. Nutr.* **2011**, *51*, 723–730. [CrossRef] [PubMed]
4. Stark, W.J.; Stoessel, P.R.; Wohlleben, W.; Hafner, A. Industrial Applications of Nanoparticles. *Chem. Soc. Rev.* **2015**, *44*, 5793–5805. [CrossRef] [PubMed]
5. Singh, R.; Nalwa, H.S. Medical applications of nanoparticles in biological imaging, cell labeling, antimicrobial agents, and anticancer nanodrugs. *J. Biomed. Nanotechnol.* **2011**, *7*, 489–503. [CrossRef] [PubMed]
6. Sakr, T.M.; Korany, M.; Katti, K.V. Selenium nanomaterials in biomedicine—An overview of new opportunities in nanomedicine of selenium. *J. Drug Deliv. Sci. Technol.* **2018**, *46*, 223–233. [CrossRef]
7. Brito-Silva, A.M.; Gómez, L.A.; de Araújo, C.B.; Galembeck, A. Laser Ablated Silver Nanoparticles with Nearly the Same Size in Different Carrier Media. *J. Nanomater.* **2010**, *2010*, 142897. [CrossRef]
8. Nee, C.H.; Yap, S.L.; Tou, T.Y.; Chang, H.C.; Yap, S.S. Direct synthesis of nanodiamonds by femtosecond laser irradiation of ethanol. *Sci. Rep.* **2016**, *6*, 33966. [CrossRef]
9. Anbarasu, M.; Anandan, M.; Chinnasamy, E.; Gopinath, V.; Balamurugan, K. Synthesis and characterization of polyethylene glycol (PEG) coated Fe_3O_4 nanoparticles by chemical co-precipitation method for biomedical applications. *Spectrochim. Acta A Mol. Biomol. Spectrosc.* **2015**, *135*, 536–539. [CrossRef]
10. Janjua, M.R.S.A. Synthesis of Co_3O_4 Nano Aggregates by Co-precipitation Method and its Catalytic and Fuel Additive Applications. *Open Chem.* **2019**, *17*, 865–873. [CrossRef]
11. Li, M.; Gu, L.; Li, T.; Hao, S.; Tan, F.; Chen, D.; Zhu, D.; Xu, Y.; Sun, C.; Yang, Z. TiO_2-Seeded Hydrothermal Growth of Spherical $BaTiO_3$ Nanocrystals for Capacitor Energy-Storage Application. *Crystals* **2020**, *10*, 202. [CrossRef]
12. Clark, I.; Smith, J.; Gomes, R.L.; Lester, E. Towards the Continuous Hydrothermal Synthesis of $ZnO@Mg_2Al-CO_3$ Core-Shell Composite Nanomaterials. *Nanomaterials* **2020**, *10*. [CrossRef] [PubMed]
13. Cheng, G.; Yang, H.; Rong, K.; Lu, Z.; Yu, X.; Chen, R. Shape-controlled solvothermal synthesis of bismuth subcarbonate nanomaterials. *J. Solid State Chem.* **2010**, *183*, 1878–1883. [CrossRef]
14. Yang, S.; Zhou, X.; Zhang, J.; Liu, Z. Morphology-controlled solvothermal synthesis of $LiFePO_4$ as a cathode material for lithium-ion batteries. *J. Mater. Chem.* **2010**, *20*, 8086–8091. [CrossRef]
15. Riccò, R.; Nizzero, S.; Penna, E.; Meneghello, A.; Cretaio, E.; Enrichi, F. Ultra-small dye-doped silica nanoparticles via modified sol-gel technique. *J. Nanopart. Res.* **2018**, *20*. [CrossRef] [PubMed]
16. Adnan, R.; Razana, N.A.; Abdul Rahman, I.; Farrukh, M.A. Synthesis and Characterization of High Surface Area Tin Oxide Nanoparticles via the Sol-Gel Method as a Catalyst for the Hydrogenation of Styrene. *J. Chin. Chem. Soc.* **2010**, *57*, 222–229. [CrossRef]
17. Brayner, R.; Dahoumane, S.A.; Nguyen, J.N.; Yepremian, C.; Djediat, C.; Coute, A.; Fievet, F. Ecotoxicological studies of CdS nanoparticles on photosynthetic microorganisms. *J. Nanosci. Nanotechnol.* **2011**, *11*, 1852–1858. [CrossRef]
18. Brayner, R.; Dahoumane, S.A.; Yepremian, C.; Djediat, C.; Meyer, M.; Coute, A.; Fievet, F. ZnO nanoparticles: Synthesis, characterization, and ecotoxicological studies. *Langmuir* **2010**, *26*, 6522–6528. [CrossRef]
19. Khaydarov, R.A.; Khaydarov, R.R.; Gapurova, O.; Estrin, Y.; Scheper, T. Electrochemical method for the synthesis of silver nanoparticles. *J. Nanopart. Res.* **2008**, *11*, 1193–1200. [CrossRef]
20. Starowicz, M.; Starowicz, P.; Zukrowski, J.; Przewoznik, J.; Lemanski, A.; Kapusta, C.; Banas, J. Electrochemical synthesis of magnetic iron oxide nanoparticles with controlled size. *J. Nanopart. Res.* **2011**, *13*, 7167–7176. [CrossRef]

21. Armijo Garcia, D.; Mendoza, L.; Vizuete, K.; Debut, A.; Arias, M.T.; Gavilanes, A.; Terencio, T.; Avila, E.; Jeffryes, C.; Dahoumane, S.A. Sugar-Mediated Green Synthesis of Silver Selenide Semiconductor Nanocrystals under Ultrasound Irradiation. *Molecules* **2020**, *25*. [CrossRef]

22. Yan, Q.; Qiu, M.; Chen, X.; Fan, Y. Ultrasound Assisted Synthesis of Size-Controlled Aqueous Colloids for the Fabrication of Nanoporous Zirconia Membrane. *Front. Chem.* **2019**, *7*, 337. [CrossRef] [PubMed]

23. Kumar, S.V.; Bafana, A.P.; Pawar, P.; Faltane, M.; Rahman, A.; Dahoumane, S.A.; Kucknoor, A.; Jeffryes, C.S. Optimized production of antibacterial copper oxide nanoparticles in a microwave-assisted synthesis reaction using response surface methodology. *Colloid Surf. A Physicochem. Eng. Aspect* **2019**, *573*, 170–178. [CrossRef]

24. Bafana, A.; Kumar, S.V.; Temizel-Sekeryan, S.; Dahoumane, S.A.; Haselbach, L.; Jeffryes, C.S. Evaluating microwave-synthesized silver nanoparticles from silver nitrate with life cycle assessment techniques. *Sci. Total Environ.* **2018**, *636*, 936–943. [CrossRef] [PubMed]

25. Gomez-Gomez, B.; Arregui, L.; Serrano, S.; Santos, A.; Perez-Corona, T.; Madrid, Y. Selenium and tellurium-based nanoparticles as interfering factors in quorum sensing-regulated processes: Violacein production and bacterial biofilm formation. *Metallomics* **2019**, *11*, 1104–1114. [CrossRef]

26. Dahoumane, S.A.; Jeffryes, C.; Mechouet, M.; Agathos, S.N. Biosynthesis of Inorganic Nanoparticles: A Fresh Look at the Control of Shape, Size and Composition. *Bioengineering* **2017**, *4*, 14. [CrossRef]

27. Rahman, A.; Lin, J.; Jaramillo, F.E.; Bazylinski, D.A.; Jeffryes, C.; Dahoumane, S.A. In Vivo Biosynthesis of Inorganic Nanomaterials Using Eukaryotes-A Review. *Molecules* **2020**, *25*. [CrossRef]

28. Kumar, S.V.; Bafana, A.P.; Pawar, P.; Rahman, A.; Dahoumane, S.A.; Jeffryes, C.S. High conversion synthesis of <10 nm starch-stabilized silver nanoparticles using microwave technology. *Sci. Rep.* **2018**, *8*, 5106. [CrossRef]

29. Rahman, A.; Kumar, S.; Bafana, A.; Lin, J.; Dahoumane, S.A.; Jeffryes, C. A Mechanistic View of the Light-Induced Synthesis of Silver Nanoparticles Using Extracellular Polymeric Substances of *Chlamydomonas reinhardtii*. *Molecules* **2019**, *24*. [CrossRef]

30. Ramani, M.; Ponnusamy, S.; Muthamizhchelvan, C.; Marsili, E. Amino acid-mediated synthesis of zinc oxide nanostructures and evaluation of their facet-dependent antimicrobial activity. *Colloid Surf. B Biointerfaces* **2014**, *117*, 233–239. [CrossRef]

31. Manivasagan, P.; Oh, J. Marine polysaccharide-based nanomaterials as a novel source of nanobiotechnological applications. *Int. J. Biol. Macromol.* **2016**, *82*, 315–327. [CrossRef] [PubMed]

32. Sharma, G.; Sharma, A.R.; Bhavesh, R.; Park, J.; Ganbold, B.; Nam, J.S.; Lee, S.S. Biomolecule-mediated synthesis of selenium nanoparticles using dried *Vitis vinifera* (raisin) extract. *Molecules* **2014**, *19*, 2761–2770. [CrossRef] [PubMed]

33. Kuppusamy, P.; Yusoff, M.M.; Maniam, G.P.; Govindan, N. Biosynthesis of metallic nanoparticles using plant derivatives and their new avenues in pharmacological applications—An updated report. *Saudi Pharm. J.* **2016**, *24*, 473–484. [CrossRef] [PubMed]

34. Singh, P.; Kim, Y.J.; Zhang, D.; Yang, D.C. Biological Synthesis of Nanoparticles from Plants and Microorganisms. *Trend Biotechnol.* **2016**, *34*, 588–599. [CrossRef]

35. Jeffryes, C.; Agathos, S.N.; Rorrer, G. Biogenic nanomaterials from photosynthetic microorganisms. *Curr. Opin. Biotechnol.* **2015**, *33*, 23–31. [CrossRef]

36. Dahoumane, S.A.; Mechouet, M.; Wijesekera, K.; Filipe, C.D.M.; Sicard, C.; Bazylinski, D.A.; Jeffryes, C. Algae-mediated biosynthesis of inorganic nanomaterials as a promising route in nanobiotechnology—A review. *Green Chem.* **2017**, *19*, 552–587. [CrossRef]

37. Dahoumane, S.A.; Wujcik, E.K.; Jeffryes, C. Noble metal, oxide and chalcogenide-based nanomaterials from scalable phototrophic culture systems. *Enzyme Microb. Technol.* **2016**, *95*, 13–27. [CrossRef]

38. Dahoumane, S.A.; Djediat, C.; Yéprémian, C.; Couté, A.; Fiévet, F.; Brayner, R. Design of magnetic akaganeite-cyanobacteria hybrid biofilms. *Thin Solid Film* **2010**, *518*, 5432–5436. [CrossRef]

39. Dahoumane, S.A.; Yéprémian, C.; Djédiat, C.; Couté, A.; Fiévet, F.; Coradin, T.; Brayner, R. A global approach of the mechanism involved in the biosynthesis of gold colloids using micro-algae. *J. Nanopart. Res.* **2014**, *16*. [CrossRef]

40. Dahoumane, S.A.; Yéprémian, C.; Djédiat, C.; Couté, A.; Fiévet, F.; Coradin, T.; Brayner, R. Improvement of kinetics, yield, and colloidal stability of biogenic gold nanoparticles using living cells of *Euglena gracilis* microalga. *J. Nanopart. Res.* **2016**, *18*. [CrossRef]

41. Das, S.K.; Marsili, E. A green chemical approach for the synthesis of gold nanoparticles: Characterization and mechanistic aspect. *Rev. Environ. Sci. Biotechnol.* **2010**, *9*, 199–204. [CrossRef]

42. Gahlawat, G.; Roy Choudhury, A. A review on the biosynthesis of metal and metal salt nanoparticles by microbes. *RSC Adv.* **2019**, *9*, 12944–12967. [CrossRef]

43. Pantidos, N.; Horsfall, L.E. Biological Synthesis of Metallic Nanoparticles by Bacteria, Fungi and Plants. *J. Nanomed. Nanotechnol.* **2014**, *5*. [CrossRef]

44. Kitching, M.; Ramani, M.; Marsili, E. Fungal biosynthesis of gold nanoparticles: Mechanism and scale up. *Microb. Biotechnol.* **2015**, *8*, 904–917. [CrossRef] [PubMed]

45. Rahman, A.; Kumar, S.; Bafana, A.; Dahoumane, S.A.; Jeffryes, C. Biosynthetic Conversion of Ag(+) to highly Stable Ag(0) Nanoparticles by Wild Type and Cell Wall Deficient Strains of *Chlamydomonas reinhardtii*. *Molecules* **2019**, *24*, 98. [CrossRef]

46. Wang, D.; Xia, X.; Wu, S.; Zheng, S.; Wang, G. The essentialness of glutathione reductase GorA for biosynthesis of Se(0)-nanoparticles and GSH for CdSe quantum dot formation in *Pseudomonas stutzeri* TS44. *J. Hazard. Mater.* **2019**, *366*, 301–310. [CrossRef]

47. Chaudhary, S.; Umar, A.; Mehta, S.K. Selenium nanomaterials: An overview of recent developments in synthesis, properties and potential applications. *Prog. Mater. Sci.* **2016**, *83*, 270–329. [CrossRef]

48. Rasouli, M. Biosynthesis of selenium nanoparticles using yeast *Nematospora coryli* and examination of their anti-candida and anti-oxidant activities. *IET Nanobiotechnol.* **2019**, *13*, 214–218. [CrossRef]
49. Piacenza, E.; Presentato, A.; Zonaro, E.; Lampis, S.; Vallini, G.; Turner, R.J. Selenium and tellurium nanomaterials. *Phys. Sci. Rev.* **2018**, *3*. [CrossRef]
50. Elahian, F.; Reiisi, S.; Shahidi, A.; Mirzaei, S.A. High-throughput bioaccumulation, biotransformation, and production of silver and selenium nanoparticles using genetically engineered *Pichia pastoris*. *Nanomed. Nanobiotechnol.* **2017**, *13*, 853–861. [CrossRef]
51. Liang, X.; Perez, M.A.M.; Nwoko, K.C.; Egbers, P.; Feldmann, J.; Csetenyi, L.; Gadd, G.M. Fungal formation of selenium and tellurium nanoparticles. *Appl. Microbiol. Biotechnol.* **2019**, *103*, 7241–7259. [CrossRef] [PubMed]
52. Ramanujam, J.; Singh, U.P. Copper indium gallium selenide based solar cells—A review. *Energy Environ. Sci.* **2017**, *10*, 1306–1319. [CrossRef]
53. Tan, L.C.; Nancharaiah, Y.V.; Lu, S.; van Hullebusch, E.D.; Gerlach, R.; Lens, P.N.L. Biological treatment of selenium-laden wastewater containing nitrate and sulfate in an upflow anaerobic sludge bed reactor at pH 5.0. *Chemosphere* **2018**, *211*, 684–693. [CrossRef] [PubMed]
54. Schwarz, K.; Foltz, C.M. Selenium as an integral part of factor 3 against dietary necrotic liver degeneration. *J. Am. Chem. Soc.* **1957**, *79*, 3292–3293. [CrossRef]
55. Shoeibi, S.; Mashreghi, M. Biosynthesis of selenium nanoparticles using *Enterococcus faecalis* and evaluation of their antibacterial activities. *J. Trace Elem. Med. Biol.* **2017**, *39*, 135–139. [CrossRef] [PubMed]
56. Dhanjal, S.; Singh Cameotra, S. Aerobic biogenesis of selenium nanospheres by *Bacillus cereus* isolated from coalmine soil. *Microbial. Cell Fact.* **2010**, *9*, 52. [CrossRef]
57. Mohammed, E.T.; Safwat, G.M. Assessment of the ameliorative role of selenium nanoparticles on the oxidative stress of acetaminophen in some tissues of male albino rats. *Beni-Suef Univ. J. Basic Appl. Sci.* **2013**, *2*, 80–85. [CrossRef]
58. Wadhwani, S.A.; Gorain, M.; Banerjee, P.; Shedbalkar, U.U.; Singh, R.; Kundu, G.C.; Chopade, B.A. Green synthesis of selenium nanoparticles using *Acinetobacter* sp. SW30: Optimization, characterization and its anticancer activity in breast cancer cells. *Int. J. Nanomed.* **2017**, *12*, 6841–6855. [CrossRef]
59. Vaigankar, D.C.; Dubey, S.K.; Mujawar, S.Y.; D'Costa, A.; Shyama, S.K. Tellurite biotransformation and detoxification by *Shewanella baltica* with simultaneous synthesis of tellurium nanorods exhibiting photo-catalytic and anti-biofilm activity. *Ecotoxicol. Environ. Saf.* **2018**, *165*, 516–526. [CrossRef]
60. Huang, W.; Wu, H.; Li, X.; Chen, T. Facile One-Pot Synthesis of Tellurium Nanorods as Antioxidant and Anticancer Agents. *Chem. Asian J.* **2016**, *11*, 2301–2311. [CrossRef]
61. Medina Cruz, D.; Mi, G.; Webster, T.J. Synthesis and characterization of biogenic selenium nanoparticles with antimicrobial properties made by *Staphylococcus aureus*, methicillin-resistant *Staphylococcus aureus* (MRSA), *Escherichia coli*, and *Pseudomonas aeruginosa*. *J. Biomed. Mater. Res. A* **2018**, *106*, 1400–1412. [CrossRef] [PubMed]
62. Medina Cruz, D.; Tien-Street, W.; Zhang, B.; Huang, X.; Vernet Crua, A.; Nieto-Arguello, A.; Cholula-Diaz, J.L.; Martinez, L.; Huttel, Y.; Ujue Gonzalez, M.; et al. Citric Juice-mediated Synthesis of Tellurium Nanoparticles with Antimicrobial and Anticancer Properties. *Green Chem.* **2019**, *21*, 1982–1988. [CrossRef] [PubMed]
63. Vahidi, H.; Kobarfard, F.; Alizadeh, A.; Saravanan, M.; Barabadi, H. Green nanotechnology-based tellurium nanoparticles: Exploration of their antioxidant, antibacterial, antifungal and cytotoxic potentials against cancerous and normal cells compared to potassium tellurite. *Inorg. Chem. Commun.* **2021**, *124*, 108385. [CrossRef]
64. Mirjani, R.; Faramarzi, M.A.; Sharifzadeh, M.; Setayesh, N.; Khoshayand, M.R.; Shahverdi, A.R. Biosynthesis of tellurium nanoparticles by *Lactobacillus plantarum* and the effect of nanoparticle-enriched probiotics on the lipid profiles of mice. *IET Nanobiotechnol.* **2015**, *9*, 300–305. [CrossRef]
65. Espinosa-Ortiz, E.J.; Rene, E.R.; Guyot, F.; van Hullebusch, E.D.; Lens, P.N.L. Biomineralization of tellurium and selenium-tellurium nanoparticles by the white-rot fungus *Phanerochaete chrysosporium*. *Int. Biodeter. Biodegr.* **2017**, *124*, 258–266. [CrossRef]
66. Espinosa-Ortiz, E.J.; Rene, E.R.; van Hullebusch, E.D.; Lens, P.N.L. Removal of selenite from wastewater in a *Phanerochaete chrysosporium* pellet based fungal bioreactor. *Int. Biodeter. Biodegr.* **2015**, *102*, 361–369. [CrossRef]
67. Espinosa-Ortiz, E.J.; Pechaud, Y.; Lauchnor, E.; Rene, E.R.; Gerlach, R.; Peyton, B.M.; van Hullebusch, E.D.; Lens, P.N. Effect of selenite on the morphology and respiratory activity of *Phanerochaete chrysosporium* biofilms. *Bioresour. Technol.* **2016**, *210*, 138–145. [CrossRef]
68. Espinosa-Ortiz, E.J.; Gonzalez-Gil, G.; Saikaly, P.E.; van Hullebusch, E.D.; Lens, P.N. Effects of selenium oxyanions on the white-rot fungus *Phanerochaete chrysosporium*. *Appl. Microbiol. Biotechnol.* **2015**, *99*, 2405–2418. [CrossRef]
69. Gharieb, M.M.; Gadd, G.M. The kinetics of ^{75}Se-selenite uptake by *Saccharomyces cerevisiae* and the vacuolization response to high concentrations. *Mycol. Res.* **2004**, *108*, 1415–1422. [CrossRef]
70. Verma, A.; Gautam, S.P.; Bansal, K.K.; Prabhakar, N.; Rosenholm, J.M. Green Nanotechnology: Advancement in Phytoformulation Research. *Medicines* **2019**, *6*. [CrossRef]
71. Zonaro, E.; Lampis, S.; Turner, R.J.; Qazi, S.J.; Vallini, G. Biogenic selenium and tellurium nanoparticles synthesized by environmental microbial isolates efficaciously inhibit bacterial planktonic cultures and biofilms. *Front. Microbiol.* **2015**, *6*, 584. [CrossRef] [PubMed]

72. Pearce, C.I.; Pattrick, R.A.; Law, N.; Charnock, J.M.; Coker, V.S.; Fellowes, J.W.; Oremland, R.S.; Lloyd, J.R. Investigating different mechanisms for biogenic selenite transformations: *Geobacter sulfurreducens*, *Shewanella oneidensis* and *Veillonella atypica*. *Environ. Technol.* **2009**, *30*, 1313–1326. [CrossRef]
73. Pasula, R.R.; Lim, S. Engineering nanoparticle synthesis using microbial factories. *Eng. Biol.* **2017**, *1*, 12–17. [CrossRef]
74. Ruiz Fresneda, M.A.; Delgado Martín, J.; Gómez Bolívar, J.; Fernández Cantos, M.V.; Bosch-Estévez, G.; Martínez Moreno, M.F.; Merroun, M.L. Green synthesis and biotransformation of amorphous Se nanospheres to trigonal 1D Se nanostructures: Impact on Se mobility within the concept of radioactive waste disposal. *Environ. Sci. Nano* **2018**, *5*, 2103–2116. [CrossRef]
75. Xia, X.; Wu, S.; Li, N.; Wang, D.; Zheng, S.; Wang, G. Novel bacterial selenite reductase CsrF responsible for Se(IV) and Cr(VI) reduction that produces nanoparticles in *Alishewanella* sp. WH16-1. *J. Hazard. Mater.* **2018**, *342*, 499–509. [CrossRef]
76. Rajasree, R.S.R.; Gayathri, S. Extracellular biosynthesis of Selenium nanoparticles using some species of *Lactobacillus*. *Indian J. Geo-Marine Sci.* **2015**, *43*, 766–775.
77. Dahoumane, S.A.; Djediat, C.; Yéprémian, C.; Couté, A.; Fiévet, F.; Coradin, T.; Brayner, R. Species selection for the design of gold nanobioreactor by photosynthetic organisms. *J. Nanopart. Res.* **2012**, *14*. [CrossRef]
78. Dahoumane, S.A.; Mechouet, M.; Alvarez, F.J.; Agathos, S.N.; Jeffryes, C. Microalgae: An outstanding tool in nanotechnology. *Bionatura* **2016**, *1*. [CrossRef]
79. Dahoumane, S.A.; Wijesekera, K.; Filipe, C.D.; Brennan, J.D. Stoichiometrically controlled production of bimetallic Gold-Silver alloy colloids using micro-alga cultures. *J. Colloid Interface Sci.* **2014**, *416*, 67–72. [CrossRef] [PubMed]
80. Alqadi, M.K.; Abo Noqtah, O.A.; Alzoubi, F.Y.; Alzouby, J.; Aljarrah, K. pH effect on the aggregation of silver nanoparticles synthesized by chemical reduction. *Mater. Sci. Pol.* **2014**, *32*, 107–111. [CrossRef]
81. Castro, L.; Blázquez, M.L.; González, F.; Muñoz, J.A.; Ballester, A. Extracellular biosynthesis of gold nanoparticles using sugar beet pulp. *Chem. Eng. J.* **2010**, *164*, 92–97. [CrossRef]
82. Wu, S.; Sun, K.; Wang, X.; Wang, D.; Wan, X.; Zhang, J. Protonation of epigallocatechin-3-gallate (EGCG) results in massive aggregation and reduced oral bioavailability of EGCG-dispersed selenium nanoparticles. *J. Agric. Food. Chem.* **2013**, *61*, 7268–7275. [CrossRef] [PubMed]
83. Akcay, F.A.; Avci, A. Effects of process conditions and yeast extract on the synthesis of selenium nanoparticles by a novel indigenous isolate *Bacillus* sp. EKT1 and characterization of nanoparticles. *Arch. Microbiol.* **2020**, *202*, 2233–2243. [CrossRef]
84. Kuroda, M.; Notaguchi, E.; Sato, A.; Yoshioka, M.; Hasegawa, A.; Kagami, T.; Narita, T.; Yamashita, M.; Sei, K.; Soda, S.; et al. Characterization of *Pseudomonas stutzeri* NT-I capable of removing soluble selenium from the aqueous phase under aerobic conditions. *J. Biosci. Bioeng.* **2011**, *112*, 259–264. [CrossRef] [PubMed]
85. Ge, J.P.; Xu, S.; Liu, L.P.; Li, Y.D. A positive-microemulsion method for preparing nearly uniform Ag$_2$Se nanoparticles at low temperature. *Chem. Eur. J.* **2006**, *12*, 3672–3677. [CrossRef]
86. Mollania, N.; Tayebee, R.; Narenji-Sani, F. An environmentally benign method for the biosynthesis of stable selenium nanoparticles. *Res. Chem. Intermed.* **2015**, *42*, 4253–4271. [CrossRef]
87. Ashour, A.H.; El-Batal, A.I.; Maksoud, M.I.A.A.; El-Sayyad, G.S.; Labib, S.; Abdeltwab, E.; El-Okr, M.M. Antimicrobial activity of metal-substituted cobalt ferrite nanoparticles synthesized by sol–gel technique. *Particuology* **2018**, *40*, 141–151. [CrossRef]
88. Wong, C.W.; Chan, Y.S.; Jeevanandam, J.; Pal, K.; Bechelany, M.; Abd Elkodous, M.; El-Sayyad, G.S. Response Surface Methodology Optimization of Mono-dispersed MgO Nanoparticles Fabricated by Ultrasonic-Assisted Sol–Gel Method for Outstanding Antimicrobial and Antibiofilm Activities. *J. Clust. Sci.* **2020**, *31*, 367–389. [CrossRef]
89. Baer, D.R. Surface Characterization of Nanoparticles: Critical needs and significant challenges. *J. Surf. Anal.* **2011**, *17*, 163–169. [CrossRef]
90. Sadtler, B.; Burgos, S.P.; Batara, N.A.; Beardslee, J.A.; Atwater, H.A.; Lewis, N.S. Phototropic growth control of nanoscale pattern formation in photoelectrodeposited Se-Te films. *Proc. Natl. Acad. Sci. USA* **2013**, *110*, 19707–19712. [CrossRef]
91. Vijayanandan, A.S.; Balakrishnan, R.M. Impact of precursor concentration on biological synthesis of cobalt oxide nanoparticles. *Data Brief* **2018**, *19*, 1941–1947. [CrossRef] [PubMed]
92. Sharifi Dehsari, H.; Halda Ribeiro, A.; Ersöz, B.; Tremel, W.; Jakob, G.; Asadi, K. Effect of precursor concentration on size evolution of iron oxide nanoparticles. *CrystEngComm* **2017**, *19*, 6694–6702. [CrossRef]
93. Moloto, N.; Revaprasadu, N.; Musetha, P.L.; Moloto, M.J. The effect of precursor concentration, temperature and capping group on the morphology of CdS nanoparticles. *J. Nanosci. Nanotechnol.* **2009**, *9*, 4760–4766. [CrossRef] [PubMed]
94. Dahoumane, S.A.; Djediat, C.; Yepremian, C.; Coute, A.; Fievet, F.; Coradin, T.; Brayner, R. Recycling and adaptation of *Klebsormidium flaccidum* microalgae for the sustained production of gold nanoparticles. *Biotechnol. Bioeng.* **2012**, *109*, 284–288. [CrossRef] [PubMed]
95. Rahman, A.; Kumar, S.; Bafana, A.; Dahoumane, S.A.; Jeffryes, C. Individual and Combined Effects of Extracellular Polymeric Substances and Whole Cell Components of *Chlamydomonas reinhardtii* on Silver Nanoparticle Synthesis and Stability. *Molecules* **2019**, *24*. [CrossRef]
96. Chaudhary, S.; Mehta, S.K. Selenium nanomaterials: Applications in electronics, catalysis and sensors. *J. Nanosci. Nanotechnol.* **2014**, *14*, 1658–1674. [CrossRef]
97. Wang, X.; Liu, G.; Zhou, J.; Wang, J.; Jin, R.; Lv, H. Quinone-mediated reduction of selenite and tellurite by *Escherichia coli*. *Bioresour. Technol.* **2011**, *102*, 3268–3271. [CrossRef]

98. Presentato, A.; Piacenza, E.; Darbandi, A.; Anikovskiy, M.; Cappelletti, M.; Zannoni, D.; Turner, R.J. Assembly, growth and conductive properties of tellurium nanorods produced by *Rhodococcus aetherivorans* BCP1. *Sci. Rep.* **2018**, *8*, 3923. [CrossRef]

99. Presentato, A.; Piacenza, E.; Anikovskiy, M.; Cappelletti, M.; Zannoni, D.; Turner, R.J. Biosynthesis of selenium-nanoparticles and -nanorods as a product of selenite bioconversion by the aerobic bacterium *Rhodococcus aetherivorans* BCP1. *New Biotechnol.* **2018**, *41*, 1–8. [CrossRef]

100. Prange, A.; Birzele, B.; Hormes, J.; Modrow, H. Investigation of different human pathogenic and food contaminating bacteria and moulds grown on selenite/selenate and tellurite/tellurate by X-ray absorption spectroscopy. *Food Control.* **2005**, *16*, 723–728. [CrossRef]

101. Wang, T.; Yang, L.; Zhang, B.; Liu, J. Extracellular biosynthesis and transformation of selenium nanoparticles and application in H_2O_2 biosensor. *Colloid Surf. B Biointerface.* **2010**, *80*, 94–102. [CrossRef] [PubMed]

102. Menon, S.; Shanmugam, V. Cytotoxicity Analysis of Biosynthesized Selenium Nanoparticles Towards A549 Lung Cancer Cell Line. *J. Inorg. Organometal. Polym. Mater.* **2019**, *30*, 1852–1864. [CrossRef]

103. Rajkumar, K.; Mvs, S.; Koganti, S.; Burgula, S. Selenium Nanoparticles Synthesized Using *Pseudomonas stutzeri* (MH191156) Show Antiproliferative and Anti-angiogenic Activity Against Cervical Cancer Cells. *Int. J. Nanomed.* **2020**, *15*, 4523–4540. [CrossRef] [PubMed]

104. Sayes, C.M.; Warheit, D.B. Characterization of nanomaterials for toxicity assessment. *Adv. Rev.* **2009**, *1*, 960–970. [CrossRef] [PubMed]

105. Warheit, D.B. Nanoparticles. *Mater. Today* **2004**, *7*, 32–35. [CrossRef]

106. Warheit, D.B.; Sayes, C.M.; Reed, K.L.; Swain, K.A. Health effects related to nanoparticle exposures: Environmental, health and safety considerations for assessing hazards and risks. *Pharmacol. Ther.* **2008**, *120*, 35–42. [CrossRef]

107. Powers, K.W.; Palazuelos, M.; Moudgil, B.M.; Roberts, S.M. Characterization of the size, shape, and state of dispersion of nanoparticles for toxicological studies. *Nanotoxicology* **2009**, *1*, 42–51. [CrossRef]

108. Kapur, M.; Soni, K.; Kohli, K. Green Synthesis of Selenium Nanoparticles from Broccoli, Characterization, Application and Toxicity. *Adv. Tech. Biol. Med.* **2017**, *5*. [CrossRef]

109. Estevam, E.C.; Griffin, S.; Nasim, M.J.; Denezhkin, P.; Schneider, R.; Lilischkis, R.; Dominguez-Alvarez, E.; Witek, K.; Latacz, G.; Keck, C.; et al. Natural selenium particles from *Staphylococcus carnosus*: Hazards or particles with particular promise? *J. Hazard. Mater.* **2017**, *324*, 22–30. [CrossRef] [PubMed]

110. Bajaj, M.; Schmidt, S.; Winter, J. Formation of Se (0) Nanoparticles by *Duganella* sp. and *Agrobacterium* sp. isolated from Se-laden soil of North-East Punjab, India. *Microb. Cell Fact.* **2012**, *11*, 64. [CrossRef] [PubMed]

111. Combs, G.F.J.; Garbisu, C.; Yee, B.C.; Yee, A.; Carlson, D.E.; Smith, N.R.; Magyarosy, A.C.; Leighton, T.; Buchanan, B.B. Bioavailability of Selenium Accumulated by Selenite-reducing Bacteria. *Biol. Trace Elem. Res.* **1996**, *52*, 209–225. [CrossRef]

112. Kenward, P.A.; Fowle, D.A.; Yee, N. Microbial Selenate Sorption and Reduction in Nutrient Limited Systems. *Environ. Sci. Technol.* **2006**, *40*, 3782–3786. [CrossRef] [PubMed]

113. Lampis, S.; Zonaro, E.; Bertolini, C.; Cecconi, D.; Monti, F.; Micaroni, M.; Turner, R.J.; Butler, C.S.; Vallini, G. Selenite biotransformation and detoxification by *Stenotrophomonas maltophilia* SeITE02: Novel clues on the route to bacterial biogenesis of selenium nanoparticles. *J. Hazard. Mater.* **2017**, *324*, 3–14. [CrossRef] [PubMed]

114. Cremonini, E.; Zonaro, E.; Donini, M.; Lampis, S.; Boaretti, M.; Dusi, S.; Melotti, P.; Lleo, M.M.; Vallini, G. Biogenic selenium nanoparticles: Characterization, antimicrobial activity and effects on human dendritic cells and fibroblasts. *Microb. Biotechnol.* **2016**, *9*, 758–771. [CrossRef] [PubMed]

115. Figueroa, M.; Fernandez, V.; Arenas-Salinas, M.; Ahumada, D.; Munoz-Villagran, C.; Cornejo, F.; Vargas, E.; Latorre, M.; Morales, E.; Vasquez, C.; et al. Synthesis and Antibacterial Activity of Metal(loid) Nanostructures by Environmental Multi-Metal(loid) Resistant Bacteria and Metal(loid)-Reducing Flavoproteins. *Front. Microbiol.* **2018**, *9*, 959. [CrossRef]

116. Tam, K.; Ho, C.T.; Lee, J.H.; Lai, M.; Chang, C.H.; Rheem, Y.; Chen, W.; Hur, H.G.; Myung, N.V. Growth mechanism of amorphous selenium nanoparticles synthesized by *Shewanella* sp. HN-41. *Biosci. Biotechnol. Biochem.* **2010**, *74*, 696–700. [CrossRef]

117. Shakibaie, M.; Khorramizadeh, M.R.; Faramarzi, M.A.; Sabzevari, O.; Shahverdi, A.R. Biosynthesis and recovery of selenium nanoparticles and the effects on matrix metalloproteinase-2 expression. *Biotechnol. Appl. Biochem.* **2010**, *56*, 7–15. [CrossRef] [PubMed]

118. Shakibaie, M.; Forootanfar, H.; Golkari, Y.; Mohammadi-Khorsand, T.; Shakibaie, M.R. Anti-biofilm activity of biogenic selenium nanoparticles and selenium dioxide against clinical isolates of *Staphylococcus aureus*, *Pseudomonas aeruginosa*, and *Proteus mirabilis*. *J. Trace Elem. Med. Biol.* **2015**, *29*, 235–241. [CrossRef] [PubMed]

119. Beheshti, N.; Soflaei, S.; Shakibaie, M.; Yazdi, M.H.; Ghaffarifar, F.; Dalimi, A.; Shahverdi, A.R. Efficacy of biogenic selenium nanoparticles against Leishmania major: In vitro and in vivo studies. *J. Trace Elem. Med. Biol.* **2013**, *27*, 203–207. [CrossRef] [PubMed]

120. Forootanfar, H.; Adeli-Sardou, M.; Nikkhoo, M.; Mehrabani, M.; Amir-Heidari, B.; Shahverdi, A.R.; Shakibaie, M. Antioxidant and cytotoxic effect of biologically synthesized selenium nanoparticles in comparison to selenium dioxide. *J. Trace Elem. Med. Biol.* **2014**, *28*, 75–79. [CrossRef]

121. Fernandez-Llamosas, H.; Castro, L.; Blazquez, M.L.; Diaz, E.; Carmona, M. Biosynthesis of selenium nanoparticles by *Azoarcus* sp. CIB. *Microb. Cell Fact.* **2016**, *15*, 109. [CrossRef] [PubMed]

122. Khoei, N.S.; Lampis, S.; Zonaro, E.; Yrjala, K.; Bernardi, P.; Vallini, G. Insights into selenite reduction and biogenesis of elemental selenium nanoparticles by two environmental isolates of *Burkholderia fungorum*. *New Biotechnol.* **2017**, *34*, 1–11. [CrossRef] [PubMed]

123. Tan, Y.; Wang, Y.; Wang, Y.; Xu, D.; Huang, Y.; Wang, D.; Wang, G.; Rensing, C.; Zheng, S. Novel mechanisms of selenate and selenite reduction in the obligate aerobic bacterium *Comamonas testosteroni* S44. *J. Hazard. Mater.* **2018**, *359*, 129–138. [CrossRef] [PubMed]

124. Kamnev, A.A.; Mamchenkova, P.V.; Dyatlova, Y.A.; Tugarova, A.V. FTIR spectroscopic studies of selenite reduction by cells of the rhizobacterium *Azospirillum brasilense* Sp7 and the formation of selenium nanoparticles. *J. Mol. Struct.* **2017**, *1140*, 106–112. [CrossRef]

125. Vogel, M.; Fischer, S.; Maffert, A.; Hubner, R.; Scheinost, A.C.; Franzen, C.; Steudtner, R. Biotransformation and detoxification of selenite by microbial biogenesis of selenium-sulfur nanoparticles. *J. Hazard. Mater.* **2018**, *344*, 749–757. [CrossRef]

126. Kora, A.J.; Rastogi, L. Biomimetic synthesis of selenium nanoparticles by *Pseudomonas aeruginosa* ATCC 27853: An approach for conversion of selenite. *J. Environ. Manag.* **2016**, *181*, 231–236. [CrossRef]

127. Pouri, S.; Motamedi, H.; Honary, S.; Kazeminezhad, I. Biological Synthesis of Selenium Nanoparticles and Evaluation of their Bioavailability. *Braz. Arch. Biol. Techn.* **2018**, *60*. [CrossRef]

128. Srivastava, N.; Mukhopadhyay, M. Biosynthesis and structural characterization of selenium nanoparticles mediated by *Zooglea ramigera*. *Powder Technol.* **2013**, *244*, 26–29. [CrossRef]

129. Subedi, G.; Taylor, J.; Hatam, I.; Baldwin, S.A. Simultaneous selenate reduction and denitrification by a consortium of enriched mine site bacteria. *Chemosphere* **2017**, *183*, 536–545. [CrossRef]

130. Wadhwani, S.A.; Shedbalkar, U.U.; Singh, R.; Chopade, B.A. Biosynthesis of gold and selenium nanoparticles by purified protein from *Acinetobacter* sp. SW 30. *Enzyme Microb. Technol.* **2018**, *111*, 81–86. [CrossRef]

131. Ramya, S.; Shanmugasundaram, T.; Balagurunathan, R. Biomedical potential of actinobacterially synthesized selenium nanoparticles with special reference to anti-biofilm, anti-oxidant, wound healing, cytotoxic and anti-viral activities. *J. Trace Elem. Med. Biol.* **2015**, *32*, 30–39. [CrossRef] [PubMed]

132. Ameri, A.; Shakibaie, M.; Ameri, A.; Faramarzi, M.A.; Amir-Heidari, B.; Forootanfar, H. Photocatalytic decolorization of bromothymol blue using biogenic selenium nanoparticles synthesized by terrestrial actinomycete *Streptomyces griseobrunneus* strain FSHH12. *Desalin. Water Treat.* **2015**, *57*, 21552–21563. [CrossRef]

133. Fernandez-Llamosas, H.; Castro, L.; Blazquez, M.L.; Diaz, E.; Carmona, M. Speeding up bioproduction of selenium nanoparticles by using *Vibrio natriegens* as microbial factory. *Sci. Rep.* **2017**, *7*, 16046. [CrossRef] [PubMed]

134. Xu, C.; Guo, Y.; Qiao, L.; Ma, L.; Cheng, Y.; Roman, A. Biogenic Synthesis of Novel Functionalized Selenium Nanoparticles by *Lactobacillus casei* ATCC 393 and Its Protective Effects on Intestinal Barrier Dysfunction Caused by Enterotoxigenic *Escherichia coli* K88. *Front. Microbiol.* **2018**, *9*, 1129. [CrossRef] [PubMed]

135. Shaaban, M.; El-Mahdy, A.M. Biosynthesis of Ag, Se, and ZnO nanoparticles with antimicrobial activities against resistant pathogens using waste isolate *Streptomyces enissocaesilis*. *IET Nanobiotechnol.* **2018**, *12*, 741–747. [CrossRef] [PubMed]

136. Ramya, S.; Shanmugasundaram, T.; Balagurunathan, R. Actinobacterial enzyme mediated synthesis of selenium nanoparticles for antibacterial, mosquito larvicidal and anthelminthic applications. *Particul. Sci. Technol.* **2019**, *38*, 63–72. [CrossRef]

137. San Keskin, N.O.; Akbal Vural, O.; Abaci, S. Biosynthesis of Noble Selenium Nanoparticles from *Lysinibacillus* sp. NOSK for Antimicrobial, Antibiofilm Activity, and Biocompatibility. *Geomicrobiol. J.* **2020**, *37*, 919–928. [CrossRef]

138. Srivastava, P.; Kowshik, M. Anti-neoplastic selenium nanoparticles from *Idiomarina* sp. PR58-8. *Enzyme Microb. Technol.* **2016**, *95*, 192–200. [CrossRef]

139. Srivastava, N.; Mukhopadhyay, M. Green synthesis and structural characterization of selenium nanoparticles and assessment of their antimicrobial property. *Bioprocess. Biosyst. Eng.* **2015**, *38*, 1723–1730. [CrossRef]

140. Song, D.; Li, X.; Cheng, Y.; Xiao, X.; Lu, Z.; Wang, Y.; Wang, F. Aerobic biogenesis of selenium nanoparticles by *Enterobacter cloacae* Z0206 as a consequence of fumarate reductase mediated selenite reduction. *Sci. Rep.* **2017**, *7*. [CrossRef]

141. Li, D.B.; Cheng, Y.Y.; Wu, C.; Li, W.W.; Li, N.; Yang, Z.C.; Tong, Z.H.; Yu, H.Q. Selenite reduction by *Shewanella oneidensis* MR-1 is mediated by fumarate reductase in periplasm. *Sci. Rep.* **2014**, *4*. [CrossRef] [PubMed]

142. Hnain, A.; Brooks, J.; Lefebvre, D.D. The synthesis of elemental selenium particles by *Synechococcus leopoliensis*. *Appl. Microbiol. Biotechnol.* **2013**, *97*, 10511–10519. [CrossRef] [PubMed]

143. Zheng, S.; Su, J.; Wang, L.; Yao, R.; Wang, D.; Deng, Y.; Wang, R.; Wang, G.; Rensing, C. Selenite reduction by the obligate aerobic bacterium *Comamonas testosteroni* S44 isolated from a metal-contaminated soil. *BMC Microbiol.* **2014**, *14*, 204. [CrossRef] [PubMed]

144. Tugarova, A.V.; Mamchenkova, P.V.; Khanadeev, V.A.; Kamnev, A.A. Selenite reduction by the rhizobacterium *Azospirillum brasilense*, synthesis of extracellular selenium nanoparticles and their characterisation. *N. Biotechnol.* **2020**, *58*, 17–24. [CrossRef] [PubMed]

145. Kora, A.J. *Bacillus cereus*, selenite-reducing bacterium from contaminated lake of an industrial area: A renewable nanofactory for the synthesis of selenium nanoparticles. *Bioresour. Bioprocess.* **2018**, *5*. [CrossRef]

146. Eszenyi, P.; Sztrik, A.; Babka, B.; Prokisch, J. Elemental, Nano-Sized (100-500 nm) Selenium Production by Probiotic Lactic Acid Bacteria. *Int. J. Biosci. Biochem. Bioinform.* **2011**, *1*, 148–152. [CrossRef]

147. Shakibaie, M.; Shahverdi, A.R.; Faramarzi, M.A.; Hassanzadeh, G.R.; Rahimi, H.R.; Sabzevari, O. Acute and subacute toxicity of novel biogenic selenium nanoparticles in mice. *Pharm. Biol.* **2013**, *51*, 58–63. [CrossRef]

148. Xu, C.; Qiao, L.; Guo, Y.; Ma, L.; Cheng, Y. Preparation, characteristics and antioxidant activity of polysaccharides and proteins-capped selenium nanoparticles synthesized by *Lactobacillus casei* ATCC 393. *Carbohydr. Polym.* **2018**, *195*, 576–585. [CrossRef]

149. Wadgaonkar, S.L.; Mal, J.; Nancharaiah, Y.V.; Maheshwari, N.O.; Esposito, G.; Lens, P.N.L. Formation of Se(0), Te(0), and Se(0)-Te(0) nanostructures during simultaneous bioreduction of selenite and tellurite in a UASB reactor. *Appl. Microbiol. Biotechnol.* **2018**, *102*, 2899–2911. [CrossRef]

150. Staicu, L.C.; van Hullebusch, E.D.; Oturan, M.A.; Ackerson, C.J.; Lens, P.N. Removal of colloidal biogenic selenium from wastewater. *Chemosphere* **2015**, *125*, 130–138. [CrossRef]
151. Dessi, P.; Jain, R.; Singh, S.; Seder-Colomina, M.; van Hullebusch, E.D.; Rene, E.R.; Ahammad, S.Z.; Carucci, A.; Lens, P.N.L. Effect of temperature on selenium removal from wastewater by UASB reactors. *Water Res.* **2016**, *94*, 146–154. [CrossRef] [PubMed]
152. Liu, W.; Golshan, N.H.; Deng, X.; Hickey, D.J.; Zeimer, K.; Li, H.; Webster, T.J. Selenium nanoparticles incorporated into titania nanotubes inhibit bacterial growth and macrophage proliferation. *Nanoscale* **2016**, *8*, 15783–15794. [CrossRef] [PubMed]
153. Eswayah, A.S.; Smith, T.J.; Gardiner, P.H. Microbial Transformations of Selenium Species of Relevance to Bioremediation. *Appl. Environ. Microbiol.* **2016**, *82*, 4848–4859. [CrossRef] [PubMed]
154. Shoeibi, S.; Mozdziak, P.; Golkar-Narenji, A. Biogenesis of Selenium Nanoparticles Using Green Chemistry. *Top. Curr. Chem.* **2017**, *375*, 88. [CrossRef] [PubMed]
155. Nancharaiah, Y.V.; Lens, P.N. Ecology and biotechnology of selenium-respiring bacteria. *Microbiol. Mol. Biol. Rev.* **2015**, *79*, 61–80. [CrossRef]
156. Tan, Y.; Yao, R.; Wang, R.; Wang, D.; Wang, G.; Zheng, S. Reduction of selenite to Se(0) nanoparticles by filamentous bacterium *Streptomyces* sp. ES2-5 isolated from a selenium mining soil. *Microb. Cell Fact.* **2016**, *15*, 157. [CrossRef]
157. Tomei, F.A.; Barton, L.L.; Lemanski, C.L.; Zocco, T.G.; Fink, N.H.; Sillerud, L.O. Transformation of selenate and selenite to elemental selenium by *Desulfovibrio desulfuricans*. *J. Indus. Microbiol.* **1995**, *14*, 329–336. [CrossRef]
158. Losi, M.E.; Frankenberger, W.T., Jr. Reduction of Selenium Oxyanions by *Enterobacter cloacae* SLD1a-1: Isolation and Growth of the Bacterium and Its Expulsion of Selenium Particles. *Appl. Environ. Microbiol.* **1997**, *63*, 3079–3084. [CrossRef]
159. Kessi, J.; Ramuz, M.; Wehrli, E.; Spycher, M.; Bachofen, R. Reduction of Selenite and Detoxification of Elemental Selenium by the Phototrophic Bacterium *Rhodospirillum rubrum*. *Appl. Environ. Microbiol.* **1999**, *65*, 4734–4740. [CrossRef]
160. Tan, L.C.; Nancharaiah, Y.V.; van Hullebusch, E.D.; Lens, P.N.L. Selenium: Environmental significance, pollution, and biological treatment technologies. *Biotechnol. Adv.* **2016**, *34*, 886–907. [CrossRef]
161. Kessi, J.; Hanselmann, K.W. Similarities between the abiotic reduction of selenite with glutathione and the dissimilatory reaction mediated by *Rhodospirillum rubrum* and *Escherichia coli*. *J. Biol. Chem.* **2004**, *279*, 50662–50669. [CrossRef] [PubMed]
162. Hunter, W.J. *Pseudomonas seleniipraecipitans* proteins potentially involved in selenite reduction. *Curr. Microbiol.* **2014**, *69*, 69–74. [CrossRef] [PubMed]
163. Chung, S.; Zhou, R.; Webster, T.J. Green Synthesized BSA-Coated Selenium Nanoparticles Inhibit Bacterial Growth While Promoting Mammalian Cell Growth. *Int. J. Nanomed.* **2020**, *15*, 115–124. [CrossRef] [PubMed]
164. Zhao, G.; Wu, X.; Chen, P.; Zhang, L.; Yang, C.S.; Zhang, J. Selenium nanoparticles are more efficient than sodium selenite in producing reactive oxygen species and hyper-accumulation of selenium nanoparticles in cancer cells generates potent therapeutic effects. *Free Radic. Biol. Med.* **2018**, *126*, 55–66. [CrossRef] [PubMed]
165. Krafft, T.; Bowen, A.; Theis, F.; Macy, J.M. Cloning and Sequencing of the Genes Encoding the Periplasmic-Cytochrome B-Containing Selenate Reductase of *Thauera selenatis*. *DNA Seq.* **2000**, *10*, 365–377. [CrossRef] [PubMed]
166. Hunter, W.J. A *Rhizobium selenitireducens* protein showing selenite reductase activity. *Curr. Microbiol.* **2014**, *68*, 311–316. [CrossRef]
167. Kuroda, M.; Yamashita, M.; Miwa, E.; Imao, K.; Fujimoto, N.; Ono, H.; Nagano, K.; Sei, K.; Ike, M. Molecular cloning and characterization of the srdBCA operon, encoding the respiratory selenate reductase complex, from the selenate-reducing bacterium *Bacillus selenatarsenatis* SF-1. *J. Bacteriol.* **2011**, *193*, 2141–2148. [CrossRef]
168. Afkar, E.; Lisak, J.; Saltikov, C.; Basu, P.; Oremland, R.S.; Stolz, J.F. The respiratory arsenate reductase from *Bacillus selenitireducens* strain MLS10. *FEMS Microbiol. Lett.* **2003**, *226*, 107–112. [CrossRef]
169. Harrison, G.; Curie, C.; Laishley, E.J. Purification and characterization of an inducible dissimilatory type sulfite reductase from *Clostridium pasteurianum*. *Arch. Microbiol.* **1984**, *138*, 72–78. [CrossRef]
170. Yanke, L.J.; Bryant, R.D.; Laishley, E.J. Hydrogenase I of *Clostridium pasteurianum* functions as a novel selenite reductase. *Anaerobe* **1995**, *1*, 61–67. [CrossRef]
171. DeMoll-Decker, H.; Macy, J.M. The periplasmic nitrite reductase of *Thauera selenatis* may catalyze the reduction of selenite to elemental selenium. *Arch. Microbiol.* **1993**, *160*, 241–247.
172. Ridley, H.; Watts, C.A.; Richardson, D.J.; Butler, C.S. Resolution of distinct membrane-bound enzymes from *Enterobacter cloacae* SLD1a-1 that are responsible for selective reduction of nitrate and selenate oxyanions. *Appl. Environ. Microbiol.* **2006**, *72*, 5173–5180. [CrossRef] [PubMed]
173. Kessi, J. Enzymic systems proposed to be involved in the dissimilatory reduction of selenite in the purple non-sulfur bacteria *Rhodospirillum rubrum* and *Rhodobacter capsulatus*. *Microbiology* **2006**, *152*, 731–743. [CrossRef] [PubMed]
174. Sonkusre, P.; Nanduri, R.; Gupta, P.; Singh Cameotra, S. Improved Extraction of Intracellular Biogenic Selenium Nanoparticles and their Specificity for Cancer Chemoprevention. *Nanomed. Nanotechnol.* **2014**, *5*. [CrossRef]
175. Lenz, M.; Kolvenbach, B.; Gygax, B.; Moes, S.; Corvini, P.F. Shedding light on selenium biomineralization: Proteins associated with bionanominerals. *Appl. Environ. Microbiol.* **2011**, *77*, 4676–4680. [CrossRef]
176. Lynch, I.; Dawson, K.A. Protein-nanoparticle interactions. *Nano Today* **2008**, *3*, 40–47. [CrossRef]
177. Xue, L.; Greisler, H.P. Biomaterials in the development and future of vascular grafts. *J. Vasc. Surg.* **2003**, *37*, 472–480. [CrossRef]
178. Dhillon, G.S.; Brar, S.K.; Kaur, S.; Verma, M. Green approach for nanoparticle biosynthesis by fungi: Current trends and applications. *Crit. Rev. Biotechnol.* **2012**, *32*, 49–73. [CrossRef]

179. Boroumand Moghaddam, A.; Namvar, F.; Moniri, M.; Md Tahir, P.; Azizi, S.; Mohamad, R. Nanoparticles Biosynthesized by Fungi and Yeast: A Review of Their Preparation, Properties, and Medical Applications. *Molecules* **2015**, *20*, 16540–16565. [CrossRef]
180. Sarkar, J.; Dey, P.; Saha, S.; Acharya, K. Mycosynthesis of selenium nanoparticles. *Micro Nano Lett.* **2011**, *6*, 599. [CrossRef]
181. Joshi, S.M.; De Britto, S.; Jogaiah, S.; Ito, S.I. Mycogenic Selenium Nanoparticles as Potential New Generation Broad Spectrum Antifungal Molecules. *Biomolecules* **2019**, *9*. [CrossRef] [PubMed]
182. Diko, C.S.; Zhang, H.; Lian, S.; Fan, S.; Li, Z.; Qu, Y. Optimal synthesis conditions and characterization of selenium nanoparticles in *Trichoderma* sp. WL-Go culture broth. *Mater. Chem. Phys.* **2020**, *246*, 122583. [CrossRef]
183. Liang, T.; Qiu, X.; Ye, X.; Liu, Y.; Li, Z.; Tian, B.; Yan, D. Biosynthesis of selenium nanoparticles and their effect on changes in urinary nanocrystallites in calcium oxalate stone formation. *3 Biotech.* **2020**, *10*, 23. [CrossRef] [PubMed]
184. Mosallam, F.M.; El-Sayyad, G.S.; Fathy, R.M.; El-Batal, A.I. Biomolecules-mediated synthesis of selenium nanoparticles using *Aspergillus oryzae* fermented Lupin extract and gamma radiation for hindering the growth of some multidrug-resistant bacteria and pathogenic fungi. *Microb. Pathog.* **2018**, *122*, 108–116. [CrossRef] [PubMed]
185. El-Batal, A.I.; Mosallam, F.M.; Ghorab, M.M.; Hanora, A.; Gobara, M.; Baraka, A.; Elsayed, M.A.; Pal, K.; Fathy, R.M.; Abd Elkodous, M.; et al. Factorial design-optimized and gamma irradiation-assisted fabrication of selenium nanoparticles by chitosan and Pleurotus ostreatus fermented fenugreek for a vigorous in vitro effect against carcinoma cells. *Int. J. Biol. Macromol.* **2020**, *156*, 1584–1599. [CrossRef] [PubMed]
186. El-Sayyad, G.S.; El-Bastawisy, H.S.; Gobara, M.; El-Batal, A.I. Gentamicin-Assisted Mycogenic Selenium Nanoparticles Synthesized Under Gamma Irradiation for Robust Reluctance of Resistant Urinary Tract Infection-Causing Pathogens. *Biol. Trace Elem. Res.* **2020**, *195*, 323–342. [CrossRef]
187. Gao, X.; Li, X.; Mu, J.; Ho, C.T.; Su, J.; Zhang, Y.; Lin, X.; Chen, Z.; Li, B.; Xie, Y. Preparation, physicochemical characterization, and anti-proliferation of selenium nanoparticles stabilized by *Polyporus umbellatus* polysaccharide. *Int. J. Biol. Macromol.* **2020**, *152*, 605–615. [CrossRef]
188. Jin, Y.; Cai, L.; Yang, Q.; Luo, Z.; Liang, L.; Liang, Y.; Wu, B.; Ding, L.; Zhang, D.; Xu, X.; et al. Anti-leukemia activities of selenium nanoparticles embedded in nanotube consisted of triple-helix β-D-glucan. *Carbohydr. Polym.* **2020**, *240*, 116329. [CrossRef]
189. Liao, W.; Yu, Z.; Lin, Z.; Lei, Z.; Ning, Z.; Regenstein, J.M.; Yang, J.; Ren, J. Biofunctionalization of Selenium Nanoparticle with *Dictyophora Indusiata* Polysaccharide and Its Antiproliferative Activity through Death-Receptor and Mitochondria-Mediated Apoptotic Pathways. *Sci. Rep.* **2015**, *5*, 18629. [CrossRef]
190. Liu, Y.; Zeng, S.; Liu, Y.; Wu, W.; Shen, Y.; Zhang, L.; Li, C.; Chen, H.; Liu, A.; Shen, L.; et al. Synthesis and antidiabetic activity of selenium nanoparticles in the presence of polysaccharides from *Catathelasma ventricosum*. *Int. J. Biol. Macromol.* **2018**, *114*, 632–639. [CrossRef]
191. Rosenfeld, C.E.; Kenyon, J.A.; James, B.R.; Santelli, C.M. Selenium (IV,VI) reduction and tolerance by fungi in an oxic environment. *Geobiology* **2017**, *15*, 441–452. [CrossRef] [PubMed]
192. Vetchinkina, E.; Loshchinina, E.; Kupryashina, M.; Burov, A.; Pylaev, T.; Nikitina, V. Green synthesis of nanoparticles with extracellular and intracellular extracts of basidiomycetes. *PeerJ* **2018**, *6*, e5237. [CrossRef] [PubMed]
193. Vetchinkina, E.; Loshchinina, E.; Kursky, V.; Nikitina, V. Reduction of organic and inorganic selenium compounds by the edible medicinal basidiomycete *Lentinula edodes* and the accumulation of elemental selenium nanoparticles in its mycelium. *J. Microbiol.* **2013**, *51*, 829–835. [CrossRef] [PubMed]
194. Vetchinkina, E.; Loshchinina, E.; Kupryashina, M.; Burov, A.; Nikitina, V. Shape and Size Diversity of Gold, Silver, Selenium, and Silica Nanoparticles Prepared by Green Synthesis Using Fungi and Bacteria. *Indus. Eng. Chem. Res.* **2019**, *58*, 17207–17218. [CrossRef]
195. Xiao, Y.; Huang, Q.; Zheng, Z.; Guan, H.; Liu, S. Construction of a *Cordyceps sinensis* exopolysaccharide-conjugated selenium nanoparticles and enhancement of their antioxidant activities. *Int. J. Biol. Macromol.* **2017**, *99*, 483–491. [CrossRef] [PubMed]
196. Zhang, H.; Zhou, H.; Bai, J.; Li, Y.; Yang, J.; Ma, Q.; Qu, Y. Biosynthesis of selenium nanoparticles mediated by fungus *Mariannaea* sp. HJ and their characterization. *Colloid Surf. A Physicochem. Eng. Aspect* **2019**, *571*, 9–16. [CrossRef]
197. Srivastava, N.; Mukhopadhyay, M. Biosynthesis and Structural Characterization of Selenium Nanoparticles Using *Gliocladium roseum*. *J. Cluster Sci.* **2015**, *26*, 1473–1482. [CrossRef]
198. Gharieb, M.M.; Wilkinson, S.C.; Gadd, G.M. Reduction of selenium oxyanions by unicellular, polymorphic and filamentous fungi: Cellular location of reduced selenium and implications for tolerance. *J. Indud. Microbiol.* **1995**, *14*, 300–311. [CrossRef]
199. Das, S.K.; Dickinson, C.; Lafir, F.; Brougham, D.F.; Marsili, E. Synthesis, characterization and catalytic activity of gold nanoparticles biosynthesized with *Rhizopus oryzae* protein extract. *Green Chem.* **2012**, *14*, 1322–1334. [CrossRef]
200. Sarkar, J.; Acharya, K. *Alternaria alternata* culture filtrate mediated bioreduction of chloroplatinate to platinum nanoparticles. *Inorg. Nano Metal. Chem.* **2017**, *47*, 365–369. [CrossRef]
201. Sathishkumar, Y.; Devarayan, K.; Ki, C.; Rajagopal, K.; Soo Lee, Y. Shape-controlled extracellular synthesis of silver nanocubes by *Mucor circinelloides*. *Mater. Lett.* **2015**, *159*, 481–483. [CrossRef]
202. Herrero, E.; Wellinger, R.E. Yeast as a model system to study metabolic impact of selenium compounds. *Microb. Cell* **2015**, *2*, 139–149. [CrossRef] [PubMed]
203. Schrauzer, G.N. Selenium yeast: Composition, quality, analysis, and safety. *Pure Appl. Chem.* **2006**, *78*, 105–109. [CrossRef]
204. Rayman, M.P. The use of high-selenium yeast to raise selenium status: How does it measure up? *Br. J. Nutr.* **2004**, *92*, 557–573. [CrossRef] [PubMed]

205. Jiménez-Lamana, J.; Abad-Álvaro, I.; Bierla, K.; Laborda, F.; Szpunar, J.; Lobinski, R. Detection and characterization of biogenic selenium nanoparticles in selenium-rich yeast by single particle ICPMS. *J. Anal. Atom. Spectrom.* **2018**, *33*, 452–460. [CrossRef]

206. Bartosiak, M.; Giersz, J.; Jankowski, K. Analytical monitoring of selenium nanoparticles green synthesis using photochemical vapor generation coupled with MIP-OES and UV–Vis spectrophotometry. *Microchem. J.* **2019**, *145*, 1169–1175. [CrossRef]

207. Lian, S.; Diko, C.S.; Yan, Y.; Li, Z.; Zhang, H.; Ma, Q.; Qu, Y. Characterization of biogenic selenium nanoparticles derived from cell-free extracts of a novel yeast *Magnusiomyces ingens*. *3 Biotech.* **2019**, *9*. [CrossRef]

208. Faramarzi, S.; Anzabi, Y.; Jafarizadeh-Malmiri, H. Nanobiotechnology approach in intracellular selenium nanoparticle synthesis using *Saccharomyces cerevisiae*-fabrication and characterization. *Arch. Microbiol.* **2020**, *202*, 1203–1209. [CrossRef]

209. Asghari-Paskiabi, F.; Imani, M.; Rafii-Tabar, H.; Razzaghi-Abyaneh, M. Physicochemical properties, antifungal activity and cytotoxicity of selenium sulfide nanoparticles green synthesized by *Saccharomyces cerevisiae*. *Biochem. Biophys. Res. Commun.* **2019**, *516*, 1078–1084. [CrossRef]

210. Khakpour, H.; Younesi, H.; Mohammadhosseini, M. Two-stage biosorption of selenium from aqueous solution using dried biomass of the baker's yeast *Saccharomyces cerevisiae*. *J. Environ. Chem. Eng.* **2014**, *2*, 532–542. [CrossRef]

211. Zhang, L.; Li, D.; Gao, P. Expulsion of selenium/protein nanoparticles through vesicle-like structures by *Saccharomyces cerevisiae* under microaerophilic environment. *World J. Microbiol. Biotechnol.* **2012**, *28*, 3381–3386. [CrossRef] [PubMed]

212. Turner, R.J.; Borghese, R.; Zannoni, D. Microbial processing of tellurium as a tool in biotechnology. *Biotechnol. Adv.* **2012**, *30*, 954–963. [CrossRef] [PubMed]

213. Borghese, R.; Brucale, M.; Fortunato, G.; Lanzi, M.; Mezzi, A.; Valle, F.; Cavallini, M.; Zannoni, D. Extracellular production of tellurium nanoparticles by the photosynthetic bacterium *Rhodobacter capsulatus*. *J. Hazard. Mater.* **2016**, *309*, 202–209. [CrossRef] [PubMed]

214. Maltman, C.; Donald, L.J.; Yurkov, V. Tellurite and Tellurate Reduction by the Aerobic Anoxygenic Phototroph *Erythromonas ursincola*, Strain KR99 Is Carried out by a Novel Membrane Associated Enzyme. *Microorganisms* **2017**, *5*. [CrossRef] [PubMed]

215. Borghese, R.; Baccolini, C.; Francia, F.; Sabatino, P.; Turner, R.J.; Zannoni, D. Reduction of chalcogen oxyanions and generation of nanoprecipitates by the photosynthetic bacterium *Rhodobacter capsulatus*. *J. Hazard. Mater.* **2014**, *269*, 24–30. [CrossRef]

216. Presentato, A.; Piacenza, E.; Anikovskiy, M.; Cappelletti, M.; Zannoni, D.; Turner, R.J. *Rhodococcus aetherivorans* BCP1 as cell factory for the production of intracellular tellurium nanorods under aerobic conditions. *Microb. Cell Fact.* **2016**, *15*, 204. [CrossRef]

217. Wang, K.; Zhang, X.; Kislyakov, I.M.; Dong, N.; Zhang, S.; Wang, G.; Fan, J.; Zou, X.; Du, J.; Leng, Y.; et al. Bacterially synthesized tellurium nanostructures for broadband ultrafast nonlinear optical applications. *Nat. Commun.* **2019**, *10*, 3985. [CrossRef]

218. Pugin, B.; Cornejo, F.A.; Muñoz-Díaz, P.; Muñoz-Villagrán, C.M.; Vargas-Pérez, J.I.; Arenas, F.A.; Vásquez, C.C. Glutathione Reductase-Mediated Synthesis of Tellurium-Containing Nanostructures Exhibiting Antibacterial Properties. *Appl. Environ. Microbiol.* **2014**, *80*, 7061–7070. [CrossRef]

219. Nguyen, V.K.; Choi, W.; Ha, Y.; Gu, Y.; Lee, C.; Park, J.; Jang, G.; Shin, C.; Cho, S. Microbial tellurite reduction and production of elemental tellurium nanoparticles by novel bacteria isolated from wastewater. *J. Indus. Eng. Chem.* **2019**, *78*, 246–256. [CrossRef]

220. Castro, L.; Li, J.; González, F.; Muñoz, J.A.; Blázquez, M.L. Green synthesis of tellurium nanoparticles by tellurate and tellurite reduction using *Aeromonas hydrophila* under different aeration conditions. *Hydrometallurgy* **2020**, *196*, 105415. [CrossRef]

221. Zare, B.; Faramarzi, M.A.; Sepehrizadeh, Z.; Shakibaie, M.; Rezaie, S.; Shahverdi, A.R. Biosynthesis and recovery of rod-shaped tellurium nanoparticles and their bactericidal activities. *Mater. Res. Bull.* **2012**, *47*, 3719–3725. [CrossRef]

222. Ramos-Ruiz, A.; Field, J.A.; Wilkening, J.V.; Sierra-Alvarez, R. Recovery of Elemental Tellurium Nanoparticles by the Reduction of Tellurium Oxyanions in a Methanogenic Microbial Consortium. *Environ. Sci. Technol.* **2016**, *50*, 1492–1500. [CrossRef] [PubMed]

223. Ottosson, L.G.; Logg, K.; Ibstedt, S.; Sunnerhagen, P.; Kall, M.; Blomberg, A.; Warringer, J. Sulfate assimilation mediates tellurite reduction and toxicity in *Saccharomyces cerevisiae*. *Eukaryot. Cell* **2010**, *9*, 1635–1647. [CrossRef] [PubMed]

224. Yang, F.; Wong, K.H.; Yang, Y.; Li, X.; Jiang, J.; Zheng, W.; Wu, H.; Chen, T. Purification and in vitro antioxidant activities of tellurium-containing phycobiliproteins from tellurium-enriched *Spirulina platensis*. *Drug Des. Devel. Ther.* **2014**, *8*, 1789–1800. [CrossRef] [PubMed]

225. Arenas-Salinas, M.; Vargas-Perez, J.I.; Morales, W.; Pinto, C.; Munoz-Diaz, P.; Cornejo, F.A.; Pugin, B.; Sandoval, J.M.; Diaz-Vasquez, W.A.; Munoz-Villagran, C.; et al. Flavoprotein-Mediated Tellurite Reduction: Structural Basis and Applications to the Synthesis of Tellurium-Containing Nanostructures. *Front. Microbiol.* **2016**, *7*, 1160. [CrossRef]

226. Klonowska, A.; Heulin, T.; Vermeglio, A. Selenite and tellurite reduction by *Shewanella oneidensis*. *Appl. Environ. Microbiol.* **2005**, *71*, 5607–5609. [CrossRef]

227. Alam, H.; Khatoon, N.; Raza, M.; Ghosh, P.C.; Sardar, M. Synthesis and Characterization of Nano Selenium Using Plant Biomolecules and Their Potential Applications. *BioNanoScience* **2019**, *9*, 96–104. [CrossRef]

228. Gunti, L.; Dass, R.S.; Kalagatur, N.K. Phytofabrication of Selenium Nanoparticles From *Emblica officinalis* Fruit Extract and Exploring Its Biopotential Applications: Antioxidant, Antimicrobial, and Biocompatibility. *Front. Microbiol.* **2019**, *10*, 931. [CrossRef]

229. Sowndarya, P.; Ramkumar, G.; Shivakumar, M.S. Green synthesis of selenium nanoparticles conjugated *Clausena dentata* plant leaf extract and their insecticidal potential against mosquito vectors. *Artif. Cells Nanomed. Biotechnol.* **2017**, *45*, 1490–1495. [CrossRef] [PubMed]

230. Marslin, G.; Siram, K.; Maqbool, Q.; Selvakesavan, R.K.; Kruszka, D.; Kachlicki, P.; Franklin, G. Secondary Metabolites in the Green Synthesis of Metallic Nanoparticles. *Materials* **2018**, *11*. [CrossRef]

231. Mittal, A.K.; Kumar, S.; Banerjee, U.C. Quercetin and gallic acid mediated synthesis of bimetallic (silver and selenium) nanoparticles and their antitumor and antimicrobial potential. *J. Colloid Interface Sci.* **2014**, *431*, 194–199. [CrossRef]

232. Cobbett, C.S. Heavy Metal Detoxification in Plants: Phytochelatin Biosynthesis and Function. *IUBMB Life* **2001**, *51*, 183–188. [CrossRef]

233. Branco-Neves, S.; Soares, C.; de Sousa, A.; Martins, V.; Azenha, M.; Gerós, H.; Fidalgo, F. An efficient antioxidant system and heavy metal exclusion from leaves make *Solanum cheesmaniae* more tolerant to Cu than its cultivated counterpart. *Food Energ. Secur.* **2017**, *6*, 123–133. [CrossRef]

234. Schiavon, M.; Pilon-Smits, E.A. Selenium Biofortification and Phytoremediation Phytotechnologies: A Review. *J. Environ. Qual.* **2017**, *46*, 10–19. [CrossRef]

235. Vogrincic, M.; Cuderman, P.; Kreft, I.; Stibilj, V. Selenium and Its Species Distribution in Above-ground Plant Parts of Selenium Enriched Buckwheat (*Fagopyrum esculentum* Moench). *Anal. Sci.* **2009**, *25*, 1357–1363. [CrossRef]

236. Hu, T.; Li, H.; Li, J.; Zhao, G.; Wu, W.; Liu, L.; Wang, Q.; Guo, Y. Absorption and Bio-Transformation of Selenium Nanoparticles by Wheat Seedlings (*Triticum aestivum* L.). *Front. Plant. Sci.* **2018**, *9*, 597. [CrossRef]

237. Anu, K.; Devanesan, S.; Prasanth, R.; AlSalhi, M.S.; Ajithkumar, S.; Singaravelu, G. Biogenesis of selenium nanoparticles and their anti-leukemia activity. *J. King Saud Univ. Sci.* **2020**, *32*, 2520–2526. [CrossRef]

238. Fan, D.; Li, L.; Li, Z.; Zhang, Y.; Ma, X.; Wu, L.; Zhang, H.; Guo, F. Biosynthesis of selenium nanoparticles and their protective, antioxidative effects in streptozotocin induced diabetic rats. *Sci. Technol. Adv. Mater.* **2020**, *21*, 505–514. [CrossRef]

239. Mulla, N.A.; Otari, S.V.; Bohara, R.A.; Yadav, H.M.; Pawar, S.H. Rapid and size-controlled biosynthesis of cytocompatible selenium nanoparticles by *Azadirachta indica* leaves extract for antibacterial activity. *Mater. Lett.* **2020**, *264*, 127353. [CrossRef]

240. Alagesan, V.; Venugopal, S. Green Synthesis of Selenium Nanoparticle Using Leaves Extract of *Withania somnifera* and Its Biological Applications and Photocatalytic Activities. *BioNanoScience* **2019**, *9*, 105–116. [CrossRef]

241. Anu, K.; Singaravelu, G.; Murugan, K.; Benelli, G. Green-Synthesis of Selenium Nanoparticles Using Garlic Cloves (*Allium sativum*): Biophysical Characterization and Cytotoxicity on Vero Cells. *J. Clust. Sci.* **2016**, *28*, 551–563. [CrossRef]

242. Li, S.; Shen, Y.; Xie, A.; Yu, X.; Zhang, X.; Yang, L.; Li, C. Rapid, room-temperature synthesis of amorphous selenium/protein composites using *Capsicum annuum* L extract. *Nanotechnology* **2007**, *18*, 405101. [CrossRef]

243. Ramamurthy, C.; Sampath, K.S.; Arunkumar, P.; Kumar, M.S.; Sujatha, V.; Premkumar, K.; Thirunavukkarasu, C. Green synthesis and characterization of selenium nanoparticles and its augmented cytotoxicity with doxorubicin on cancer cells. *Bioprocess. Biosyst. Eng.* **2013**, *36*, 1131–1139. [CrossRef]

244. Vennila, K.; Chitra, L.; Balagurunathan, R.; Palvannan, T. Comparison of biological activities of selenium and silver nanoparticles attached with bioactive phytoconstituents: Green synthesized using *Spermacoce hispida* extract. *Adv. Nat. Sci. Nanosci. Nanotechnol.* **2018**, *9*, 015005. [CrossRef]

245. Kokila, K.; Elavarasan, N.; Sujatha, V. *Diospyros montana* leaf extract-mediated synthesis of selenium nanoparticles and their biological applications. *New J. Chem.* **2017**, *41*, 7481–7490. [CrossRef]

246. Rajaee Behbahani, S.; Iranbakhsh, A.; Ebadi, M.; Majd, A.; Ardebili, Z.O. Red elemental selenium nanoparticles mediated substantial variations in growth, tissue differentiation, metabolism, gene transcription, epigenetic cytosine DNA methylation, and callogenesis in bittermelon (*Momordica charantia*); an in vitro experiment. *PLoS ONE* **2020**, *15*, e0235556. [CrossRef]

247. Cui, D.; Liang, T.; Sun, L.; Meng, L.; Yang, C.; Wang, L.; Liang, T.; Li, Q. Green synthesis of selenium nanoparticles with extract of hawthorn fruit induced HepG2 cells apoptosis. *Pharm. Biol.* **2018**, *56*, 528–534. [CrossRef]

248. Fardsadegh, B.; Vaghari, H.; Mohammad-Jafari, R.; Najian, Y.; Jafarizadeh-Malmiri, H. Biosynthesis, characterization and antimicrobial activities assessment of fabricated selenium nanoparticles using *Pelargonium zonale* leaf extract. *Green Process. Synth.* **2019**, *8*, 191–198. [CrossRef]

249. Fardsadegh, B.; Jafarizadeh-Malmiri, H. *Aloe vera* leaf extract mediated green synthesis of selenium nanoparticles and assessment of their in vitro antimicrobial activity against spoilage fungi and pathogenic bacteria strains. *Green Process. Synth.* **2019**, *8*, 399–409. [CrossRef]

250. Hassanien, R.; Abed-Elmageed, A.A.I.; Husein, D.Z. Eco-Friendly Approach to Synthesize Selenium Nanoparticles: Photocatalytic Degradation of Sunset Yellow Azo Dye and Anticancer Activity. *Chem. Select* **2019**, *4*, 9018–9026. [CrossRef]

251. Mellinas, C.; Jimenez, A.; Garrigos, M.D.C. Microwave-Assisted Green Synthesis and Antioxidant Activity of Selenium Nanoparticles Using *Theobroma Cacao* L. Bean Shell Extract. *Molecules* **2019**, *24*, 4048. [CrossRef]

252. Menon, S.; Devi, K.S.S.; Agarwal, H.; Shanmugam, V.K. Efficacy of Biogenic Selenium Nanoparticles from an Extract of Ginger towards Evaluation on Anti-Microbial and Anti-Oxidant Activities. *Colloid Interface Sci. Commun.* **2019**, *29*, 1–8. [CrossRef]

253. Wang, L.; Li, C.; Huang, Q.; Fu, X. Biofunctionalization of selenium nanoparticles with a polysaccharide from *Rosa roxburghii* fruit and their protective effect against H_2O_2-induced apoptosis in INS-1 cells. *Food Funct.* **2019**, *10*, 539–553. [CrossRef]

254. Zhang, W.; Zhang, J.; Ding, D.; Zhang, L.; Muehlmann, L.A.; Deng, S.E.; Wang, X.; Li, W.; Zhang, W. Synthesis and antioxidant properties of *Lycium barbarum* polysaccharides capped selenium nanoparticles using tea extract. *Artif. Cell Nanomed. Biotechnol.* **2017**, *46*, 1463–1470. [CrossRef]

255. Ezhuthupurakkal, P.B.; Polaki, L.R.; Suyavaran, A.; Subastri, A.; Sujatha, V.; Thirunavukkarasu, C. Selenium nanoparticles synthesized in aqueous extract of *Allium sativum* perturbs the structural integrity of Calf thymus DNA through intercalation and groove binding. *Mater. Sci. Eng. C Mater. Biol. Appl.* **2017**, *74*, 597–608. [CrossRef]

256. Tanaka, Y.K.; Takada, S.; Kumagai, K.; Kobayashi, K.; Hokura, A.; Ogra, Y. Elucidation of tellurium biogenic nanoparticles in garlic, *Allium sativum*, by inductively coupled plasma-mass spectrometry. *J. Trace Elem. Med. Biol.* **2020**, *62*, 126628. [CrossRef]
257. Anan, Y.; Yoshida, M.; Hasegawa, S.; Katai, R.; Tokumoto, M.; Ouerdane, L.; Lobinski, R.; Ogra, Y. Speciation and identification of tellurium-containing metabolites in garlic, *Allium sativum*. *Metallomics* **2013**, *5*, 1215–1224. [CrossRef]
258. Cowgill, U.M. The Tellurium Content of Vegetation. *Biol. Trace Elem. Res.* **1988**, *17*, 43–67. [CrossRef]
259. Yurkov, V.; Jappé, J.; Verméglio, A. Tellurite Resistance and Reduction by Obligately Aerobic Photosynthetic Bacteria. *Appl. Environ. Microbiol.* **1996**, *62*, 4195–4198. [CrossRef]
260. Sridharan, K.; Ollakkan, M.S.; Philip, R.; Park, T.J. Non-hydrothermal synthesis and optical limiting properties of one-dimensional Se/C, Te/C and Se–Te/C core–shell nanostructures. *Carbon* **2013**, *63*, 263–273. [CrossRef]
261. Fu, S.; Cai, K.; Wu, L.; Han, H. One-step synthesis of high-quality homogenous Te/Se alloy nanorods with various morphologies. *CrystEngComm* **2015**, *17*, 3243–3250. [CrossRef]
262. Baesman, S.M.; Stolz, J.F.; Kulp, T.R.; Oremland, R.S. Enrichment and isolation of *Bacillus beveridgei* sp. nov., a facultative anaerobic haloalkaliphile from Mono Lake, California, that respires oxyanions of tellurium, selenium, and arsenic. *Extremophiles* **2009**, *13*, 695–705. [CrossRef]
263. Bajaj, M.; Winter, J. Se (IV) triggers faster Te (IV) reduction by soil isolates of heterotrophic aerobic bacteria: Formation of extracellular SeTe nanospheres. *Microb. Cell Fact.* **2014**, *13*, 168. [CrossRef]
264. Thanh, N.T.K.; Green, L.A.W. Functionalisation of nanoparticles for biomedical applications. *Nano Today* **2010**, *5*, 213–230. [CrossRef]
265. Salata, O.V. Applications of nanoparticles in biology and medicine. *J. Nanobiotechnol.* **2004**, *2*, 3. [CrossRef]
266. Zamora-Ledezma, C.; Clavijo, C.D.F.; Medina, E.; Sinche, F.; Santiago Vispo, N.; Dahoumane, S.A.; Alexis, F. Biomedical Science to Tackle the COVID-19 Pandemic: Current Status and Future Perspectives. *Molecules* **2020**, *25*, 4620. [CrossRef]
267. Mout, R.; Moyano, D.F.; Rana, S.; Rotello, V.M. Surface functionalization of nanoparticles for nanomedicine. *Chem. Soc. Rev.* **2012**, *41*, 2539–2544. [CrossRef]
268. Soloviev, M. Nanobiotechnology today: Focus on nanoparticles. *J. Nanobiotechnol.* **2007**, *5*, 11. [CrossRef]
269. Zhang, J.; Wang, X.; Xu, T. Elemental selenium at nano size (Nano-Se) as a potential chemopreventive agent with reduced risk of selenium toxicity: Comparison with se-methylselenocysteine in mice. *Toxicol. Sci.* **2008**, *101*, 22–31. [CrossRef]
270. Wang, H.; Zhang, J.; Yu, H. Elemental selenium at nano size possesses lower toxicity without compromising the fundamental effect on selenoenzymes: Comparison with selenomethionine in mice. *Free Radic. Biol. Med.* **2007**, *42*, 1524–1533. [CrossRef]
271. Kheradmand, E.; Rafii, F.; Yazdi, M.H.; Sepahi, A.A.; Shahverdi, A.R.; Oveisi, M.R. The antimicrobial effects of selenium nanoparticle-enriched probiotics and their fermented broth against *Candida albicans*. *DARU J. Pharm. Sci.* **2014**, *22*, 48. [CrossRef] [PubMed]
272. Shakibaie, M.; Adeli-Sardou, M.; Mohammadi-Khorsand, T.; ZeydabadiNejad, M.; Amirafzali, E.; Amirpour-Rostami, S.; Ameri, A.; Forootanfar, H. Antimicrobial and Antioxidant Activity of the Biologically Synthesized Tellurium Nanorods; A Preliminary In vitro Study. *Iran. J. Biotechnol.* **2017**, *15*, 268–276. [CrossRef] [PubMed]
273. Chang, C.; Gershwin, M.E. Drugs and autoimmunity -a contemporary review and mechanistic approach. *J. Autoimmun.* **2010**, *34*, J266–J275. [CrossRef] [PubMed]
274. Inoue, K.; Takano, H.; Yanagisawa, R.; Ichinose, T.; Sakurai, M.; Yoshikawa, T. Effects of nano particles on cytokine expression in murine lung in the absence or presence of allergen. *Arch. Toxicol.* **2006**, *80*, 614–619. [CrossRef]
275. Liao, C.; Li, Y.; Tjong, S.C. Bactericidal and Cytotoxic Properties of Silver Nanoparticles. *Int. J. Mol. Sci.* **2019**, *20*. [CrossRef]
276. Ahmad, M.S.; Yasser, M.M.; Sholkamy, E.N.; Ali, A.M.; Mehanni, M.M. Anticancer activity of biostabilized selenium nanorods synthesized by *Streptomyces bikiniensis* strain Ess_amA-1. *Int. J. Nanomed.* **2015**, *10*, 3389–3401. [CrossRef]
277. Filipe, V.; Hawe, A.; Jiskoot, W. Critical evaluation of Nanoparticle Tracking Analysis (NTA) by NanoSight for the measurement of nanoparticles and protein aggregates. *Pharm. Res.* **2010**, *27*, 796–810. [CrossRef]
278. Prasad, K.S.; Vyas, P.; Prajapati, V.; Patel, P.; Selvaraj, K. Biomimetic synthesis of selenium nanoparticles using cell-free extract of *Microbacterium* sp. ARB05. *Micro. Nano Lett.* **2012**, *7*, 1–4. [CrossRef]
279. Selenius, M.; Rundlöf, A.-K.; Olm, E.; Fernandes, A.P.; Björnstedt, M. Selenium and the Selenoprotein Thioredoxin Reductase in the Prevention, Treatment and Diagnostics of Cancer. *Antioxid. Redox Sign.* **2010**, *12*, 867–880. [CrossRef]
280. Bao, P.; Xiao, K.Q.; Wang, H.J.; Xu, H.; Xu, P.P.; Jia, Y.; Haggblom, M.M.; Zhu, Y.G. Characterization and Potential Applications of a Selenium Nanoparticle Producing and Nitrate Reducing Bacterium *Bacillus oryziterrae* sp. nov. *Sci. Rep.* **2016**, *6*, 34054. [CrossRef]
281. Schrauzer, G.N. Anticarcinogenic effects of selenium. *CMLS Cell Mol. Life Sci.* **2000**, *57*, 1864–1873. [CrossRef]
282. Zeng, H.; Combs, G.F., Jr. Selenium as an anticancer nutrient: Roles in cell proliferation and tumor cell invasion. *J. Nutr. Biochem.* **2008**, *19*, 1–7. [CrossRef] [PubMed]
283. Brown, C.D.; Cruz, D.M.; Roy, A.K.; Webster, T.J. Synthesis and characterization of PVP-coated tellurium nanorods and their antibacterial and anticancer properties. *J. Nanopart. Res.* **2018**, *20*, 254. [CrossRef]
284. Mohanty, A.; Kathawala, M.H.; Zhang, J.; Chen, W.N.; Chye Loo, J.S.; Kjelleberg, S.; Yang, L.; Cao, B. Biogenic Tellurium Nanorods as a Novel Antivirulence Agent Inhibiting Pyoverdine Production in *Pseudomonas aeruginosa*. *Biotechnol. Bioeng.* **2013**, *111*, 858–865. [CrossRef] [PubMed]
285. Aydin, N.; Arslan, M.E.; Sonmez Sonmez, E.; Turkez, H. Cytotoxicity analysis of tellurium dioxide nanoparticles on cultured human pulmonary alveolar epithelial and peripheral blood cell cultures. *Biomed. Res. India* **2017**, *28*, 3300–3304.

286. Forootanfar, H.; Amirpour-Rostami, S.; Jafari, M.; Forootanfar, A.; Yousefizadeh, Z.; Shakibaie, M. Microbial-assisted synthesis and evaluation the cytotoxic effect of tellurium nanorods. *Mater. Sci. Eng. C Mater. Biol. Appl.* **2015**, *49*, 183–189. [CrossRef] [PubMed]
287. Wen, H.; Zhong, J.; Shen, B.; Gan, T.; Fu, C.; Zhu, Z.; Li, R.; Yang, X. Comparative study of the cytotoxicity of the nanosized and microsized tellurium powders on HeLa cells. *Front. Biol.* **2013**, *8*, 444–450. [CrossRef]

International Journal of
Molecular Sciences

Review

Prussian Blue: A Nanozyme with Versatile Catalytic Properties

Joan Estelrich [1,2,*] and M. Antònia Busquets [1,2]

1   Department of Pharmacy and Pharmaceutical Technology and Physical Chemistry, Faculty of Pharmacy and Food Sciences, University of Barcelona, Avda Joan XXIII, 27-31, 08028 Barcelona, Catalonia, Spain; mabusquetsvinas@ub.edu
2   Institute of Nanoscience and Nanotechnology (IN2UB), University of Barcelona, Avda. Diagonal 645, 08028 Barcelona, Catalonia, Spain
*   Correspondence: joanestelrich@ub.edu

Abstract: Nanozymes, nanomaterials with enzyme-like activities, are becoming powerful competitors and potential substitutes for natural enzymes because of their excellent performance. Nanozymes offer better structural stability over their respective natural enzymes. In consequence, nanozymes exhibit promising applications in different fields such as the biomedical sector (in vivo diagnostics/and therapeutics) and the environmental sector (detection and remediation of inorganic and organic pollutants). Prussian blue nanoparticles and their analogues are metal–organic frameworks (MOF) composed of alternating ferric and ferrous irons coordinated with cyanides. Such nanoparticles benefit from excellent biocompatibility and biosafety. Besides other important properties, such as a highly porous structure, Prussian blue nanoparticles show catalytic activities due to the iron atom that acts as metal sites for the catalysis. The different states of oxidation are responsible for the multicatalytic activities of such nanoparticles, namely peroxidase-like, catalase-like, and superoxide dismutase-like activities. Depending on the catalytic performance, these nanoparticles can generate or scavenge reactive oxygen species (ROS).

Keywords: Prussian blue analogues; hybrid nanoparticles; multicatalytic activity; ROS-scavenge; ROS generation; superoxide dismutase-like activity; catalase-like activity; redox potential

check for
updates

Citation: Estelrich, J.; Busquets, M.A. Prussian Blue: A Nanozyme with Versatile Catalytic Properties. *Int. J. Mol. Sci.* **2021**, 22, 5993. https://doi.org/10.3390/ijms22115993

Academic Editor: Raghvendra Singh Yadav

Received: 14 May 2021
Accepted: 28 May 2021
Published: 1 June 2021

Publisher's Note: MDPI stays neutral with regard to jurisdictional claims in published maps and institutional affiliations.

1. Introduction

Nanozymes are one kind of materials with nanoscale sizes (1–100 nm) and enzymatic catalytic properties. The term "nanozyme" can encompass two different types of materials: (1) enzymes or enzymatic catalytic groups which are immobilized on nanomaterials, and (2) nanomaterials that possess inherent enzymatic catalytic properties [1]. At the present, nanozyme is specifically referred to as nanomaterials with intrinsic catalytic properties [1–3]. The first nanomaterial found with enzymatic properties were the iron oxide nanoparticles in 2007 [4]. Since this discovery, the significant advances that nanotechnology, biotechnology, and nanomaterials science have experimented, have allowed emerging of a large number of studies on nanomaterial-based artificial enzymes.

Nanozymes overcome the drawbacks that characterize the natural enzymes, i.e., high cost for preparation and purification, low operational stability, the sensitivity of catalytic activity to environmental conditions, and difficulties in recycling and reusing. Compared with natural enzymes and conventional artificial enzymes, nanozymes have the advantages such as low cost, easy large-scale production, and high stability and durability. However, the most important advantage of nanozymes is their size/composition-dependent activity. It appears that the enzymatic activities of nanozymes are closely related to their size, surface lattice, surface modification, and composition [5]. This allows the design of materials with a broad range of catalytic activity simply by varying shape, structure, and composition. Given these advantages, it is easy to understand the widespread interest that nanozymes have generated and that are becoming powerful competitors and potential substitutes for natural enzymes. In this way, nanozymes have been utilized for practical use in scientific research, biotechnology and food industries, agriculture, environmental treatment, biosensing, disease diagnosis and treatment, antibacterial agents, and cryoprotection against biomolecules in the cell.

According to Huang et al. [1], nanozymes can be divided into two categories: (1) oxidoreductases (oxidase, peroxidase, catalase, superoxide dismutase, and nitrate reductase), and (2) hydrolases (nuclease, esterase, phosphatase, protease, and silicatein).

To date, many nanomaterials have demonstrated to possess remarkable enzyme-like activities. For example, noble and transient metals nanoparticles, carbon nanoparticles (graphene oxide nanosheets, carbon nanotubes, and fullerene), metal oxides, metal sulfides and tellurides, carbon nitride, quantum dots, 2D-nanomaterials with confined single metal and nonmetal atoms, polypyrrole nanoparticles, hemin micelles, and Prussian blue (PB) nanoparticles (PBNPs) [3].

In this review, we will focus on the enzyme-like properties of PBNPs and their applications in this field for biomedicine development. PBNPs have the potential to mimic multiple enzymes in a natural cascade-like system (peroxidase (POD); catalase (CAT); and superoxide dismutase (SOD)). Moreover, the ability to mimic the three indicated antioxidant enzymes makes PBNPs excellent scavengers of reactive oxygen species (ROS), and, contrarily, in various determined conditions, PBNPs can generate ROS by means of a Fenton reaction.

2. Prussian Blue Nanoparticles

PB is a kind of metal–organic framework (MOF) composed of alternating ferric and ferrous irons coordinated with cyanides [6]. PB is an aggregated form of ~ 10 nm nanoparticles (PBNPs) [7]. A usual formulation of PB (the insoluble form) is $Fe^{III}_4(Fe^{II}(CN)_6)_3 \cdot x\ H_2O$, where the extent of hydration varies from 10 to 16. PB analogues (PBA) are coordination polymers with the general formula $AM_a(M_b(CN)_6)^{2/3} \cdot x\ H_2O$, where A is an ion (K^+, Na^+, NH_4^+) alkali ion, and M_a and M_b are cations in oxidation state +2 or +3. PB is a commonly used dye (it was the first of the modern pigments, discovered by Diesbach in 1706 [8]) but its applications are not limited to the artistic field but affect many other fields [9]. The development of PBNPs nanoparticles has involved an upturn of their use in a myriad of biomedical applications [10]. Some of them are based on the photothermal effects of these nanoparticles. Recently, interesting antibacterial features of PBNPs coming from this issue have been described [11]. Such nanoparticles benefit from excellent biocompatibility and biosafety since the insoluble form of PB was approved by the Food and Drug Administration (FDA) for the treatment of exposure with cesium and thallium ions. Moreover, it is found in the List of Essential Medicines of the World Health Organization [12]. Apart from the excellent adsorptive properties [13], facilitated by the mesoporous structure, PBNPs have shown catalytic activities, due to the iron atoms that act as metal active sites for catalysis. The different states of oxidation are responsible for the multicatalytic properties of these nanoparticles: one can find a fully oxidized form of Prussian blue (the so-called Prussian yellow, PY), a partially oxidized form (Berlin green, BG), and a reduced form (Prussian white, PW) (Figure 1).

The conversion of one form into other occurs by means of the reactions:

$$PB + 4\,e^- + 4\,K^+ \rightarrow PW \tag{1}$$

$$PB \rightarrow BG + 3\,e^- + 3\,K^+ \tag{2}$$

$$PB \rightarrow PY + e^- + K^+ \tag{3}$$

Each oxidation state affords a different color. In the visible spectrum, PB and BG ($K_{1/3}Fe(Fe^{III}(CN)_6)_{2/3}(Fe^{II}(CN)_6)_{1/3}$) have two absorption bands at 700 and 480 nm; PW ($K_2Fe^{II}Fe^{II}(CN)_6$) has no obvious band, and PY ($Fe^{III}Fe^{III}(CN)_6$) presents a band at 420 nm. The multienzyme-like activities of PBNPs (SOD, POD, and CAT) [2] are caused by the abundant redox potentials of their different forms (Figure 2). In this aspect, these nanoparticles are unique since can exhibit simultaneously the indicated three enzyme mimetic activities. Moreover, PBNPs present some advantages that are absent in other nanozymes. In this way, PBNPs can protect ischemia-injured neurons in a rat model due to the antioxidative activity of PBNPs that can eliminate excessive reactive oxygen species (ROS) [15].

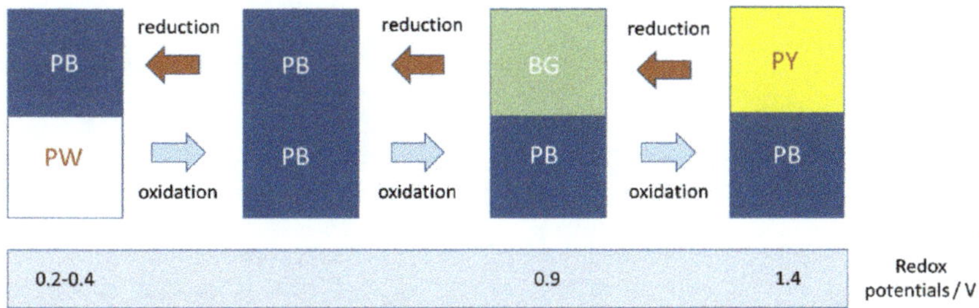

Figure 1. Different forms of Prussian blue depending on the state of oxidation. Each form possesses a characteristic redox potential, which affects the catalytic reaction mechanisms. Reproduced, after modification, with permission from [14].

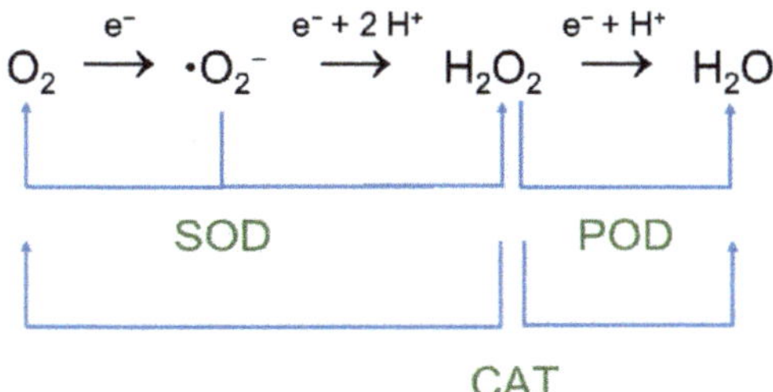

Figure 2. Schematic diagram of the reactions catalyzed by oxidoreductases mimicked by PB and PBA. SOD alternatively catalyzes the dismutation of superoxide into oxygen and hydrogen peroxide; POD breakdown hydrogen peroxide into the water; and CAT breakdown hydrogen peroxide into oxygen and water.

There are several ways to obtain PBNPs [16–18]. Some of them are easy, relatively cheap, and environmentally friendly. As indicated, the size of such nanoparticles is usually reported to be over 10 nm [7]. Recently, ultrasmall PBNPs ~3.4 nm in diameter have been prepared by using an ethanol/water mixture as the solvent and polyvinyl pyrrolidone (PVP) as the surface capping agent [19]. The stability of the PB or PBA nanoparticles plays a key role in determining the suitability for potential utilization. In most applications, PBNPs are dispersed in a liquid medium such as blood and cellular cytoplasm in living organisms. The colloidal stability is of special importance for any kind of nanoparticle, but in the case of the enzymatic properties of PBNPs, it is crucial, since particle aggregation can lead to significant loss of the catalytic activity. A way to achieve stable dispersions is by modification of the nanoparticle surface. The surface modification of PBNPs has been performed with polyethylene glycol, poly(diallyldimethylammonium chloride), chitosan, native proteins, and poly(vinylpyrrolidone) macromolecules [20]. A simple one-step aqueous solution route for preparing biocompatible PBNPs is coating their surface with a carboxylic acid, namely citric or oxalic acid [21]. The ferric ions are complexed by these anions and form a precursor that reduces the rate of nucleation. Citrate-coated PBNPs did not undergo any change in particle size, size distribution, and dispersibility over a period of more than one year.

At the present, no inherent inhibitor has been reported for PB/PBA nanozymes. However, in other nanozymes such as iron oxide nanozymes, the peroxidase-like activity decreased in the presence of guanidine chloride (GuHCl) [22]. In this case, GuHCl is supposed to interact directly with the iron atoms of the iron oxide resulting in a change in the electron density of iron. Since PB/PBA also contains iron atoms, the existence of inhibitors cannot be excluded.

3. Peroxidase-Like Activity

Peroxidases are a group of enzymes (E.C. 1.11.1.x) that typically catalyze a reaction of the form $ROOR' + 2e^- + 2H^+ \rightarrow ROH + R'OH$. For instance, in the presence of H_2O_2, POD can effectively catalyze the peroxidase substrates into their oxidized substrates. Moreover, POD can decompose reactive oxygen species (ROS), which are involved in multiple key oxidative reactions [23]. PB and PBA show POD-like activity in acidic media because of the large area of ferric and ferrous iron available. These compounds exhibit good H_2O_2-induced POD mimicking activity including the oxidation of multiple biomolecules such as 3,5,3',5'-tetramethylbenzidine (TMB), 2,2'-azino-di(3-ethylbenzthiazoline-6-sulfonic acid) (ABTS), O-phenylenediamine (OPD), glucose, dopamine, luminol, guaiacol, and NADH [24]. The generally utilized natural POD substrate is TMB which, when oxidized in the presence of H_2O_2, presents two absorption peaks at 369 nm and 652 nm. The intensity of the peaks is proportional to the substrate, and enzyme concentration. Similar to the natural enzymes, the catalytic kinetic behavior of PBNPs followed the Michaelis–Menten equation and depended on the substrate concentration, temperature, and pH. The nanomaterial activity is often compared to the natural enzyme, horseradish peroxidase (HRP).

The Michaelis–Menten kinetics is one of the best-known models of enzyme kinetics. The model takes the form of an equation that relates reaction rate (v) (rate of formation of product) to the concentration of substrate, [S]

$$v = v_{max}\frac{[S]}{K_M + [S]} \tag{4}$$

where v_{max} represents the maximum rate achieved by the system, happening at saturating substrate concentration. The value of K_M is numerically equal to the substrate concentration at which the reaction rate is half of v_{max}. K_M approximates the affinity of the enzyme for the substrate, and smaller the K_M, the larger the affinity.

PB exhibits high POD-like activity for the reduction of hydrogen peroxide due to high electroactivity. As said above, the oxidized TMB absorbs at 652 nm. In the presence of PB, such absorption could be masked by the huge band of PB (absorption peak at ~700 nm). In this case, the catalytic activity can be monitored at 450 nm, the absorption band of the fully oxidized form of the mediator TMB^{2+} [24]. The enzymatic-like activity has been observed in PB (and PBA) alone or combined with other substances forming hybrid composites.

3.1. Prussian Blue and Prussian Blue Analogues

Although the group led by Gu and Zhang reported for the first time that PB-modified γ-Fe_2O_3 nanoparticles had POD-like activity [25], the first study about the use of PB alone as nanozyme was performed by Zhang et al. [26], which observed a catalytic activity towards the hydrogen peroxide-mediated oxidation of classical peroxidase substrate ABTS to produce a colored product. The catalysis followed Michaelis–Menten kinetics and showed a strong affinity for H_2O_2. Čunderlová et al. [27] determined that cubic crystals of PB of 15 nm diameter presented a high catalytic POD-like activity. The estimated value of K_M for TMB was 0.76 mmol L^{-1}, and for H_2O_2 was 840 mmol L^{-1} (for the same substrates, HRP presented values of 0.147 and 790 mmol L^{-1}, respectively). Moreover, in this study, the surface of PBNPs was biotinylated. These modified nanocrystals showed their use in assays based on the competitive affinity of biotin and human serum albumin. Importantly, biotin-PBNPs retained their catalytic activity towards TMB and H_2O_2. As a differential trend, the dependence of the rate of reduction of H_2O_2 (v_{H2O2}) on the concentration of TMB did not fit with the model of Michaelis–Menten. An explanation could be the limiting rate of diffusion of TMB through the biotinylated surface.

Vázquez-González compared the POD-like activity of four inorganic clusters formed by PB and PBA. They observed the generation of chemiluminescence in the presence of luminol and H_2O_2, the catalyzed oxidation of dopamine to aminochrome, and the catalyzed oxidation of NDPH to NAD^+ by H_2O_2 [28]. The catalytic activities were different in each compound showing the importance of the chemical structure on the intrinsic properties of nanozymes. PBNPs were most efficient for the oxidation of dopamine and NADH by H_2O_2, whereas for the generation of chemiluminescence, the most efficient was the PBA formed by M_a = Fe, M_b = Co (PBCoFe). Komkova et al. synthesized PBNPs at the highest catalytic activity by reducing the mixture of the reagents forming PB using either hydrogen peroxide or a conducting polymer, such as polyaniline [24]. The authors observed that the initial reaction rate of H_2O_2 reduction, catalyzed by PBNPs, was linearly dependent on hydrogen peroxide concentration, which had never been described either for POD-like nanozymes or for natural POD. This fact pointed out that H_2O_2 activation by the nanoparticles occurred much faster even compared to the natural enzyme. On the other hand, in the absence of H_2O_2, no oxidation of TMB was registered. Hence, the employed PBNPs did not display oxidase-like activity (reduction of molecular oxygen). Contrarily, Wang et al. have used PBA nanocages that mimic the oxidase-like activity to the continuous detection of hydrogen sulfide [14]. A PBA described previously (PBCoFe) [28] was employed to catalyze the oxidation of L-tyrosine into dopachrome in the presence of L-ascorbic acid/H_2O_2. [29]. In the first step, L-tyrosine is hydroxylated to form L-DOPA, and then, is subsequently oxidized to dopachrome. The mixture L-ascorbic acid/H_2O_2 is basic to provoke the hydroxylation of L-tyrosine. The PBA also catalyzed the oxidation of L-phenylalanine to dopachrome. A study conducted by Farka et al. conjugated directly PBNPs with antibodies and applied these nanoparticles in nanozyme-linked immunosorbent assay (NLISA) [30]. The method consisted of the colorimetric readout of the color generated by the oxidized TMB (Figure 3). This technique was used to detect human serum albumin in urine for the diagnosis of albuminuria, as well as for the detection of *Salmonella typhimurirum* in powdered milk, as an example of microbial antigen. When the same antibodies were used in standard sandwich ELISA formats with natural HRP enzyme, similar assay parameters were obtained. This suggested that the assay depended mainly on the affinity of the antibodies for the target analyte and not by the detection step. The easy synthesis from cheap precursors and higher stability of PBNPs in comparison with natural enzymes confirmed the suitability of PBNPs to be used as an enzyme replacement.

Many properties of nanomaterials are dimensionally dependent, which is considered as size effects. The catalytic activity of nanozymes exhibits a size-dependent manner since, for the same volume, nanozymes with a small size expose more active sites. According to this, ultrasmall PBNPs have been demonstrated to possess the highest POD-like activity [19]. The catalytic activity of these nanoparticles was one order of magnitude higher than that of PBNPs obtained by the conventional methods. The values v_{max} and K_M values of such nanoparticles were 2.5 μM s^{-1} and 0.22 mM, respectively, which were much larger than those of PBNPs reported in the literature (i.e., 0.22 μM s^{-1} and 0.34 mM [31]).

3.2. Hybrid Nanoparticles Formed by PB/PBA

As indicated above, the group led by Gu and Zhang reported for the first time that a hybrid composite formed by PB-modified γ-Fe_2O_3 nanoparticles had POD-like activity [25]. It was observed that POD-like activity was increased as the PB proportion increased. This result may reason from that to the more PB providing more ferrous ions as catalysis centers to interact with substrates. The catalytic activity was high, three orders of magnitudes larger than that for magnetite nanoparticles of similar size. With such nanocomposites, an enzyme immunoassay model was established. In this assay, staphylococcal protein A was conjugated on the nanoparticle's surface and the nanoprobe served to detect IgG immobilized to 96-well plates. Later, this group studied the applications of PB-modified ferritin nanoparticles in biological detection [32]. The same group synthesized PB and gold-modified polystyrene nanocomposites (PS@Au@PB) [33]. In a later study, they remarked that PBNPs showed weak POD activity under neutral conditions [31]. Moreover, in an alkaline environment, they were nearly inactive. More recently, this group inferred that PBNPs may play as ascorbic acid-related nanozyme, which can catalyze the oxidation of ascorbic acid without producing H_2O_2 [34]. As the oxidation of ascorbic acid by PBNPs can also be performed in presence of H_2O_2, the PBNPs possess ascorbic acid oxidase-like activity besides ascorbic acid peroxidase-like activity.

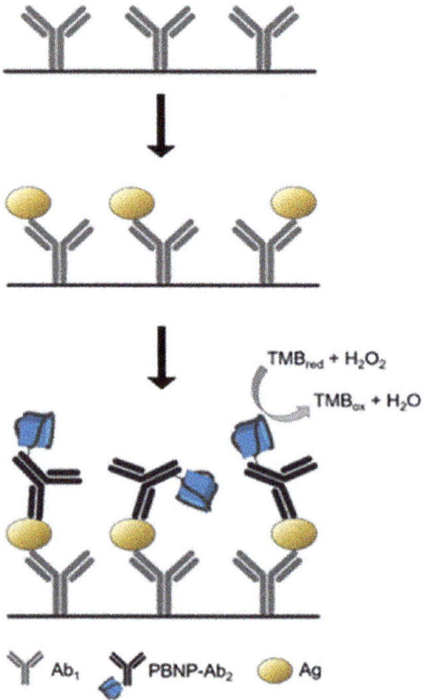

Figure 3. Scheme of PBNPs-based sandwich NLISA. The microtiter plate is coated by capture antibody (Ab$_1$), followed by specific binding of antigen (Ag) and formation of sandwich with the detection PBNPs-Ab$_2$ label. The colorimetric readout is based on PBNPs-catalyzed oxidation of TMB in the presence of H_2O_2. Reproduced with permission from [30].

Cui et al. prepared a nanocomposite by growing PB on the microporous metalorganic framework MIL-101(Fe) [35]. The K_M values of the composite, PB/MIL101(Fe) with respect to TMB and H_2O_2 were 0.127 and 0.0058 mM, lower than those of MIL-101(Fe) for the same substrates (0.490 and 0.620 mM, respectively). The difference was attributed to the fact that the presence of PB in the composite involves more active sites for peroxidase substrates.

Yang et al. modified PB nanocubes with Co_3O_4, a kind of transition metal oxide that exhibits a catalytic activity towards HRP substrates [36]. Such nanoparticles (PB@Co_3O_4) exhibited both intrinsic oxidase- and peroxidase-like activities, namely, they could rapidly oxidize TMB with or without H_2O_2 at acidic conditions. In this way, a colorimetric method for the detection of glutathione (GSH) by using PB@Co_3O_4 nanoparticles as oxidase mimics was developed. GSH produced the fading of the color generated by oxidized TMB (reduction of the absorbance at 652 nm).

He et al prepared hybrid composites formed by PB and $Ti_3C_2T_x$, in which this latter acted as a reducing agent, and the multilayer 2D nanostructure could effectively keep the PB nanoparticles away from aggregation. The composite was used as a colorimetric sensor for H_2O_2, dopamine, and glucose detection [37].

In a study conducted by Sahar et al., a hybrid organic/inorganic hydrogel, which has the potential to mimic multiple enzymes (POD, SOD, and CAT) in a natural cascade-like system was prepared [38]. PB was partially oxidized to Berlin green (BG) and this was mixed with sodium trioxovanadate ($NaVO_3$) in the presence of polyvinylpyrrolidone (PVP). After stirring and heating, a hybrid hydrogel (VO_xBG analogue) was formed. This hydrogel was used as a cascade-like system in the detection of glucose, hydrogel photolithography, oxidation of dopamine, and photocatalytic oxidative degradation of recalcitrant substrates in aqueous media.

An increase in the enzymatic activity was achieved in PBA nanocages doped with molybdenum [14]. This doping successfully tailored the size, morphology, composition, and complex structure of the nanocage. The associate POD-like activity was enhanced by over 37 times compared with pristine PBA.

The POD-like activity of PB/PBA has been used for the detection of several substances such as glucose, glutathione, hydrogen peroxide, or ethanol. In some PBA, which additionally possess oxidase-like activity, this is used to detect glutathione or hydrogen sulfide (Table 1). In the majority of cases, these assays are based on the oxidation of a substrate by hydrogen peroxide. For instance, the aerobic oxidation of glucose in the presence of glucose oxidase (GOx) yields gluconic acid and H_2O_2. The oxidation of a substrate by H_2O_2 gives a product usually readable by spectrophotometry. In the research by Vázquez-González et al. [28], after the oxidation of glucose, the oxidation of luminol as substrate generates a light of ~428 nm, providing a quantitative readout signal for the concentration of glucose (Figure 4).

Table 1. Prussian blue nanoparticles used as sensors. LOD = limit of detection.

Nanoparticles	Size/nm	Analyte	Linear Range	LOD	Ref
PB	<50	Glucose	0.1–50 µM	0.03 µM	[26]
PB	<50	Hydrogen peroxide	0.05–50 µM	0.031 µM	[26]
Biotinylated-PB	15	Human serum albumin	0.35–???	0.27 µg/mL	[27]
PBA (FeCo)	100–200	Glucose	Not indicated	Not indicated	[28]
PBA nanocages	60	Hydrogen sulfide	0.1–15 µM	33 nM	[14]
Antibody modified PB	24	Human serum albumin	1.2–1000 ng/mL	1.2 ng/mL	[30]
PS@Au@PB	20	Glucose	15.6–250.0 µM	3.9 µM	[33]
Co_3O_4-modified PB	~200	Glutathione	0.1–10 µM	0.021 µM	[36]
PB-T13C2Tx		Hydrogen peroxide	2–240 µM	0.4667 µM	[37]
PB		Dopamine	Not indicated	Not indicated	[37]
PB-T13C2Tx		Dopamine	5–120 µM	3.36 µM	[37]
PB-$T_{13}C_2T_x$		Glucose	10–350 µM	6.52 µM	[37]
VO_xBG hydrogel	250	Glucose	1–50 µM	0.046 µM	[38]
Hollow PB	~20	Glucose	0.05–7.3 mM	40 µM	[39]
Hollow PB nanocubes	80	Ethanol	2–500 µg/mL	1.41 µg/mL	[40]
PB/Polypyrrole		Hydrogen peroxide	5–2775 µM	1.6 µM	[41]

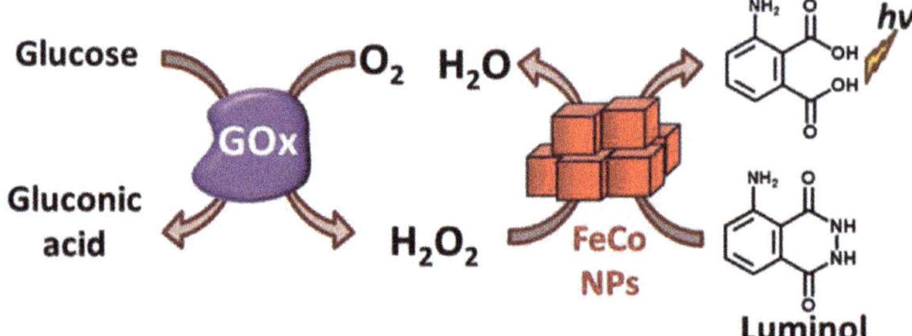

Figure 4. Coupling of the PBFeCo-catalyzed chemiluminescence generation system to the GOx-mediated aerobic oxidation of glucose to gluconic acid and H_2O_2 for the development of a glucose sensor. Reproduced with permission from [28].

Figure 5 illustrates the main reactions implied in the use of PBNPs as POD-like nanozymes. Despite many nanocomposites based on PB/PBA with POD-like activity have been reported until now, their catalytic mechanisms have rarely been investigated. Chen et al. have proved for the first time, by comparing their POD-like activity with that of a series of PBA, in which Fe atoms were replaced by Ni, Cu, and Co [42]. They have demonstrated that the catalytic-like properties can be ascribed to the FeN_x (x = 4–6). Inspired by the plane quadrilateral structure of heme, they have proposed that the catalytic mechanism was also similar to that of the heme enzymes (HRP, cytochrome P450).

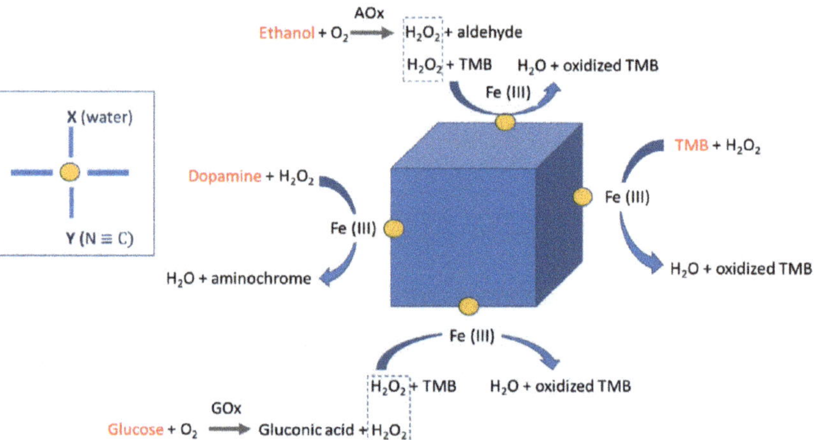

Figure 5. Reactions of the main analytes detected by using PB/PBA as POD-like nanozymes. Inside, the structure of FeN_x band according to [42]; the blue bands represent the macrocyclic assembly of four N ≡ C in the same plane.

From the above, it is evident that PBNPs, as a POD-like nanozyme, can be used as a universal vehicle to build cascades of catalytic reactions. For an efficient action of PBNPs as sensors, such nanoparticles must meet certain requirements such as high surface/volume ratio, compatibility with the other enzymes present in the reaction (in tandem reactions), optimal pH for both enzymes, and the same chemical substrates as natural enzymes.

4. Catalase-Like Activity

Catalases (EC 1.11.1.6) are common enzymes that catalyze the decomposition of hydrogen peroxide into water and oxygen ($2\,H_2O_2 \rightarrow H_2O + O_2$). In consequence, the catalytic activity of CAT can be quantitatively analyzed by measuring the concentration of oxygen produced. Due to H_2O_2 being a main indicator of inflammation, the evaluation of generated O_2 is important to evidence any inflammatory process, such as that produced after photothermal therapy. It is habitual for nanozymes to present POD-like activity at low pH and CAT-like activity at high pH. Thus, depending on the pH conditions, H_2O_2 can be adsorbed or decomposed on the surface of the nanozymes, and this fact determines their catalytic activity.

The group led by Zhang and Gu demonstrated that PBNPs could serve as promising CAT candidates in a redox-state-dependent manner under neutral conditions (optimum pH = 7.4) [31]. This group observed that after incubating PBNPs with H_2O_2 in an alkaline environment the nanoparticles were inactive, as the POD-like activity is concerned, but bubbles of oxygen appeared in the solution. The amount of gas increased at higher pH levels. The group verified that the CAT-like activity was linearly dependent on the concentration of PBNPs. This study is extremely important because it demonstrated why PB can show POD-like or CAT-like activity as a function of pH. PB, as efficient electron transporter, can present one or another enzymatic activity depending on the redox potentials. For instance, at higher pH, the redox potential of H_2O_2/O_2 is very low, and H_2O_2 can be more easily oxidized into O_2. The same group evidenced the effect of pH on the enzymatic activity of PS@Au@PB nanocomposites [33]. Whereas the POD-like activity was maximal at pH 5.2, the CAT-like activity was remarkable in an alkaline buffer.

The incorporation of a chemical element can modify the pH dependence of the CAT-like activity of PB. In this way, PB nanocubes doped with platinum prepared by in situ reduction of $PtCl_6^{2-}$ on the surface of PB showed prominent CAT-like activity at pH 6.5: the dissolved O_2 concentration rose from 4.41 mg/L for PB alone to 38.25 mg/L for a PB with a high content in Pt [43].

It is interesting to cite the study carried out by Zhou et al. [44]. They designed a tumor-targeted redox-responsive nanocomposite that may combine tumor starvation therapy and low-temperature photothermal therapy for the treatment of oxygen-deprived tumors. The nanosystem was prepared by loading porous hollow PBNPs with GOx. After this, their surface was coated with hyaluronic acid, therefore allowing the nanocarrier to bind specifically with tumor cells overexpressed in the tumor. Once the nanosystem was introduced into the cell by endocytosis, the GOx was released and oxidized the glucose. As the growth of tumors is highly dependent on glucose supply, this provokes tumor starvation. However, the oxygen concentration in most solid tumors is significantly lower than in normal tissues, which would severely limit the catalytic efficiency of GOx. Consequently, the catalase-like activity of PBNPs can be used to catalyze the intratumoral H_2O_2 for rapid reoxygenation, which may help to circumvent the tumor-hypoxia-related tissues.

The oxygen produced by PBNPs' catalysis has been utilized in some applications for the diagnosis of diseases. Yang et al. developed an oxygen bubble nanogenerator for ultrasound (US) and magnetic resonance (MR) imaging [45]. First, PBNPs decomposed H_2O_2 into O_2 in neutral conditions; then, the nucleated O_2 molecules could act as gas-bubble US contrast agents. When the oxygen concentration reached supersaturation in tissues, the bubbles could be observed, and an enhanced ultrasound image could also be detected by an acoustic measuring system. Moreover, the paramagnetic oxygen bubbles could shorten the $T1$ relaxation time, acting as an MRI contrast agent. In summary, PBNPs could be an excellent ultrasound (US) and magnetic resonance (MR) dual-modality imaging probe for in vitro and in vivo diagnosing H_2O_2 with excellent sensitivity and resolution.

The CAT-like activity has been used to enhance chemotherapy/photothermal therapy. In this way, a possible synergy between catalytic activity exploited as a photodynamic tool and photothermal effects could be spent. As indicated above, when a tumor grows, there are some portions of the tumor with regions where the oxygen concentration is significantly lower than in healthy tissues. These hypoxic microenvironments in solid tumors are a result of the consumption of available oxygen by rapidly proliferating tumor cells. Moreover, cancer cells produce large amounts of lactate, and this acidic environment promotes tumor cell invasion of adjacent non-cancerous tissue. The indicated hypoxia in solid tumors extremely limits the antitumor of photodynamic therapy. However, the tumor microenvironment also presents high levels of H_2O_2, and this can be a source of O_2. The CAT-like activity of PBNPs can deliver oxygen to the tumor. The generated O_2 can support the photodynamic therapy of tumors and reduce tumor growth [46].

Peng et al. prepared a PB/manganese dioxide (PBMn) nanoparticle coated by erythrocyte membrane to carry doxorubicin and prolong the circulation time in vivo [47]. The generated oxygen from H_2O_2 under the catalysis of PBMn relieved the hypoxia of tumors, and its expansion disrupted the erythrocyte membrane coated on the PBMn surface, which accelerated the release of doxorubicin from the nanoparticles. This made such nanoparticles an H_2O_2-responsive activated drug release nanosystem and enhanced the chemotherapy by PBMn. In addition, PBMn, as any PB/PBA, has strong absorption in the infrared region, with high photothermal conversion efficiency, and can be used as an ideal photothermal therapeutic agent [48].

Similar to the observed POD-like activity, the CAT-like activity is also size-dependent. Ultrasmall PBNPs tend to be more effective and exhibit higher enzyme-like catalytic activity than PBNPs of higher size [19]. The initial catalytic rate of the ultrasmall nanoparticles was $5.3\ \text{mg L}^{-1}\ \text{min}^{-1}$ compared to $0.77\ \text{mg L}^{-1}\ \text{min}^{-1}$ of PBNPs [31].

5. Superoxide Dismutase-Like Activity

Superoxide dismutase (SOD, EC 1.15.1.1) is an enzyme that catalyzes the dismutation of the superoxide ($O_2^{\bullet-}$) radical into ordinary molecular oxygen and hydrogen peroxide, two less damaging species. The reaction is: $2\ O_2^{\bullet-} + 2\ H^+ \rightarrow O_2 + H_2O_2$. Superoxide is produced as a by-product of oxygen metabolism and, if not regulated, causes many types of cell damage [49].

As indicated, PBNPs display SOD-like activity [14]. Moreover, the H_2O_2 produced by the disproportionation of superoxide can be transformed into H_2O or O_2 by PBNPs acting as POD- or CAT-like nanozymes. The standard redox potential values of $O_2/O_2^{\bullet-}$ (0.73 V) and $O_2^{\bullet-}/H_2O_2$ (1.5 V) indicate that the PBNPs are capable of catalyzing the following half-reactions (see Figure 1):

$$2HO_2^{\bullet} + 2H^+ + PB + 2e^- \rightarrow 2H_2O_2 + BG \tag{5}$$

$$2HO_2^{\bullet} + BG \rightarrow 2O_2 + 2H^+ + PB + 2e^- \tag{6}$$

$$2HO_2^{\bullet} + 2H^+ + BG + 2e^- \rightarrow 2H_2O_2 + PY \tag{7}$$

$$HO_2^{\bullet} + PY \rightarrow O_2 + H^+ + BG + e^- \tag{8}$$

The ability of PBNPs to undergo the dismutation of superoxide radical ions can be assessed by the Fridovich assay [50]. In this assay, the oxidation of xanthine by xanthine oxidase generates superoxide radical ions. Such radical ions reduce nitroblue tetrazolium (NBT) giving diformazan. In this way, the reduction of NBT, yellow-colored, by the generated superoxide radical ions implies the formation of the blue-colored diformazan. In the presence of PB/PBA, the generated radicals are totally or partially scavenged reducing the amount of diformazan and, thus, the intensity of blue color. This reduction can be monitored spectrophotometrically (Figure 6) [20].

Another way to verify the ability of PBNPs for scavenging superoxide anions is by using the xanthine/xanthine oxidase system with 5-tert-butoxycarbonyl 5-methyl-1-pyrroline-N-oxide (BMPO) as a spin trap to form the adduct BMPO/OOH$^{\bullet}$. In the presence of SOD or PBNPs, the signal of the electron spin resonance (ESR) spectrum declined [31,51].

In one study, PB was immobilized on amidine functionalized polystyrene latex (AL) particles [20]. The ability of PB and PB-AL materials in dismutation of superoxide radical ions was determined colorimetrically [49]. The PB did not lose the SOD activity upon immobilization on AL. However, because of the inevitable hindrance of some catalytic sites on the surfaces of PBNPs upon attachment to the AL surface, the maximum inhibition values decrease for PB-AL. In this way, $K_M = 2.19$ mM for PB in comparison to 2.92 mM for PB-AL, and $v_{max} = 6.71 \times 10^{-6}\ \text{M s}^{-1}$ for PB and $4.09 \times 10^{-6}\ \text{M s}^{-1}$ for PB-AL. As an important advantage of these hybrid nanoparticles, the authors indicate their high colloidal stability.

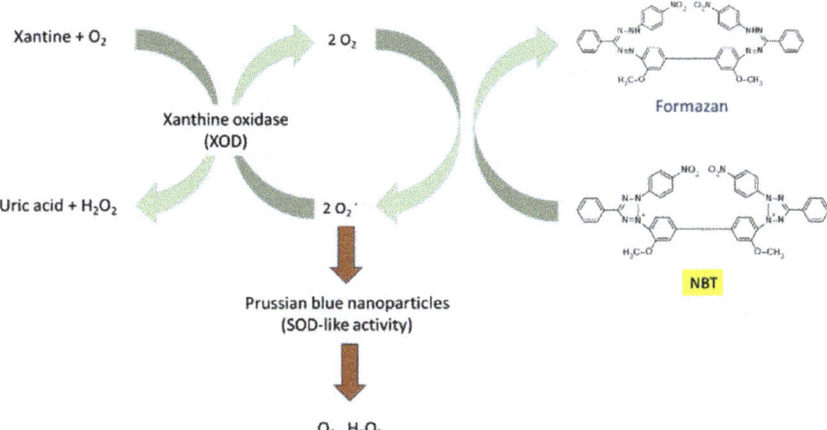

Figure 6. The conversion of xanthine and O_2 to uric acid and H_2O_2 by XOD generates superoxide radicals. The superoxide anions reduce the nitroblue tetrazolium [NBT] to a colored formazan product that absorbs radiation at 560 nm. PBNPs scavenge superoxide anions, thereby reducing the rate of formazan formation.

The VO$_x$BG analogue aforementioned [38] has multiple oxidation states and this confers it a multi-enzymatic activity. The SOD-like activity of this analogue was determined by the transformation of NBT.

The platinum-doped PB nanoparticles presented an increased enzyme-like activity in comparison to plain PB nanoparticles [42]. The activity was enhanced with increasing Pt content. Under the same condition, Pt nanoparticles presented a slight SOD-like activity, which contributed to the increased enzyme-like activity of Pt-doped PB nanoparticles.

6. ROS-Scavenging or ROS-Generating Activity

The term ROS describes reactive oxygen-derived free radicals such as hydroxyl ($OH^•$), superoxide anion ($O_2^{•-}$), nitric oxide ($NO^•$), and peroxyl ($RO_2^•$), as well as nonradical reactive oxygen derivatives such as H_2O_2. ROS is a double-edged sword because, although the biological antioxidant system controls the levels of ROS, an excess of them leads to oxidative stress damage and dysfunction related to many human diseases (Alzheimer's disease, Parkinson's disease, and diabetes). However, ROS can also play a positive role, since they can induce the activation of apoptosis in cancer cells.

PB and PBA can generate or scavenge ROS depending on their catalytic properties. In this way, PB/PBA with POD- or oxidase-like activity generates abundant ROS; PB/PBA with CAT-like and/or SOD-like activity can remove ROS. In both cases, one can find important biomedical applications, as will be indicated below.

The formation of ROS by PBNPs follows the Fenton reaction [51]. The conventional Fenton process is associated with the electron transfer from Fe^{2+} to H_2O_2 to generate highly active $OH^•$, which is capable of effectively attacking the target pollutants and decomposing them into harmless species. The reaction pathway can be illustrated by the following equations:

$$Fe^{2+} + H_2O_2 \rightarrow Fe^{3+} + OH^- + OH^• \qquad (9)$$

$$Fe^{3+} + H_2O_2 \rightarrow Fe^{2+} + OOH^• + H^+ \qquad (10)$$

$$OH^• + \text{substrates} \rightarrow \text{intermediates} \rightarrow CO_2 + H_2O \qquad (11)$$

Over the years, several approaches have been carried out to use PB/PBA as Fenton reagents to break down various organic contaminants [52–56]. For instance, Liu et al. employed PB as a photo-Fenton-like reagent and investigated its high catalytic efficiency for rhodamine B (RhB) degradation under visible irradiation in neutral conditions [52]. Li et al. performed PB/TiO_2 micro composites as a heterogeneous photo-Fenton catalyst to enhance the Fe (II) improvement in destroying organic contaminants [53]. Furthermore, the same group also prepared two kinds of PBA, based on Co and Fe, with different iron valence states and examined their heterogeneous photo-Fenton catalytic mechanism and their use in the oxidation of bisphenol [54,55]. Bu et al. used ultrathin nanosheets of PB to achieve the degradation of methylene blue (MB), one of the most used dyes in various industries [56]. In this study, it was found that, by means of the peroxidase-like activity, 73% of MB had been removed within 1 min in the presence of nanosheets. In the same study, cubic PB only removed 25% of MB. These preliminary studies elucidated that PB/PBA are promising alternatives to conventional transition metals and metal oxides for more efficient catalysis. More recently, the study by Wang et al. focused on the structural control and design of micro/nanocrystals and investigated the relationship between structure and performance of PB [57]. To this end, the authors synthesized a series of PB microcrystals with morphology evolution from microtubes to hexapod stars by adjusting the concentration of chloroplatinic acid in the reaction system (Figure 7).

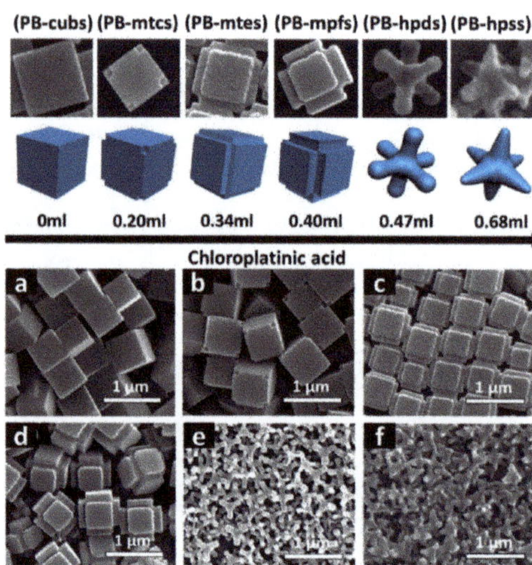

Figure 7. SEM images of PB crystals of different morphology obtained increasing the concentration of chloroplatinic acid. (**a**) PB-cubs: microcubes; (**b**) PB-mcts: PB cubs with truncated corners; (**c**) PB-mtes: PB cubs with truncated edges; (**d**) PB-mpfs: PB cubs with protrusive faces; (**e**) PB-hpds: hexapods; and (**f**) PB-hpss: hexapod stars. Reproduced with permission from [57].

Results showed that the degradation efficiency of RhB was closely related to the morphology of PB, where the pristine PB-cubs only account for 16.6% of RhB removal within 60 min and PB-hpds enhance that to 97.1% under the same conditions. PB-hpss also presented superior catalytic performance compared to PB-cub and other PB intermediates. The reason was the high specific surface areas and adequate exposure of Fe^{III}-NC coordination active sites of PB-hpds and PB-hpss.

The hybrid hydrogel constituting VO_x incorporated BG analogue complex was explored as a potential catalyst for photocatalytic degradation of organic pollutants for wastewater treatment purposes [38]. Unlike other photocatalysts, this complex hydrogel followed an unusual catalytic mechanism of $OH^\bullet$ radical generation, which in this case function as entrapped holes rather than free radicals in the reaction system. This ability accompanied by the simultaneous generation of $OO^{\bullet -}$ contributed to a superior multicatalytic performance of VO_xBG.

By modulating ROS production, PB/PBA can be exploited for disease therapeutics. For instance, by catalytically generating abundant ROS selectively in a tumor microenvironment, PB/PBA can be used for antitumor therapeutics, since excessive intracellular levels of ROS may damage lipids, proteins, and DNA. However, the antitumor effect can be reverted by PB itself in some conditions. It is known that ascorbic acid (AA) is able to induce cancer cell death by improving the intracellular ROS level [58]. Although iron is considered to be able to reinforce the anti-cancer effect of AA via Fenton reaction, PBNPs, despite being an iron-based nanomaterial, can inhibit the anti-cancer effect of AA. Zhang et al. have inferred that PBNPs may play as an AA-related nanozyme that can catalyze the oxidation of AA without producing H_2O_2 [34]. Moreover, this study showed that PBNPs could catalyze the oxidation of AA in the absence and presence of H_2O_2, indicating PBNPs possess ascorbic acid oxidase-like activity besides ascorbic acid peroxidase-like activity. Both enzymatic activities are ROS-scavengers.

PBNPs, after being taken up by cells, can be found in different intracellular microenvironments, where may exert differential enzyme-like activities. PBNPs mainly function as SOD-mimetics, high-activity CAT mimetics, and low-activity POD-mimetics when distributed in the cytosol. In lysosomes, PBNPs unlike other iron-based nanoparticles can inhibit the production of $OH^{\bullet}$ [31]. Without the Fenton reaction, PBNPs could effectively eliminate ROS such as hydrogen peroxide, hydroxyl radicals ($OH^{\bullet}$), and superoxide radicals ($OH^{\bullet}$). With these unique performances, PBNPs have great potential as an anti-inflammatory agent to protect cells from ROS damage. To demonstrate the ROS scavenging ability of PBNPs, Zhang et al. established an in vivo inflammation model [31]. Mice with an inflammation induced by lipoproteins were treated with PBNPs. Results showed that PBNPs could protect the animals from oxidative damage and slow down the inflammatory response. Zhao et al. also showed the ROS-scavenging ability of PBNPs when mice with inflammatory bowel disease were treated with PBNPs administered intravenously [59]. Zhao et al. developed a PBA (based on Mn) to treat inflammatory bowel disease in mice with induced colitis [60]. The prepared PBA accumulated at inflamed sites after oral administration. PBA significantly improved colitis via a primary effect on the toll-like receptor signaling pathway without causing adverse side effects.

Zhang et al. studied the ability of hollow PB to scavenger ROS and reactive nitrogen species, such as nitric oxide ($NO^{\bullet}$) and peroxynitrite ($ONOO^{-}$), in a rat model of ischemia stroke [15]. Apart from attenuating oxidative stress, PB also suppressed apoptosis and counteracted inflammation both in vitro and in vivo, thereby contributing to increasing brain tolerance of ischemic injury with minimal side effects.

The reduction of the size of PBNPs obtained by using a mixture of water and ethanol [17] increased the ability to break down H_2O_2 to generate oxygen, scavenging free radicals and protecting cells from oxidation. The scavenging ability was related to the results of the catalytic activity of the nanozyme. This activity increased by reducing the size of the nanoparticles due to the fact that the catalytic activity is size-dependent. The catalytic activity of PBNPs mainly emanates from the electron transfer between Fe ions with different valences [31], and since these catalytic sites are mostly situated at the FeN_x ($x = 4$–6 units), the reduction of the size of PBNPs results in the increase of their specific surface area and allows for the maximization of the exposed FeN_x sites.

7. Conclusions and Future Perspectives

PBNPs and their analogues have gathered increasing research interest since their first discovery as nanozymes just over ten years ago [25,26] because they exhibit catalytic properties as peroxidase, catalase, and superoxide dismutase. In this review, we systematically summarized the enzyme-like activities and the catalytic applications of PB and PBA. Compared to natural enzymes, PBNPs overcome their main disadvantages such as the high cost of production and purification. Moreover, PBNPs offer improved tolerance to harsh conditions and higher stability. PBNPs have an added value: their different states of oxidation can be changed by regulating external potentials (PB presents low redox potentials) and this fact is responsible for their multicatalytic properties. Depending on the type of enzymatic-like activity, PB/PBA can be used to remove excess ROS from the body to treat ROS-related diseases and decompose the H_2O_2 into oxygen to overcome the hypoxia of solid tumors. However, the same PB/PBA can generate abundant ROS and this property serves for the remediation of environmental pollutants.

Although great progress has been made, the development of PB/PBA as nanozymes with desired properties is still limited by some challenges, and an in-depth understanding of the fundamental principles of these nanoparticles as nanozymes remains limited. It is still necessary to define exactly the influence of size, how the surface modification affect the enzyme-like properties beyond acting as a stabilizer, which is the maximal density and thickness of the coating layer to avoid that the interaction between the nanoparticles and the substrate is masked, which elements are optimal to dope the chemical structure, and so on. Finally, in spite of the safety of PB, the preparation of new analogues, or new hybrid nanocomposites should be accompanied by studies about the potential nanotoxicity of these nanosystems on living beings well as the toxicity in the environment.

Author Contributions: Conceptualization, writing—original draft preparation J.E.; writing—review and editing, M.A.B. All authors have read and agreed to the published version of the manuscript.

Funding: This research received no external funding.

Institutional Review Board Statement: Not applicable.

Informed Consent Statement: Not applicable.

Data Availability Statement: Not applicable.

Conflicts of Interest: The authors declare no conflict of interest.

References

1. Huang, Y.; Ren, J.; Qu, X. Nanozymes: Classification, catalytic mechanisms, activity regulation, and applications. *Chem. Rev.* **2019**, *119*, 357–4412. [CrossRef] [PubMed]
2. Liang, M.; Yan, X. Nanozymes: From new concepts, mechanisms, and standards to applications. *Acc. Chem. Res.* **2019**, *52*, 2190–2200. [CrossRef] [PubMed]
3. Stasyuk, S.; Smutok, O.; Demkiv, O.; Prokopiv, T.; Gayda, G.; Nisnevitch, M.; Gonchar, M. Synthesis, catalytic properties and application in biosensorics of nanozymes and electronanocatalysts: A review. *Sensors* **2020**, *20*, 4509. [CrossRef] [PubMed]
4. Gao, L.; Zhuang, J.; Nie, L.; Zhang, J.; Zhang, Y.; Gu, N.; Wang, T.; Feng, J.; Yang, D.; Perrett, S.; et al. Intrinsic peroxidase-like activity of ferromagnetic nanoparticles. *Nat. Nanotechnol.* **2007**, *2*, 577–583. [CrossRef] [PubMed]
5. Wang, Z.; Zhang, R.; Yan, X.; Fan, K. Structure and activity of nanozymes: Inspirations for de novo design of nanozymes. *Mater. Today* **2020**, *41*, 81–119. [CrossRef]
6. Thompson, D.F.; Callen, E.D. Soluble or insoluble Prussian blue for radiocesium and thallium poisoning. *Ann. Pharmacother.* **2004**, *38*, 1509–1514. [CrossRef] [PubMed]
7. Ishizaki, M.; Kanaizuka, K.; Abe, M.; Sakamoto, M.; Kawamoto, T.; Tanaka, H.; Kurihara, M. Preparation of electrochromic Prussian blue nanoparticles dispersible into various solvents for realisation of printed electronics. *Green Chem.* **2012**, *14*, 1537–1544. [CrossRef]
8. Ware, M. Prussian blue: Artists' pigment and chemists' sponge. *J. Chem. Educ.* **2008**, *85*, 612–620. [CrossRef]
9. Guari, Y.; Larionova, J. (Eds.) *Prussian Blue-Type Nanoparticles and Nanocomposites. Synthesis, Devices and Applications*; Pan Stanford: Singapore, 2019.
10. Busquets, M.A.; Estelrich, J. Prussian blue nanoparticles: Synthesis, surface modification and biomedical applications. *Drug Discov. Today* **2020**, *25*, 1413–1443. [CrossRef] [PubMed]
11. Taglietti, A.; Pallavicini, P.; DaCarro, G. Prussian blue and its analogs as a novel nanostructured antibacterial materials. *Appl. Nano* **2021**, *2*, 85–97. [CrossRef]
12. *WHO Model List of Essential Medicines*, 3rd ed.; World Health Organization: Geneva, Switzerland, 2013; p. 4. Available online: https://www.who.int/groups/expert-committee-on-selection-and-use-of-essential-medicines/essential-medicines-lists (accessed on 22 April 2021).
13. Estelrich, J.; Busquets, M.A. Prussian blue: A safe pigment with zeolitic-like activity. *Int. J. Mol. Sci.* **2021**, *22*, 780. [CrossRef] [PubMed]

14. Wang, C.; Wang, M.; Zhang, W.; Liu, J.; Lu, M.; Li, K.; Lin, Y. Integrating Prussian blue analog-based nanozyme and online visible light absorption approach for continuous hydrogen sulfide monitoring in brains of living rats. *Anal. Chem.* **2020**, *92*, 662–667. [CrossRef] [PubMed]

15. Zhang, K.; Tu, M.; Gao, W.; Cai, X.; Song, F.; Chen, Z.; Zhang, Q.; Wang, J.; Jin, C.; Shi, J.; et al. Hollow Prussian blue nanozymes drive neuroprotection against ischemic stroke via attenuating oxidative stress, counteracting inflammation, and suppressing cell apoptosis. *Nano Lett.* **2019**, *19*, 2812–2823. [CrossRef]

16. Ming, H.; Torad, N.L.K.; Chiang, Y.-D.; Wu, K.C.-W.; Yamauchi, Y. Size- and shape-controlled synthesis of Prussian blue nanoparticles by a polyvinylpyrrolidone-assisted crystallization process. *CrystEngComm* **2012**, *14*, 3387–3396. [CrossRef]

17. Azhar, A.; Li, Y.; Cai, Z.; Zakaria, M.B.; Masud, M.K.; Hossain, M.S.A.; Kim, J.; Zhang, W.; Na, J.; Yamauchi, Y.; et al. Nanoarchitectonics: A new materials horizon for Prussian blue and its analogues. *Bull. Chem. Soc. Jpn.* **2019**, *92*, 875–904. [CrossRef]

18. Dacarro, G.; Taglietti, A.; Pallavicini, P. Prussian blue nanoparticles as a versatile photothermal tool. *Molecules* **2018**, *23*, 1414. [CrossRef] [PubMed]

19. Qin, Z.; Chen, B.; Mao, Y.; Shi, C.; Li, Y.; Huang, X.; Yang, F.; Gu, N. Achieving ultrasmall Prussian blue nanoparticles as high-performance biomedical agents with multifunctions. *ACS Appl. Mater. Interfaces* **2020**, *12*, 35068–35078. [CrossRef] [PubMed]

20. Alsharif, N.B.; Samu, G.F.; Sáringer, S.; Muráth, S.; Szilagyi, I. A colloid approach to decorate latex particles with Prussian blue nanozymes. *J. Mol. Liq.* **2020**, *309*, 113066. [CrossRef]

21. Shokouhimehr, M.; Soehnlen, E.S.; Khitrin, A.; Basu, S.; Huang, S.D. Biocompatible Prussian blue nanoparticles: Preparation, stability, cytotoxicity, and potential use as an MRI contrast agent. *Inorg. Chem. Commun.* **2010**, *13*, 58–61. [CrossRef]

22. Mo, W.; Yu, J.; Gao, L.; Liu, Y.; Wei, Y.; He, R. Reversible inhibition of iron oxide nanozyme by guanidine chloride. *Front. Chem.* **2020**, *8*, 491. [CrossRef]

23. Pham-Huy, L.A.; He, H.; Pham-Huy, C. Free radicals, antioxidants in disease and health. *Int. J. Biomed. Sci.* **2008**, *4*, 89–96. [PubMed]

24. Komkova, M.A.; Karyakina, E.E.; Karyakin, A.A. Catalytically synthesized Prussian blue nanoparticles defeating natural enzyme peroxidase. *J. Am. Chem. Soc.* **2018**, *140*, 11302–11307. [CrossRef] [PubMed]

25. Zhang, X.Q.; Gong, S.W.; Zhang, Y.; Yang, T.; Wang, C.Y.; Gu, N. Prussian blue modified iron oxide magnetic nanoparticles and their high peroxidase-like activity. *J. Mater. Chem.* **2010**, *20*, 5110–5116. [CrossRef]

26. Zhang, W.; Ma, D.; Du, J. Prussian blue nanoparticles as peroxidase mimetics for sensitive colorimetric detection of hydrogen peroxide and glucose. *Talanta* **2014**, *120*, 362–367. [CrossRef] [PubMed]

27. Čunderlová, V.; Hlaváček, A.; Horňaková, V.; Peterek, M.; Němeček, D.; Hampl, A.; Eyer, L.; Skládal, P. Catalytic nanocrystalline coordination polymers as an efficient peroxidase mimic for labeling and optical immunoassays. *Microchim. Acta* **2016**, *183*, 651–658. [CrossRef]

28. Vázquez-González, M.; Torrente-Rodríguez, R.M.; Kozell, A.; Liao, W.-C.; Cecconello, A.; Campuzano, S.; Pingarrón, J.M.; Willner, I. Mimicking peroxidase activities with Prussian blue nanoparticles and their cyanometallate structural analogues. *Nano Lett.* **2017**, *17*, 4958–4963. [CrossRef]

29. Hou, J.; Vázquez-González, M.; Fadeev, M.; Liu, X.; Lavi, R.; Willner, I. Catalyzed and electrocatalyzed oxidation of L-tyrosine and L-phenylalanine to dopachrome by nanozymes. *Nano Lett.* **2018**, *18*, 4015–4022. [CrossRef]

30. Farka, Z.; Čunderlová, V.; Horáčová, V.; Pastucha, M.; Mikušová, Z.; Hlaváček, A.; Skládal, P. Prussian blue nanoparticles as a catalytic label in sandwich nanozyme-linked immunosorbent assay. *Anal. Chem.* **2018**, *90*, 2348–2354. [CrossRef] [PubMed]

31. Zhang, W.; Hu, S.; Yin, J.-J.; He, W.; Lu, W.; Ma, M.; Gu, N.; Zhang, Y. Prussian blue nanoparticles as multienzyme mimetics and reactive oxygen species scavengers. *J. Am. Chem. Soc.* **2016**, *138*, 5860–5865. [CrossRef] [PubMed]

32. Zhang, W.; Zhang, Y.; Chen, Y.; Li, S.; Gu, N.; Hu, S.; Sun, Y.; Chen, X.; Li, Q. Prussian blue modified ferritin as peroxidase mimetics and its applications in biological detection. *J. Nanosci. Nanotechnol.* **2013**, *13*, 60–67. [CrossRef] [PubMed]

33. Zhang, X.-Z.; Zhou, Y.; Zhang, W.; Zhang, Y.; Gu, N. Polystyrene@Au@prussian blue nanocomposites with enzyme-like activity and their application in glucose detection. *Colloids Surfaces A* **2016**, *490*, 291–299. [CrossRef]

34. Zhang, W.; Wu, Y.; Dong, H.-J.; Yin, J.-J.; Zhang, H.; Wu, H.-A.; Song, L.-N.; Chong, Y.; Li, Z.-X.; Gu, N.; et al. Sparks fly between ascorbic acid and iron-based nanozymes: A study on Prussian blue nanoparticles. *Colloids Surfaces B* **2018**, *163*, 379–384. [CrossRef]

35. Cui, F.; Deng, Q.; Sun, L. Prussian blue modified metal-organic framework MIL-101(Fe) with intrinsic peroxidase-like catalytic activity as a colorimetric biosensing platform. *RSC Adv.* **2015**, *5*, 98215–98221. [CrossRef]

36. Yang, W.; Hao, J.; Zhang, Z.; Zhang, B. PB@Co$_3$O$_4$ nanoparticles as both oxidase and peroxidase mimics and their application for colorimetric detection of glutathione. *New J. Chem.* **2015**, *39*, 8802–8806. [CrossRef]

37. He, Y.; Zhou, X.; Zhou, L.; Zhang, X.; Ma, L.; Jiang, Y.; Gao, J. Self-reducing Prussian blue on Ti3C2Tx MXene nanosheets as a dual-functional nanohybrid for hydrogen peroxide and pesticide sensing. *Ind. Eng. Chem. Res.* **2020**, *59*, 15556–15564. [CrossRef]

38. Sahar, S.; Zeb, A.; Ling, C.; Raja, A.; Wang, G.; Ullah, N.; Lin, X.-M.; Xu, A.-W. A hybrid VO$_x$ incorporated hexacyanoferrate nanostructured hydrogel as a multienzyme mimetic *via* cascade reactions. *ACS Nano* **2020**, *14*, 3017–3031. [CrossRef]

39. Zheng, K.; Yang, M.; Liu, N.-Y.; Rassoly, A. Dual function hollow structures mesoporous Prussian blue mesocrystals for glucose biosensors. *Anal. Methods* **2018**, *10*, 3951–3958. [CrossRef]

40. Wang, S.; Yan, H.; Wang, Y.; Wang, N.; Lin, Y.; Li, M. Hollow Prussian blue nanocubes as peroxidase mimetic and enzyme carriers for colorimetric determination of ethanol. *Microchim. Acta* **2019**, *186*, 738. [CrossRef] [PubMed]

41. Yang, Z.; Zheng, X.; Zheng, J. Facile synthesis of Prussian blue/hollow polypyrrole nanocomposites for enhanced hydrogen peroxide sensing. *Ind. Eng. Chem. Res.* **2016**, *55*, 12161–12166. [CrossRef]
42. Chen, J.; Wang, Q.; Huang, L.; Zhang, H.; Rong, K.; Zhang, H.; Dong, S. Prussian blue with intrinsec heme-like structure as peroxidase enzyme. *Nano Res.* **2018**, *11*, 4905–4913. [CrossRef]
43. Li, Z.-H.; Chen, Y.; Zhang, X.-Z. Platinum-doped Prussian blue nanozymes for multiwavelength bioimaging guided photothermal therapy of tumor and anti-inflammation. *ACS Nano* **2021**, *15*, 5189–5200. [CrossRef] [PubMed]
44. Zhou, J.; Li, M.; Hou, Y.; Luo, Z.; Chen, Q.; Cao, H.; Huo, R.; Xue, C.; Sutrisno, L.; Hao, L.; et al. Engineering of a sized biocatalyst for combined tumor starvation and low-temperature photothermal therapy. *ACS Nano* **2018**, *12*, 2858–2872. [CrossRef] [PubMed]
45. Yang, F.; Hu, S.; Zhang, Y.; Cai, X.; Huang, Y.; Wang, F.; Wen, S.; Teng, G.; Gu, N. A hydrogen peroxide-responsive O$_2$ nanogenerator for ultrasound and magnetic resonance dual modality imaging. *Adv. Mater.* **2012**, *24*, 5205–5211. [CrossRef] [PubMed]
46. Gao, X.; Wang, Q.; Cheng, C.; Lin, S.; Lin, T.; Liu, C.; Han, X. The application of Prussian blue nanoparticles in tumor diagnosis and treatment. *Sensors* **2020**, *20*, 6905. [CrossRef] [PubMed]
47. Peng, J.; Yang, Q.; Li, W.; Tan, L.; Xiao, Y.; Chen, L.; Hao, Y.; Qian, Z. Erythrocyte-membrane-coated Prussian blue/manganese dioxide nanoparticles as H$_2$O$_2$-responsive oxygen generators to enhance cancer chemotherapy/photothermal therapy. *ACS Appl. Mater. Interfaces* **2017**, *9*, 44410–44422. [CrossRef]
48. Peng, J.; Dong, M.; Ran, B.; Li, W.; Hao, Y.; Yang, Q.; Tan, L.; Shi, K.; Qian, Z. "One-for-all"-type, biodegradable Prussian blue/manganese dioxide hybrid nanocrystal for trimodal imaging-guided photothermal therapy and oxygen regulation of breast cancer. *ACS Appl. Mater. Interfaces* **2017**, *9*, 13875–13886. [CrossRef] [PubMed]
49. McCord, J.M.; Fridovich, I. Superoxide dismutase. An enzymic function for erythrocuprein (hemocuprein). *J. Biol. Chem.* **1969**, *244*, 6049–6055. [CrossRef]
50. Beaucham, C.; Fridovich, I. Superoxide dismutase—Improved assays and an assay applicable to acrylamide gels. *Anal. Biochem.* **1971**, *44*, 276–287. [CrossRef]
51. Fenton, H.J.H. Oxidation of tartaric acid in presence of iron. *J. Chem. Soc. Trans.* **1894**, *65*, 899–910. [CrossRef]
52. Liu, S.-Q.; Cheng, S.; Luo, L.; Cheng, H.-Y.; Wang, S.-J.; Lou, S. Degradation of dye rhodamine B under visible irradiation with Prussian blue as a photo-Fenton reagent. *Environm. Chem. Lett.* **2009**, *9*, 31–35. [CrossRef]
53. Li, X.; Wang, J.; Rykov, A.I.; Sharma, V.K.; Wei, H.; Jin, C.; Liu, X.; Li, M.; Yu, S.; Sun, C.; et al. Prussian blue/TiO$_2$ nanocomposites as a heterogeneous photo-Fenton catalyst for degradation of organic pollutants in water. *Catal. Sci. Technol.* **2015**, *5*, 504–514. [CrossRef]
54. Li, X.; Liu, J.; Rykov, A.I.; Han, H.; Jin, C.; Liu, X.; Wang, J. Excellent photo-Fenton catalysts of Fe-Co Prussian blue analogues and their reaction mechanism study. *Appl. Catal. B* **2015**, *179*, 196–205. [CrossRef]
55. Li, X.; Rykov, A.I.; Wang, J. Hydrazine drastically promoted Fenton oxidation of bisphenol A catalysed by a Fe III-Co Prussian blue analogue. *Catal. Commun.* **2016**, *77*, 32–36. [CrossRef]
56. Bu, F.-X.; Hu, M.; Zhang, W.; Meng, Q.; Xu, L.; Jiang, D.-M.; Jiang, J.-S. Three-dimensional hierarchical Prussian blue composed of ultrathin nanosheets: Enhanced hetero-catalytic and adsorption properties. *Chem. Commun.* **2015**, *51*, 17568–17571. [CrossRef] [PubMed]
57. Wang, N.; Ma, W.; Du, Y.; Ren, Z.; Han, B.; Zhang, L.; Sun, B.; Xu, P.; Han, X. Prussian blue microcrystals with morphology evolution as a high-performance photo-Fenton catalyst for degradation of organic pollutants. *ACS Appl. Mater. Interfaces* **2019**, *11*, 1174–1184. [CrossRef] [PubMed]
58. Doskey, C.M.; Buranasudja, V.; Wagner, B.A.; Wilkes, J.G.; Du, J.; Cullen, J.J.; Buettner, G.R. Tumor cells have decreased ability to metabolize H$_2$O$_2$: Implications for pharmacological ascorbate to cancer therapy. *Redox Biol.* **2016**, *10*, 274–284. [CrossRef] [PubMed]
59. Zhao, J.; Cai, X.; Gao, W.; Zhang, L.; Zou, D.; Zheng, Y.; Li, Z.; Chen, H. Prussian blue nanozyme with multienzyme activity reduces colitis in mice. *ACS Appl. Mater. Interfaces* **2018**, *10*, 26108–26117. [CrossRef] [PubMed]
60. Zhao, J.; Goo, W.; Cai, X.; Xu, J.; Zou, D.; Li, Z.; Hu, B.; Zheng, Y. Nanozyme-mediated catalytic nanotherapy for inflammatory bowel disease. *Theranostics* **2019**, *9*, 2843–2855. [CrossRef]

International Journal of
Molecular Sciences

Article

Feasibility of Monitoring Tumor Response by Tracking Nanoparticle-Labelled T Cells Using X-ray Fluorescence Imaging—A Numerical Study

Henrik Kahl [1,2], Theresa Staufer [2], Christian Körnig [2], Oliver Schmutzler [2], Kai Rothkamm [1] and Florian Grüner [2,*]

[1] University Medical Center Hamburg-Eppendorf, Department of Radiotherapy and Radiation Oncology, Medical Faculty, University of Hamburg, Martinistraße 52, 20246 Hamburg, Germany; henrik.kahl@stud.uke.uni-hamburg.de (H.K.); k.rothkamm@uke.de (K.R.)

[2] Center for Free-Electron Laser Science (CFEL), Universität Hamburg, Luruper Chaussee 149, 22761 Hamburg, Germany; theresa.staufer@desy.de (T.S.); ckoernig@mail.desy.de (C.K.); oliver.schmutzler@desy.de (O.S.)

* Correspondence: florian.gruener@uni-hamburg.de

Citation: Kahl, H.; Staufer, T.; Körnig, C.; Schmutzler, O.; Rothkamm, K.; Grüner, F. Feasibility of Monitoring Tumor Response by Tracking Nanoparticle-Labelled T Cells Using X-ray Fluorescence Imaging—A Numerical Study. *Int. J. Mol. Sci.* **2021**, *22*, 8736. https://doi.org/10.3390/ijms22168736

Academic Editors: Raghvendra Singh Yadav and Donald J. Buchsbaum

Received: 25 June 2021
Accepted: 10 August 2021
Published: 14 August 2021

Publisher's Note: MDPI stays neutral with regard to jurisdictional claims in published maps and institutional affiliations.

Abstract: Immunotherapy has been a breakthrough in cancer treatment, yet only a subgroup of patients responds to these novel drugs. Parameters such as cytotoxic T-cell infiltration into the tumor have been proposed for the early evaluation and prediction of therapeutic response, demanded for non-invasive, sensitive and longitudinal imaging. We have evaluated the feasibility of X-ray fluorescence imaging (XFI) to track immune cells and thus monitor the immune response. For that, we have performed Monte Carlo simulations using a mouse voxel model. Spherical targets, enriched with gold or palladium fluorescence agents, were positioned within the model and imaged using a monochromatic photon beam of 53 or 85 keV. Based on our simulation results, XFI may detect as few as 730 to 2400 T cells labelled with 195 pg gold each when imaging subcutaneous tumors in mice, with a spatial resolution of 1 mm. However, the detection threshold is influenced by the depth of the tumor as surrounding tissue increases scattering and absorption, especially when utilizing palladium imaging agents with low-energy characteristic fluorescence photons. Further evaluation and conduction of in vivo animal experiments will be required to validate and advance these promising results.

Keywords: XFI; X-ray fluorescence imaging; T cell; immunotherapy; nanoparticles; gold; palladium; simulation

1. Introduction

Over the past decades the field of cancer therapy has seen major breakthroughs by the implementation of novel therapeutic approaches. Most prominently, the idea of not targeting tumor cells directly but rather harnessing the body's immune response has revolutionized the way neoplastic diseases are looked upon [1]. While this concept itself is not particularly new, it has been majorly restrained by malignant cells' capabilities of evading immune response. This immune evasion is based on multiple mechanisms, mostly either inhibiting effector cells or disguising themselves by downregulating surface proteins [2]. Three major approaches regarding immunotherapy have been pursued, namely cancer vaccines [3], adoptive T cell transfer [4] and checkpoint inhibition [5]. The latter, working through inhibiting the activation of cytotoxic T lymphocyte-associated protein 4 (CTLA-4) or programmed cell death protein 1 (PD-1), T cell membrane proteins suppressing immune response, have shown unprecedented anti-tumor response in clinical trials [6,7], but only in a subgroup of patients. With a multitude of immune checkpoint inhibitors entering the market and multimodal therapeutic concepts being introduced, early evaluation and even prediction of treatment response is indispensable to improve the patients' outcome [8]. Nevertheless, traditional methods of assessing tumor response mostly relying on change in tumor size were shown to be inadequate in this context due to the presence of atypical tumor responses [9]. As one of multiple markers being proposed, cytotoxic T cells (CTL) infiltrating the tumor were discovered to correlate with tumor response under these novel therapeutics and could thus be used to quickly assess and monitor the immune reaction [10]. While early investigations of T-cell distribution were limited to highly invasive biopsies not suited for routine clinical implementation [11], the emergence of molecular imaging has paved the way for non-invasive monitoring of cellular targets [12]. Four imaging modalities have been of particular interest for observing CTLs in the context of immunotherapy, either imaging injected cells or the endogenous T-cell population. The majority of the research groups apply PET and SPECT imaging due to their high sensitivity. However, as a significant efflux of radionucleotides has been observed when directly labelling T cells, the majority of PET studies has subsequently utilized radioactively labelled CD8 antibody fragments [13], with the first human trials being conducted [14]. Nevertheless, imaging of in vitro-labelled, injected T cells has also been studied using magnetic resonance imaging (MRI) [15], computed tomography (CT) [16] and bioluminescence imaging (BLI) [17]. Whereas all four approaches show promising results in either sensitivity or spatial resolution and some even allow for longitudinal imaging, none has been able to combine all aspects, with each modality having its specific drawbacks. Due to this reason, neither approach has been implemented in a clinical setting as of yet, albeit being urgently needed. Thus, we want to introduce a further imaging modality—X-ray fluorescence imaging (XFI)—and discuss the expected limits.

Ever since their discovery in 1895 [18], X-rays have been extensively studied as a way to non-invasively visualize body structures and processes. However, the possible applications of medical X-rays are far beyond the scope of established attenuation-based imaging [19]. When a photon in the X-ray energy range collides with matter, multiple interactions with the atom are possible. Regarding XFI, two effects are of major importance, namely the photoelectric effect and Compton scattering. Photons above an element-dependent energy (the so-called "absorption-edge") have the ability to ionize atoms by removing an electron from one of their shells [20]. While the primary photon is absorbed in this process, the resulting empty position in the atom's shell is filled by an outer shell electron, releasing energy isotropically in the form of a secondary photon. Such emissions of high energy are called X-ray fluorescence (like an "X-ray echo") and have a characteristic energy based on the interacting atomic element. Materials with a higher atomic number Z provide higher photon emission energies [21], increasing the probability of traversing even larger targets [19]. However, detected fluorescence signal derived from these interactions is impeded by other photons, mainly originating from multiple Compton scattering, which is dependent on the target size but even applies to objects as small as a mouse. Compton scattering similarly describes the excitation of an electron through photon–atom interactions, without absorbing the primary photon but rather diverting its path. During this process, the photon loses a fraction of its energy by passing it on to an electron, depending on its energy prior to scattering and the scattering angle.

The challenge for XFI lies in differentiating the fluorescence signal from detected background photons. As multiple Compton scattering results in a shift of the incident photon energy down into the fluorescence signal region, the detection of fluorescence signals of medically feasible tracer concentrations in large targets becomes nearly impossible. However, the intrinsic problem of high Compton background has been partially overcome through the so-called spatial filtering as described in previous work by our group [22], showing the principal feasibility of XFI even for human-sized objects.

In contrast to established imaging modalities, XFI has the potential to provide both high sensitivity and excellent spatial resolution while simultaneously allowing for serial imaging to monitor targets longitudinally.

For the visualization of certain structures, specific materials with known fluorescence photon energies are used as imaging agents, commonly delivered in the form of molecular tracers or metallic nanoparticles (NPs). Nanoparticles are chosen because of their high customizability regarding size, shape and surface characteristics as well as possible functionalization, each influencing cell labelling efficiency [23,24]. NPs have been extensively studied in the context of immunotherapy, not only for imaging cellular targets [25] but also as an approach of treatment [26] and a way to specifically deliver drugs [27]. While NPs are passively accumulated in tumors due to enhanced permeability and retention (EPR) effects [28], more specific targeting is required when utilizing them in diagnostics and therapy. This can be achieved by functionalizing them through coating, change of surface structures or binding to ligands, depending on the target to be investigated [29].

Most prominently, gold nanoparticles (GNP) have been thoroughly evaluated due to their high atomic number Z (Z = 79) resulting in high X-ray absorption levels, making them well suited as a CT contrast agent [30]. GNPs have gained high popularity due to their simplicity in production and excellent customizability as well as being regarded to be of little toxicity [31]. Immune cell labelling with GNPs is usually performed ex vivo, such that only injected cells can be monitored. While cellular uptake was evaluated for a multitude of immune cells such as macrophages, monocytes and dendritic cells, T-cell labelling has rarely been performed [30]. Whereas in first studies utilizing GNPs for T-cell imaging, nanoparticles were used to deliver radionucleotides for PET imaging rather than imaging GNPs themselves [32], imaging of GNP-labelled T cells using CT was shown to be feasible [16,33]. Furthermore, the effects of GNP size, labelling duration and Au concentration on cell viability and labelling efficiency were investigated by Meir et al., achieving up to 195 pg gold per cell [23,34].

Another element that is of particular interest in cancer research is palladium nanoparticles (PdNPs). Whereas T-cell labelling has, to our knowledge, not been performed as of yet, PdNPs have been extensively researched in the context of tumor therapy, imaging and drug delivery [35]. However, there are concerns about potential toxicological and immunomodulatory effects in applications of PdNPs; hence, further investigation is needed [36]. While PdNPs have been deployed in various imaging modalities including photoacoustic imaging, SPECT and MRI are nonetheless not ideally suited for CT imaging in large objects due to the comparatively low atomic number (Z = 46) of palladium [37]. However, hybrid NPs consisting of gold-coated PdNPs have been successfully utilized in CT imaging. Regarding XFI, a reduction in tissue penetration depth is to be expected, yet it can be compensated through differing background characteristics in comparison to GNPs, as discussed in our work. Whereas gold nanoparticles have been frequently used in X-ray fluorescence studies [22,38,39], palladium is not commonly considered as an X-ray fluorescence agent.

In this work, we thus investigate the feasibility of imaging GNPs and, as a complementary approach, also PdNPs using XFI in the context of immunotherapy. For this purpose, we conducted multiple simulations using the software toolkit Geant4 v.10.5.1 [40], implementing a tumor-bearing mouse voxel model [41] irradiated by a monochromatic X-ray beam. The tumors were placed either subcutaneously, in the kidney, or in the center of the abdomen and enriched with gold or palladium nanoparticles. The software Geant4 is designed to simulate interactions of particles passing through matter using Monte Carlo methods [42,43]. Geant4-based simulations were shown to be consistent with first ex vivo and in situ experiments in previous work of our group [22,44,45] and can thus be used to explore diverse applications of XFI without the need of extensive animal research. As a result of our simulations, we intend to illuminate both prospects and challenges of implementing X-ray fluorescence in the highly topical area of cancer research to lay the foundation for future in vivo experiments.

2. Results

2.1. Subcutaneous Targets

In a first series of simulations, we mimicked imaging conditions found in several small animal imaging studies, in particular referencing a study by Meir et al. [16] utilizing

GNP-labelled cytotoxic T cells for CT imaging. A tumor of 5.5 mm in diameter was placed at a subcutaneous position on the dorsolateral abdomen of the mouse voxel model, as many small animal studies investigating tumor imaging make use of artificially grown subcutaneous tumors for convenient handling. In our study, the beam enters the mouse either through the front, the back, or at a 45° angle to minimize tissue along the beam. This is achieved by rotating and repositioning the mouse, thereby influencing the amount of tissue the photons have to pass prior to and after hitting the target; however, the same effect can be achieved by repositioning of the X-ray source and detectors.

In each of the three scenarios, the significance Z was observed to be highly influenced by the detector angle. While fluorescence photons show isotropic emission, background photons predominantly composed of multiple Compton scattering are considerably influenced by the detector position. While K_α significance, with a signal region below the initial Compton peak, is increasingly impaired at higher detector angles through rising background, K_β significance does thrive with Compton energy being shifted below its signal range, as displayed in Figure 1. Moreover, the target position within the mouse has to be considered because of the influence of additional tissue between target and detector, increasing the probability of additional scattering to occur, which further reduces photon energy. When rotating the phantom to keep the tissue depth along the beam propagation axis as low as possible, the significance is further improved through an overall reduction of the Compton background. In this scenario, K_α-fluorescence yields substantially higher significance values than K_β as the Compton background is less shifted towards the K_α region. In addition, L-shell fluorescence was examined, which will not be further discussed because of consistently performing below K-fluorescence regions. For all scenarios and signal regions, it can be observed that the significance values for gold do scale with the agent concentration, as can be seen in Figure 2. The detection thresholds for Au imaging were extrapolated to be 0.1 mg/mL for the front and back scenario and between 0.1 to 0.033 mg/mL for optimized rotation.

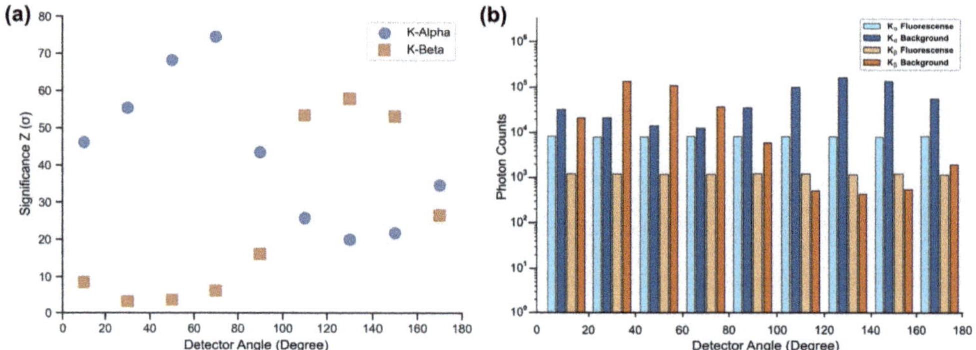

Figure 1. Au fluorescence significance and photon counts as simulated for an agent concentration of 1mg/mL. (**a**) Significance values for both K_α and K_β plotted for each of the 9 detector angles. (**b**) Box chart displaying the detected photon counts for both K_α and K_β fluorescence photons and background photons at the different detector angles.

As can be seen in Figure 2, the Pd simulations conducted herein show superior significance across all scenarios when compared to equally enriched Au targets. However, signal intensity is substantially limited by the penetration depth of fluorescence photons emitted by Pd atoms, as can be assessed for detector angles in which the fluorescence photons have to pass the mouse's body. When lifting these restrictions through phantom rotation, the best significance can be achieved with the detectors orthogonal (90°) to the beam direction where Compton scattering is suppressed. Derived from the dilution series, the detection threshold for Pd is estimated to be around 5 µg Pd/mL for front/back beam direction, and <3.3 µg Pd/mL for optimized rotation. The effect of Pd fluorescence being able to achieve noticeably higher significance than their Au equivalents is most prominent when the tissue thickness is as small as possible.

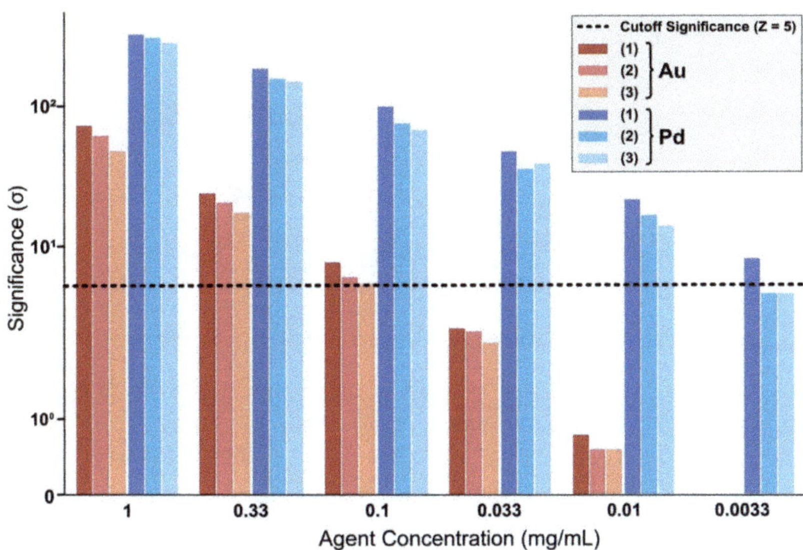

Figure 2. Significance comparison for a subcutaneous target at (1) optimized rotation, (2) beam hitting the mouse from the front or (3) beam entering the mouse from the back. Gold fluorescence significances are displayed in red, palladium fluorescence significances in blue.

2.2. Kidney and Central Target

It was further evaluated whether this imaging technique is feasible for deep tumor imaging in small animals. This challenge was addressed by both simulating a spherical kidney lesion of 5.5 mm diameter as well as exploring the worst-case scenario, namely a similar-sized lesion in the center of the mouse with surrounding tissue of approximately 12–15 mm.

It is observed that kidney and central targets perform similarly, with the tendency of the central target position being slightly inferior. A limitation of K_α significance can be seen due to the shift in Compton background described before, thus lowering the maximum significance that can be achieved in these scenarios compared with previous simulations. However, when only one of both signal regions is to be examined, K_α remains the signal region of choice. Angular dependence is critical, with the same behavior of K_α and K_β fluorescence significance as described before. The resulting Au detection threshold for both scenarios is estimated to be between 0.33 and 0.1 mg/mL.

When utilizing Pd as an imaging agent, significance in deeper targets noticeably decreases in comparison to subcutaneous scenarios. Nevertheless, as is displayed in Figure 3, Pd does still offer higher significance than Au when comparing similar concentrations, with an observed detection limit of around 0.01 mg/mL, ergo performing at least one order of magnitude better. This is due to the different background behavior in the simulated X-ray spectra: the background in the K_α Pd-signal region is much less than in the K_β-region of Au. Angular dependencies are once again mostly influenced by the thickness of surrounding tissue, performing worst for orthogonal detector positions.

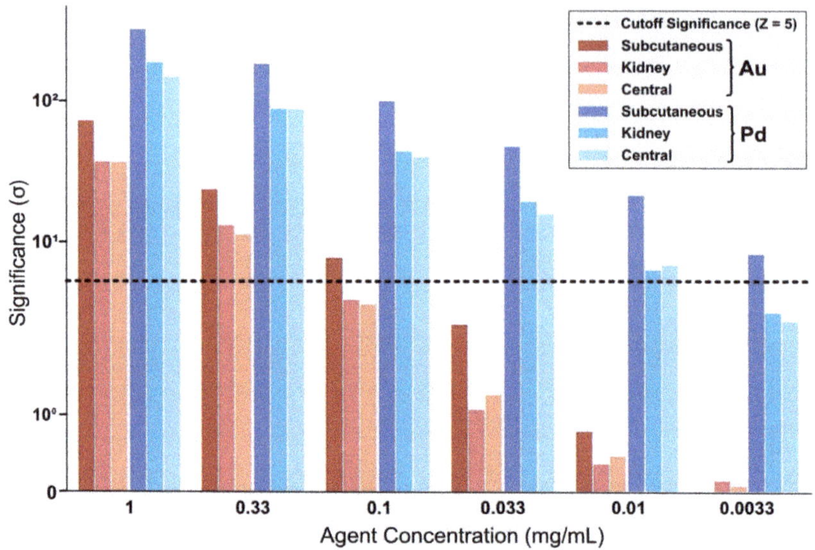

Figure 3. Comparison of a subcutaneous, a kidney and a centered target regarding Au-fluorescence (red) and Pd-fluorescence (blue) significances.

2.3. Influence of Target Size

In further simulations utilizing the same center target scenario, the target diameter was varied between sizes of 10, 5, 2.5, and 1.25 mm. For better comparability, the simulated agent concentrations were adjusted such that the amount of fluorescence marker within the beam volume is nearly consistent for different target sizes.

The agent mass is the determining factor for the fluorescence yield; hence, the signal strength and significance Z, as scenarios differing in target size and concentration, but containing the same amount of agent mass, achieve similar results (see Figure 4). Nevertheless, it is noticed that the calculated significance does also scale with target size, albeit to a much lesser extent, as larger targets do perform slightly better than smaller ones containing the same agent mass, presumably due to the decrease of surrounding tissue. However, when looking upon the smallest target diameter, a steep decrease in significance is present, with two effects most likely contributing to this finding. On the one side, changing the shape of the beam-target intersection from cylindrical in larger targets to spherical in targets with a size approaching the beam diameter does influence the amount of fluorescence marker contributing to the observed signal. On the other hand, imperfections of beam alignment only detectable for small targets cannot be excluded.

2.4. Dose

Furthermore, we evaluated the deposited dose utilizing a dose tracker specified in the methods section (see Figure 5). The dose for an ideally positioned tumor is estimated to be 25.5 mGy within the tumor and 0.1 mGy of full body dose for a single beam position. When imaging a centrally placed target 5 mm in diameter, we found the organ dose to be 28.38 mGy for Au imaging and 24.45 mGy for Pd imaging with respective full body doses of 0.34 mGy (Au) and 0.33 mGy (Pd) for a single beam position. A planar scan of this 5 mm target would require 25 scan positions utilizing a 1 mm beam; thus, the full body dose for a scan is estimated to be 8.5 mGy (Au) and 8.25 mGy (Pd). As the tumor location may not always be known a priori, we also extrapolated the dose applied through a whole-body scan. When an entire mouse is scanned by a photon beam, the full-body dose roughly approaches the local dose seen within the beam volume. Applying this rationale, we estimate the full body dose to be between 250 and 350 mGy at 53 keV and 300 to 400 mGy for 85 keV.

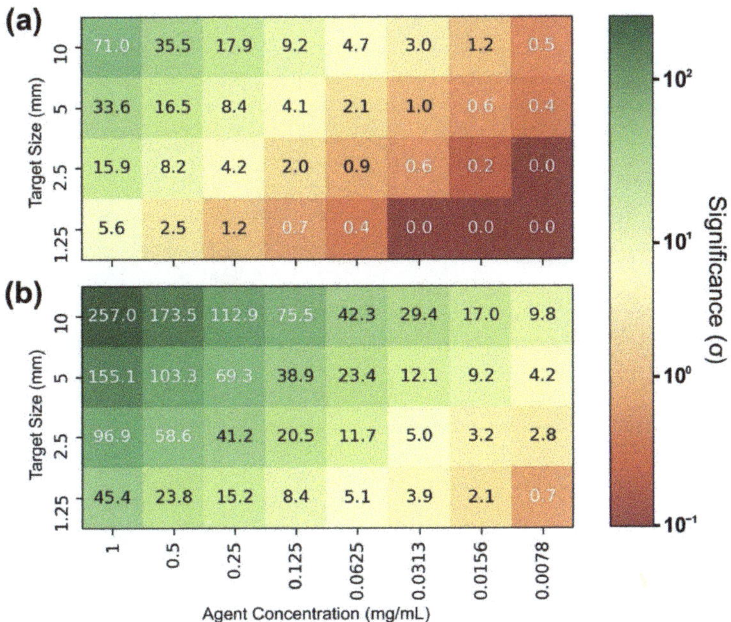

Figure 4. Significance Z at varying agent concentration and target size with Au-fluorescence (**a**) and Pd-fluorescence (**b**).

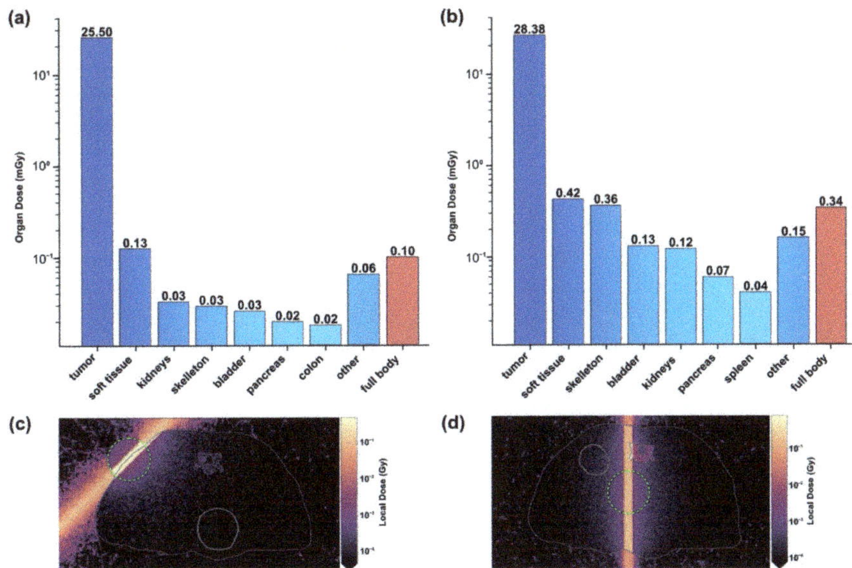

Figure 5. Dose deposition when imaging Au-labelled subcutaneous (**a–c**) and central (**b–d**) targets. (**a,b**) Organ doses and full body dose. (**c,d**) Transversal slice of the mouse voxel model showing the local dose at beam level. The target is highlighted by a green circle.

3. Discussion

Based on the determined agent concentration detection limits, possible cell detection thresholds can now be estimated. As reported in the literature, 195 pg/cell Au T-cell labelling is feasible [34], whereas for Pd labelling, no reference could be found. Therefore, only the cellular detection threshold of Au is calculated, assuming a homogenous distribution of cells within the tumor. Based on our simulations of a 5.5 mm diameter tumor, the detection threshold for a subcutaneous target is estimated to be between 1.5×10^4 and 4.5×10^4 cells/mL (Au) for a best-case scenario and 4.5×10^4 cells/mL (Au) without optimized rotation. When simulating a kidney and centered lesion, detection thresholds are extrapolated to be between 4.5×10^4 and 1.5×10^5 cells (Au).

When only considering the tumor volume intersecting with the 1 mm^2 beam, it can be estimated that as little as 730 to 2400 T cells can be detected in a subcutaneous target when using gold as a fluorescence marker. For both the kidney and central target, these values increase to 2200 to 7250 cells. As shown by our simulations, the absolute mass of fluorescence marker, correlating with the number of cells within the beam, is of supreme importance. At the same time, the volume in which these cells are distributed is of minor relevance. This fact can be explained with the intrinsic strength of XFI, as the spatial resolution that can be achieved is solely limited by the beam diameter. Hence, according to our simulation study, XFI is able to detect lesions in the sub-millimeter range for scenarios in which a high cell concentration is present, with comparable spatial resolution to computed tomography and MRI, vastly surpassing the resolution achieved by PET and SPECT imaging. While this opens up promising possibilities in molecular imaging as synchrotrons are able to generate photon beams of only a few μm in diameter or even less, the size of the beam does inversely affect image acquisition times when scanning objects and more scanning positions result in increased radiation dose; thus, it cannot be shrunk ad libitum. However, when only scanning small objects and using high photon flux X-ray sources, this drawback is of minor importance.

Based on the fundamental research performed by Tumeh et al. [46], investigating the CD8+ T-cell infiltration in melanoma patients receiving immunotherapy through serial biopsies, we extrapolated the difference in T-cell abundance between responding and non-responding patients to estimate the required cell labelling efficiency needed for detection through XFI. With the data given in Figure 3 of [46], we estimate the difference of the groups to be between 1500 and 2500 CD8+ T cells per mm^2, equaling 5.8–12.5×10^4 cells/mm^3 or 3–6.5×10^7 cells in a 1 cm diameter circular tumor, assuming homogenous cell distribution. Subsequently, we calculate the required labelling efficiency to be between 0.286 and 0.044 pg Au/T cell as well as between 0.009 and 0.004 pg Pd/T cell. In a subcutaneous target, these values increase to between 0.945 and 0.133 pg Au/cell and between 0.029 and 0.013 pg Pd/cell, highlighting the potential benefit of utilizing Pd as an imaging agent due to its reduced requirements on labelling efficiency. Moreover, while these estimations only include one tumor entity and endogenous T cells; they suggest that the labelling values lie well within currently achievable Au labelling efficiencies of up to 195 pg/cell.

Furthermore, when comparing our results with established imaging modalities, the solid angle covered by the detector has to be considered. Herein, a single detector with a detection area of 25 mm^2 (Au) or 50 mm^2 (Pd) and a detector to target distance of 6 to 6.5 cm is used, thus only covering approximately 0.06% (Au) and 0.11% (Pd) of the full ($4\,\pi$) solid angle. In comparison, modern PET, SPECT or CT detectors do feature much higher coverage, even investigating full body scanners [47]. Nevertheless, based on the observed anisotropy of the Compton background, covering the entire solid angle would not necessarily be the ideal solution for all scenarios. A detector covering 30–40% of the solid angle would be ideal for human sized targets, when targeting the K_α signal region of gold fluorescent agents [48], as it was previously investigated by our group.

Based on these considerations, we can assess our sensitivity data in the context of current literature regarding other imaging modalities. As mentioned above, our setup was designed such that it can provide results comparable with other studies monitoring

T-cell distribution. Research performed by Meir et al. utilizing gold nanoparticles as a CT contrast agent [16] was of particular interest because the GNP-labelled cells could also be imaged using XFI. The group reported successful monitoring of injected T cell abundance in a tumor containing 8×10^4 to 4.6×10^5 T cells. However, no detection threshold is provided for this particular study.

Whereas this approach of injected T cells directly labelled with nanoparticles has been proven feasible, labelling endogenous cells with antibodies or antibody fragments, as has been performed in PET imaging [14], is yet to be investigated. The PET sensitivity levels that have been achieved utilizing radionucleotide-labelled anti CD8+ antibody fragments range from 2×10^4 CD8+ T cells per milligram in lymphoid organs [13] to 1.6–4×10^6 CD8+ lymphocytes in a tumor volume of ~480 mm^3 [49], equaling 3.3–8.3×10^6 cells/mL. Based on this data, it is to be assumed that XFI could deliver comparable detection thresholds, while simultaneously providing substantially higher spatial resolution. However, even though binding of GNPs to antibodies has been conducted in literature [50], there is a lack of studies regarding T-cell labelling using GNP-conjugated antibodies.

Direct ex vivo cell labelling of T cells has been examined in MRI studies utilizing superparamagnetic iron oxide [15,51], showing sensitivity levels of <3 labelled cells/voxel in vivo [51], with the voxel size being 75 μm × 75 μm × 500 μm, equaling <1×10^6 cells/mL. However, all these publications suffered from imaging only being feasible within 48–72 h post injection, due to the fast biodegradability seen in SPIO labelling [52]. The same limitations apply for PET and SPECT imaging due to the inevitable tradeoffs made for radionuclide halftime, balancing longitudinal imaging against potential reductions in cell viability of radiosensitive lymphocytes [53]. In contrast, XFI is ideally suited for longitudinal imaging, experiencing no intrinsic decrease in signal over time. While longitudinal studies for other cell types labelled with GNPs show that imaging is feasible for at least 4 weeks [54], this number may vary for different types of cells due to variations in efflux, proliferation and cell longevity. As a fourth approach, optical imaging of T cells has achieved promising sensitivities of up to 10^4 cells [17]; however, the translation into a clinical setting has proven difficult due to strong limitations in the tissue penetration depth as well as the complexity of cell preparation due to the required genetic cell modifications being necessary. The issue of limitations in penetration depth in XFI has been discussed in this work, endorsing previous research by showing a substantial dependance on the fluorescent agent. While palladium is limited to small animal imaging studies or the examination of superficial lesions due to its low energetic fluorescence photons, gold fluorescence in the K$_\alpha$ and K$_\beta$ region was shown to be feasible in human-sized objects in previous investigations by our group [48,55]. However, most of these setups rely on a brilliant, monochromatic pencil beam X-ray source providing a high photon flux, features currently only achieved by synchrotron facilities, too large and too expensive for a clinical implementation. Aiming at narrowing the gap between traditional X-ray tubes and synchrotron facilities, inverse Compton X-ray sources (ICS) have been thoroughly studied over the last decades [56]. While offering quasi-monochromatic photon beams and a significantly increased photon flux in an affordable and compact size, they are often limited to lower energies [57]. To overcome these limits, our group also works on ultra-compact laser-driven Thomson X-ray sources [58], an approach envisaged for clinical use in the future.

Other approaches of achieving X-ray fluorescence imaging in a compact setup using a high-energy polychromatic X-ray tube, either as a cone or a pencil beam, have been surveyed to image tumor bearing mice [38,39,59]. In contrast to setups utilizing a pencil beam, the position of fluorescent agents in cone-beam imaging is determined using pinhole collimation, theoretically offering faster scanning times when using parallel signal attenuation in a detector array [38]. However, as such a setup that can provide a sufficient flux and narrow energy spectrum is not yet available, those methods currently suffer from the same limitations of long measurement times and a higher dose as presented in our approach. Nevertheless, great sensitivity has been achieved, even reaching synchrotron-like detection thresholds of 0.007 mg Au/mL; however, for ex vivo imaging of small targets [60], whereby using high radiation dose not suited for in vivo imaging. These drawbacks could be vastly improved in a study by Larsson et al. [39], pointing out the advantages of a pencil beam driven approach and indicating clinical applicability for molecular imaging in the sub-millimeter range for small targets.

The estimated full-body dose in the present study ranges from 0.1 to 0.34 mGy for a single beam position to 250 to 400 mGy for a whole-body scan. These doses lie markedly below the median lethal dose of 9.25 Gy after 30 days described for mice in the literature [61]. Furthermore, studies indicate that mice exposed with around 300 mGy can repair the damages within hours after exposure [62]. However, through improvements in labelling efficiency and detector size, the dose of our approach may be vastly reduced. Moreover, it should be feasible to initially locate primary and larger metastatic tumor sites by other imaging methods such as CT or MRI scans, which would then facilitate targeted XFI analysis of those specific locations.

While passive GNP attenuation in tumors has been thoroughly investigated, no study specifically targeting T cells through X-ray fluorescence has been reported as of yet, to the best of our knowledge. However, the findings presented here indicate that X-ray fluorescence imaging can be of great value in future applications of molecular imaging, specifically cell tracking. Providing similar sensitivity to established functional imaging modalities such as PET and SPECT imaging while at the same time achieving spatial resolutions usually only seen for morphological imaging in MRI and CT, XFI is ideally suited for the application of monitoring the intratumoral cell abundance. Moreover, a huge benefit over radioisotope-based approaches lies in improved longitudinal evaluation through serial imaging.

While our findings indicate a promising future for X-ray fluorescence-based imaging, its scope is a feasibility pre-study without animal research, but will be of help for test animal proposals. Hence, we believe that this pre-study paves the way for future XFI-based cell tracking research. When additional research in all areas of XFI, ranging from X-ray sources to novel labelling techniques to detector improvements, adds to the already promising capabilities of this method and allow for clinical implementation, it may play a decisive role in tumor imaging and a variety of other applications.

4. Materials and Methods

4.1. Geant4

The setup used herein consists of 3 major elements: the mouse model, detectors and additional elements modelling a beam line at PETRA3 at DESY [63]. The general setup is illustrated in Figure 6.

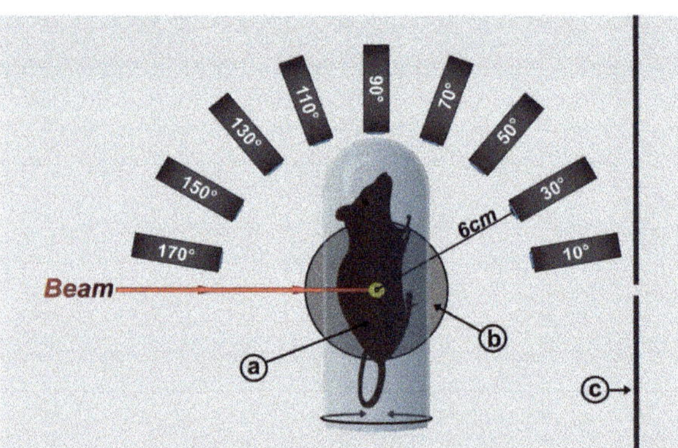

Figure 6. Illustration of the simulation setup as seen from above. (**a**) Rotational mouse voxel model enclosed in a plexiglass tube; (**b**) stage inlay + stage plate; (**c**) front plate.

4.1.1. Mouse Model

We utilized a segmented 3D-voxel model published and described by Dogdas et al. [41], derived through co-registration of CT and cryosection mouse data, dividing a 28 g nude male mouse into 78.4×10^6 cubes. Each voxel is assigned to structures/organs with defined chemical composition and properties such as density. This material data for both the mouse voxels as well as for other objects was derived from the integrated Geant4 database. The mouse model is available free of charge and can be downloaded at the website of the Biomedical Imaging Group at the University of Southern California [64]. The model was implemented using half the maximum resolution, with $104 \times 496 \times 190 = 9,800,960$ cubic voxels and a voxel size of 0.2 mm. The mouse position was adjusted based on the target position, such that the target always remained at the center of the setup.

4.1.2. Detectors

A total of 9 detectors were added horizontally at the target height, ranging from 10 to 170 degrees with respect to the beam direction with a spacing of 20 degrees, thus covering the entire semicircle. The detector-to-target distance was 6 cm in a subcutaneous scenario and 6.5 cm in a kidney and center scenario to avoid geometric interferences between the detectors and the tube.

For the simulations targeting Pd, a GEANT4 implementation of an Amptek silicon drift diode detector was modeled (70 mm^2 FAST SDD$^®$; Amptek Inc., Bedford, MA, USA). This detector offers an active detector area of 70 mm^2 collimated to 50 mm^2 with a silicon sensor thickness of 500 μm, providing a superior energy resolution to planar detectors [65]. The energy resolution in the region of Pd K-shell fluorescence is on the order of 120 eV (rms). However, for higher energy X-rays, the active area made of silicon does not offer enough stopping power; thus, Amptek only recommends detector usage for energies up to 30 keV [66].

For gold simulations, a Cadmium-Telluride Detector by Amptek (XR-100CdTe; Amptek Inc., Bedford, MA, USA) was implemented with an active region of 25 mm^2 and 1 mm thickness. CdTe detectors offer a higher quantum efficiency, with the downside of a lower energy resolution of 530 eV (rms) at 14.4 keV [67]. For both detector types, multiple detector-specific effects are considered which influence their performance [68].

4.1.3. Additional Geometry

The mouse model was embedded in a plexiglass tube of 2 mm thickness and of an outer diameter of 30 mm. Moreover, a vague display of a typical beamline setup was added, consisting of an aluminum stage mounting plate. In its center, an inlay made of iron is placed to reduce scattering contributions. A stainless-steel front plate is added to block scattering of air molecules along the beam path.

4.2. Parameters

4.2.1. Target Position

Three distinctive positions of a spherical target were simulated through specifying its center and adding agent material to each voxel containing soft tissue (density $\rho = 1035$ mg/mL) within a defined radius. In the case of the tumor protruding the model (subcutaneous position), an additional 0.5 mm thick layer of soft tissue was added to mimic skin.

4.2.2. Agent Concentration

The agent material was chosen to be either gold or palladium. A dilution series with an agent concentration of (1, 0.33, 0.1, 0.033, 0.01, 0.0033) mg/mL inside the simulated tumor volume was performed in each scenario to cover a broad range of concentrations. For the variation of the target size, a different dilution series was chosen to improve comparability between data points (1, 0.5, 0.25, 0.125, 0.0625, 0.03125, 0.015675, 0.0078375) mg/mL, as the target diameter was also altered by a factor of 2.

4.2.3. Beam

A monoenergetic photon X-ray beam of 0.5 mm radius, horizontally polarized relative to the lab frame and containing a total photon number of 10^{10}, was used to keep the applied dose low while maintaining sufficient signal intensity. Optimum beam energy and detector type depend on the agent material as a tradeoff between signal intensity and background behavior, as seen in previous work of our group examining ideal beam energy [48]. In the case of Au, a beam energy of 85 keV was chosen, and for palladium we have selected 53 keV.

4.3. Simulation and Analysis

4.3.1. Histogram Generation and Analysis

The entire spectrum is not evaluated but only predefined signal regions around the distinct fluorescence peaks. The width of these regions is defined as $(E_{Fluo} - 3\sigma; E_{Fluo} + 3\sigma)$, with well-known fluorescence line energies E_{Fluo} specific for the target material [21] and the standard deviation σ equivalent to the energy resolution at E_{Fluo} depending on the used detector model [66,67]. Due to limitations in detector resolution as of today, certain fluorescence peaks being close together such as Pd $K_{\alpha1}$ and $K_{\alpha2}$ cannot be effectively discriminated and are therefore analyzed as one peak.

Within these defined signal regions, the significance of a fluorescence signal can be calculated by analyzing both signal and background photons and performing a one-tailed hypothesis test. The null hypothesis H_0 is defined as the non-existence of fluorescence photons in an observed spectrum, stating that the observed behavior is entirely explainable through background characteristics [48]. As it is also discussed in [48], the total number of photons counted when adding a fluorescent agent is not expected to be less than for Compton background alone; thus, a one-tailed test can be performed. The resulting *p*-value states the probability that an effect; herein, an increase in photon counts, is observed, albeit H_0 being true. However, as small *p*-values are inconvenient to handle, it is converted to the significance Z, expressed as the number of standard deviations σ, stating by which probability the null hypothesis H_0 is to be discarded. It can be approximated using:

$$Z \approx \frac{\left(n_{observed} - n_{expected} \right)}{\sqrt{\left(n_{expected} \right)}} = \frac{N_{signal}}{\sqrt{N_{background}}},$$

(1)

with N_{signal} as the number of observed fluorescence photons and $N_{background}$ as the detected background photons [55]. In the context of this research, a significant signal is defined as $Z \geq 5\sigma$, often considered as the significance required for discovery in particle physics [69], as previous experiments conducted by our group indicate that this value is sufficient for the detection of simulated fluorescence in experiments, indicating that even $Z \geq 3\sigma$ could be satisfactory [48,55]. $Z \geq 5\sigma$ is equivalent to a type 1 error probability of $p \leq 2.867 \times 10^{-5}\%$.

4.3.2. Dose

A dose tracker was implemented in our simulations, storing the deposited energy for all voxels of the mouse model. Therefore, we are able to determine the dose for each organ segmented in the model as well as for each material individually.

Author Contributions: Conceptualization, H.K., F.G. and T.S.; methodology, T.S., C.K., O.S., F.G. and H.K.; software, C.K., O.S., T.S. and H.K.; validation, F.G. and T.S.; formal analysis, H.K., T.S., C.K. and F.G.; investigation, H.K.; resources, F.G., T.S.; data curation, H.K., T.S. and F.G.; writing—original draft preparation, H.K. and T.S.; writing—review and editing, H.K., T.S., F.G., C.K., K.R. and O.S.; visualization, H.K. and C.K.; supervision, F.G. and K.R.; project administration, F.G. All authors have read and agreed to the published version of the manuscript.

Funding: This research received no external funding.

Institutional Review Board Statement: Not applicable.

Data Availability Statement: The data presented in this study are available on reasonable request.

Conflicts of Interest: The authors declare no conflict of interest.

References

1. Couzin-Frankel, J. Breakthrough of the Year 2013. Cancer Immunotherapy. *Science* **2013**, *342*, 1432–1433. [CrossRef]
2. Zindl, C.L.; Chaplin, D.D. Immunology. Tumor Immune Evasion. *Science* **2010**, *328*, 697–698. [CrossRef] [PubMed]
3. Schumacher, T.N.; Schreiber, R.D. Neoantigens in Cancer Immunotherapy. *Science* **2015**, *348*, 69–74. [CrossRef] [PubMed]
4. Weber, E.W.; Maus, M.V.; Mackall, C.L. The Emerging Landscape of Immune Cell Therapies. *Cell* **2020**, *181*, 46–62. [CrossRef]
5. Sharma, P.; Allison, J.P. The Future of Immune Checkpoint Therapy. *Science* **2015**, *348*, 56–61. [CrossRef]
6. Larkin, J.; Chiarion-Sileni, V.; Gonzalez, R.; Grob, J.J.; Cowey, C.L.; Lao, C.D.; Schadendorf, D.; Dummer, R.; Smylie, M.; Rutkowski, P.; et al. Combined Nivolumab and Ipilimumab or Monotherapy in Untreated Melanoma. *N. Engl. J. Med.* **2015**, *373*, 23–34. [CrossRef]
7. Hodi, F.S.; O'Day, S.J.; McDermott, D.F.; Weber, R.W.; Sosman, J.A.; Haanen, J.B.; Gonzalez, R.; Robert, C.; Schadendorf, D.; Hassel, J.C.; et al. Improved Survival with Ipilimumab in Patients with Metastatic Melanoma. *N. Engl. J. Med.* **2010**, *363*, 711–723. [CrossRef] [PubMed]
8. Decazes, P.; Bohn, P. Immunotherapy by immune checkpoint inhibitors and nuclear medicine imaging: Current and future applications. *Cancers* **2020**, *12*, 371. [CrossRef]
9. Borcoman, E.; Kanjanapan, Y.; Champiat, S.; Kato, S.; Servois, V.; Kurzrock, R.; Goel, S.; Bedard, P.; Le Tourneau, C. Novel Patterns of Response under Immunotherapy. *Ann. Oncol.* **2019**, *30*, 385–396. [CrossRef]
10. Yamada, N.; Oizumi, S.; Kikuchi, E.; Shinagawa, N.; Konishi-Sakakibara, J.; Ishimine, A.; Aoe, K.; Gemba, K.; Kishimoto, T.; Torigoe, T.; et al. CD8+ tumor-infiltrating lymphocytes predict favorable prognosis in malignant pleural mesothelioma after resection. *Cancer Immunol. Immunother.* **2010**, *59*, 1543–1549. [CrossRef]
11. McCracken, M.N.; Tavaré, R.; Witte, O.N.; Wu, A.M. Advances in PET detection of the antitumor T cell response. *Adv. Immunol.* **2016**, *131*, 187–231.
12. Liu, Z.; Li, Z. Molecular imaging in tracking tumor-specific cytotoxic T lymphocytes (CTLs). *Theranostics* **2014**, *4*, 990–1001. [CrossRef] [PubMed]
13. Rashidian, M.; Ingram, J.R.; Dougan, M.; Dongre, A.; Whang, K.A.; Le Gall, C.; Cragnolini, J.J.; Bierie, B.; Gostissa, M.; Gorman, J.; et al. Predicting the response to CTLA-4 blockade by longitudinal noninvasive monitoring of CD8 T Cells. *J. Exp. Med.* **2017**, *214*, 2243–2255. [CrossRef] [PubMed]
14. Pandit-Taskar, N.; Postow, M.A.; Hellmann, M.D.; Harding, J.J.; Barker, C.A.; O'Donoghue, J.A.; Ziolkowska, M.; Ruan, S.; Lyashchenko, S.K.; Tsai, F.; et al. First-in-humans imaging with 89Zr-Df-IAB22M2C anti-CD8 minibody in patients with solid malignancies: Preliminary pharmacokinetics, biodistribution, and lesion targeting. *J. Nucl. Med.* **2020**, *61*, 512–519. [CrossRef]
15. Smirnov, P.; Lavergne, E.; Gazeau, F.; Lewin, M.; Boissonnas, A.; Doan, B.-T.; Gillet, B.; Combadière, C.; Combadière, B.; Clément, O. In vivo cellular imaging of lymphocyte trafficking by MRI: A tumor model approach to cell-based anticancer therapy. *Magn. Reason. Med.* **2006**, *56*, 498–508. [CrossRef] [PubMed]
16. Meir, R.; Shamalov, K.; Betzer, O.; Motiei, M.; Horovitz-Fried, M.; Yehuda, R.; Popovtzer, A.; Popovtzer, R.; Cohen, C.J. Nanomedicine for cancer immunotherapy: Tracking cancer-specific T-cells in vivo with gold nanoparticles and CT imaging. *ACS Nano* **2015**, *9*, 6363–6372. [CrossRef]
17. Kleinovink, J.W.; Mezzanotte, L.; Zambito, G.; Fransen, M.F.; Cruz, L.J.; Verbeek, J.S.; Chan, A.; Ossendorp, F.; Löwik, C. A dual-color bioluminescence reporter mouse for simultaneous in vivo imaging of T cell localization and function. *Front. Immunol.* **2019**, *9*, 3097. [CrossRef]
18. Behling, R. X-Ray sources: 125 years of developments of this intriguing technology. *Phys. Med.* **2020**, *79*, 162–187. [CrossRef]
19. Chen, H.; Rogalski, M.M.; Anker, J.N. Advances in functional X-ray imaging techniques and contrast agents. *Phys. Chem. Chem. Phys.* **2012**, *14*, 13469–13486. [CrossRef]
20. Berger, M.; Yang, Q.; Maier, A. X-ray Imaging. In *Medical Imaging Systems: An Introductory Guide*; Maier, A., Steidl, S., Christlein, V., Hornegger, J., Eds.; Springer: Cham, Switzerland, 2018.
21. Bearden, J.A. X-Ray Wavelengths. *Rev. Mod. Phys.* **1967**, *39*, 78–124. [CrossRef]
22. Grüner, F.; Blumendorf, F.; Schmutzler, O.; Staufer, T.; Bradbury, M.; Wiesner, U.; Rosentreter, T.; Loers, G.; Lutz, D.; Richter, B.; et al. Localising functionalised gold-nanoparticles in murine spinal cords by X-ray fluorescence imaging and background-reduction through spatial filtering for human-sized objects. *Sci. Rep.* **2018**, *8*, 16561. [CrossRef] [PubMed]
23. Betzer, O.; Meir, R.; Dreifuss, T.; Shamalov, K.; Motiei, M.; Shwartz, A.; Baranes, K.; Cohen, C.J.; Shraga-Heled, N.; Ofir, R.; et al. In-vitro optimization of nanoparticle-cell labeling protocols for in-vivo cell tracking applications. *Sci. Rep.* **2015**, *5*, 15400. [CrossRef] [PubMed]
24. Xie, X.; Liao, J.; Shao, X.; Li, Q.; Lin, Y. The Effect of shape on cellular uptake of gold nanoparticles in the forms of stars, rods, and triangles. *Sci. Rep.* **2017**, *7*, 3827. [CrossRef] [PubMed]
25. Ou, Y.-C.; Wen, X.; Bardhan, R. Cancer Immunoimaging with smart nanoparticles. *Trends Biotechnol.* **2020**, *38*, 388–403. [CrossRef]
26. Riley, R.S.; Day, E.S. Gold Nanoparticle-Mediated Photothermal Therapy: Applications and Opportunities for Multimodal Cancer Treatment. *Wiley Interdiscip. Rev. Nanomed. Nanobiotechnol.* **2017**, *9*, e1449. [CrossRef]

27. Riley, R.S.; June, C.H.; Langer, R.; Mitchell, M.J. Delivery technologies for cancer immunotherapy. *Nat. Rev. Drug Discov.* **2019**, *18*, 175–196. [CrossRef]
28. Prabhakar, U.; Maeda, H.; Jain, R.K.; Sevick-Muraca, E.M.; Zamboni, W.; Farokhzad, O.C.; Barry, S.T.; Gabizon, A.; Grodzinski, P.; Blakey, D.C. Challenges and key considerations of the enhanced permeability and retention effect for nanomedicine drug delivery in oncology. *Cancer Res.* **2013**, *73*, 2412–2417. [CrossRef]
29. Nicol, J.R.; Dixon, D.; Coulter, J.A. Gold nanoparticle surface functionalization: A necessary requirement in the development of novel nanotherapeutics. *Nanomedicine* **2015**, *10*, 1315–1326. [CrossRef]
30. Chandrasekaran, R.; Madheswaran, T.; Tharmalingam, N.; Bose, R.J.; Park, H.; Ha, D.-H. Labeling and tracking cells with gold nanoparticles. *Drug Discov. Today* **2021**, *26*, 94–105. [CrossRef] [PubMed]
31. Adewale, O.B.; Davids, H.; Cairncross, L.; Roux, S. Toxicological behavior of gold nanoparticles on various models: Influence of physicochemical properties and other factors. *Int. J. Toxicol.* **2019**, *38*, 357–384. [CrossRef]
32. Li, H.; Diaz, L.; Lee, D.; Cui, L.; Liang, X.; Cheng, Y. In vivo imaging of T cells loaded with gold nanoparticles: A pilot study. *Radiol. Med.* **2014**, *119*, 269–276. [CrossRef]
33. Chen, M.; Betzer, O.; Fan, Y.; Gao, Y.; Shen, M.; Sadan, T.; Popovtzer, R.; Shi, X. Multifunctional dendrimer-entrapped gold nanoparticles for labeling and tracking T cells via dual-modal computed tomography and fluorescence imaging. *Biomacromolecules* **2020**, *21*, 1587–1595. [CrossRef] [PubMed]
34. Meir, R.; Popovtzer, R. Cell Tracking Using Gold Nanoparticles and Computed Tomography Imaging: Cell Tracking Using Gold Nanoparticles and Computed Tomography Imaging. *WIREs Nanomed. Nanobiotechnol.* **2018**, *10*, e1480. [CrossRef]
35. Phan, T.T.V.; Huynh, T.-C.; Manivasagan, P.; Mondal, S.; Oh, J. An up-to-date review on biomedical applications of palladium nanoparticles. *Nanomaterials* **2019**, *10*, 66. [CrossRef] [PubMed]
36. Leso, V.; Iavicoli, I. Palladium nanoparticles: Toxicological effects and potential implications for occupational risk assessment. *Int. J. Mol. Sci.* **2018**, *19*, 503. [CrossRef] [PubMed]
37. Liu, Y.; Li, J.; Chen, M.; Chen, X.; Zheng, N. Palladium-based nanomaterials for cancer imaging and therapy. *Theranostics* **2020**, *10*, 10057–10074. [CrossRef]
38. Manohar, N.; Reynoso, F.J.; Diagaradjane, P.; Krishnan, S.; Cho, S.H. Quantitative imaging of gold nanoparticle distribution in a tumor-bearing mouse using benchtop X-ray fluorescence computed tomography. *Sci. Rep.* **2016**, *6*, 22079. [CrossRef]
39. Larsson, J.C.; Vogt, C.; Vågberg, W.; Toprak, M.S.; Dzieran, J.; Arsenian-Henriksson, M.; Hertz, H.M. High-spatial-resolution X-ray fluorescence tomography with spectrally matched nanoparticles. *Phys. Med. Biol.* **2018**, *63*, 164001. [CrossRef]
40. Agostinelli, S.; Allison, J.; Amako, K.; Apostolakis, J.; Araujo, H.; Arce, P.; Asai, M.; Axen, D.; Banerjee, S.; Barrand, G.; et al. Geant4—A Simulation Toolkit. *Nucl. Instrum. Methods Phys. Res. Sect. A Accel. Spectrometers Detect. Assoc. Equip.* **2003**, *506*, 250–303. [CrossRef]
41. Dogdas, B.; Stout, D.; Chatziioannou, A.F.; Leahy, R.M. Digimouse: A 3D whole body mouse atlas from CT and cryosection data. *Phys. Med. Biol.* **2007**, *52*, 577–587. [CrossRef]
42. Allison, J.; Amako, K.; Apostolakis, J.; Araujo, H.; Arce Dubois, P.; Asai, M.; Barrand, G.; Capra, R.; Chauvie, S.; Chytracek, R.; et al. Geant4 developments and applications. *IEEE Trans. Nucl. Sci.* **2006**, *53*, 270–278. [CrossRef]
43. Allison, J.; Amako, K.; Apostolakis, J.; Arce, P.; Asai, M.; Aso, T.; Bagli, E.; Bagulya, A.; Banerjee, S.; Barrand, G.; et al. Recent Developments in Geant4. *Nucl. Instrum. Methods Phys. Res. Sect. A Accel. Spectrometers Detect. Assoc. Equip.* **2016**, *835*, 186–225. [CrossRef]
44. Sanchez-Cano, C.; Alvarez-Puebla, R.A.; Abendroth, J.M.; Beck, T.; Blick, R.; Cao, Y.; Caruso, F.; Chakraborty, I.; Chapman, H.N.; Chen, C.; et al. X-ray-based techniques to study the nano-bio interface. *ACS Nano* **2021**, *15*, 3754–3807. [CrossRef] [PubMed]
45. Schmutzler, O.; Graf, S.; Behm, N.; Mansour, W.Y.; Blumendorf, F.; Staufer, T.; Körnig, C.; Salah, D.; Kang, Y.; Peters, J.N.; et al. X-Ray Fluorescence Uptake Measurement of Functionalized Gold Nanoparticles in Tumor Cell Microsamples. *Int. J. Mol. Sci.* **2021**, *22*, 3691. [CrossRef] [PubMed]
46. Tumeh, P.C.; Harview, C.L.; Yearley, J.H.; Shintaku, I.P.; Taylor, E.J.M.; Robert, L.; Chmielowski, B.; Spasic, M.; Henry, G.; Ciobanu, V.; et al. PD-1 Blockade Induces Responses by Inhibiting Adaptive Immune Resistance. *Nature* **2014**, *515*, 568–571. [CrossRef]
47. Vandenberghe, S.; Moskal, P.; Karp, J.S. State of the Art in Total Body PET. *EJNMMI Phys.* **2020**, *7*, 35. [CrossRef]
48. Blumendorf, F. Background Reduction for XFI with Human-Sized Phantoms. Ph.D. Thesis, University of Hamburg, Hamburg, Germany, 2019.
49. Olafsen, T.; Jiang, Z.K.; Romero, J.; Zamilpa, C.; Marchioni, F.; Zhang, G.; Torgov, M.; Satpayev, D.; Gudas, J.M. Abstract LB-188: Sensitivity of 89 Zr-labeled anti-CD8 minibody for PET imaging of infiltrating CD8+ T Cells. *Cancer Res.* **2016**, *76*, LB-188.
50. Jazayeri, M.H.; Amani, H.; Pourfatollah, A.A.; Pazoki-Toroudi, H.; Sedighimoghaddam, B. Various methods of gold nanoparticles (GNPs) conjugation to antibodies. *Sens. Bio Sens. Res.* **2016**, *9*, 17–22. [CrossRef]
51. Kircher, M.F.; Allport, J.R.; Graves, E.E.; Love, V.; Josephson, L.; Lichtman, A.H.; Weissleder, R. In vivo high resolution three-dimensional imaging of antigen-specific cytotoxic t-lymphocyte trafficking to tumors. *Cancer Res.* **2003**, *63*, 6838–6846.
52. Ahrens, E.T.; Bulte, J.W.M. Tracking immune cells in vivo using magnetic resonance imaging. *Nat. Rev. Immunol.* **2013**, *13*, 755–763. [CrossRef]

53. Tavaré, R.; Escuin-Ordinas, H.; Mok, S.; McCracken, M.N.; Zettlitz, K.A.; Salazar, F.B.; Witte, O.N.; Ribas, A.; Wu, A.M. An effective immuno-PET imaging method to monitor CD8-dependent responses to immunotherapy. *Cancer Res.* **2016**, *76*, 73–82. [CrossRef] [PubMed]
54. Meir, R.; Betzer, O.; Motiei, M.; Kronfeld, N.; Brodie, C.; Popovtzer, R. Design principles for noninvasive, longitudinal and quantitative cell tracking with nanoparticle-based CT imaging. *Nanomed. Nanotechnol. Biol. Med.* **2017**, *13*, 421–429. [CrossRef] [PubMed]
55. Schmutzler, O. Experimental and Numerical Studies for Synchrotron-Based X-Ray Fluorescence Imaging in Medium Sized Objects. Ph.D. Thesis, University of Hamburg, Hamburg, Germany, 2020.
56. Jacquet, M. Potential of Compact Compton Sources in the Medical Field. *Phys. Med.* **2016**, *32*, 1790–1794. [CrossRef] [PubMed]
57. Kulpe, S.; Dierolf, M.; Günther, B.; Brantl, J.; Busse, M.; Achterhold, K.; Pfeiffer, F.; Pfeiffer, D. Spectroscopic Imaging at Compact Inverse Compton X-Ray Sources. *Phys. Med.* **2020**, *79*, 137–144. [CrossRef]
58. Brümmer, T.; Debus, A.; Pausch, R.; Osterhoff, J.; Grüner, F. Design study for a compact laser-driven source for medical X-ray fluorescence imaging. *Phys. Rev. Accel. Beams* **2020**, *23*, 031601. [CrossRef]
59. Zhang, S.; Li, L.; Chen, J.; Chen, Z.; Zhang, W.; Lu, H. Quantitative imaging of gd nanoparticles in mice using benchtop cone-beam X-Ray fluorescence computed tomography system. *Int. J. Mol. Sci.* **2019**, *20*, 2315. [CrossRef]
60. Manohar, N.; Reynoso, F.; Jayarathna, S.; Moktan, H.; Ahmed, F.; Diagaradjane, P.; Krishnan, S.; Cho, S.H. High-sensitivity imaging and quantification of intratumoral distributions of gold nanoparticles using a benchtop X-ray fluorescence imaging system. *Opt. Lett.* **2019**, *44*, 5314. [CrossRef]
61. Sharma, N.K.; Holmes-Hampton, G.P.; Kumar, V.P.; Biswas, S.; Wuddie, K.; Stone, S.; Aschenake, Z.; Wilkins, W.L.; Fam, C.M.; Cox, G.N.; et al. Delayed effects of acute whole body lethal radiation exposure in mice pre-treated with BBT-059. *Sci. Rep.* **2020**, *10*, 6825. [CrossRef]
62. Parkins, C.S.; Fowler, J.F.; Maughan, R.L.; Roper, M.J. Repair in mouse lung for up to 20 fractions of X rays or neutrons. *Br. J. Radiol.* **1985**, *58*, 225–241.
63. Drube, W.; Bieler, M.; Caliebe, W.A.; Schulte-Schrepping, H.; Spengler, J.; Tischer, M.; Wanzenberg, R. The PETRA III Extension. In *AIP Conference Proceedings*; AIP Publishing LLC: New York, NY, USA, 2016; Volume 1741, p. 020035.
64. USC University of Southern California. Biomedical Imaging Group. Digimouse: Download. Available online: https://neuroimage.usc.edu/neuro/Digimouse_Download (accessed on 8 March 2021).
65. Amptek Silicon Drift Detectors. Available online: https://www.amptek.com/-/media/ametekamptek/documents/products/amptek-silicon-drift-detectors.pdf?dmc=1&la=en&revision=47add5cd-a5b0-4590-ba42-c32d559f6d0d (accessed on 8 March 2021).
66. Fast-Sdd-Specifications. Available online: https://www.amptek.com/-/media/ametekamptek/documents/products/fast-sdd-specifications.pdf?dmc=1&la=en&revision=c4bd8a28-f8ed-46be-a2d4-c1bd4c86dbd0 (accessed on 8 March 2021).
67. Xr100cdte. Available online: https://www.amptek.com/-/media/ametekamptek/documents/products/xr100cdte.pdf?dmc=1&la=en&revision=4fc7bfd8-4247-4abf-b9ee-d5f5e992e7ee (accessed on 8 March 2021).
68. Redus, R.H.; Pantazis, J.A.; Pantazis, T.J.; Huber, A.C.; Cross, B.J. Characterization of CdTe detectors for quantitative X-Ray spectroscopy. *IEEE Trans. Nucl. Sci.* **2009**, *56*, 2524–2532. [CrossRef]
69. Sinervo, P.K. Signal significance in particle physics. *arXiv* **2002**, arXiv:hep-ex/0208005.

International Journal of
Molecular Sciences

MDPI

Article

Zn Promoted Mg-Al Mixed Oxides-Supported Gold Nanoclusters for Direct Oxidative Esterification of Aldehyde to Ester

Jie Li [1], Shiyi Wang [1], Huayin Li [1], Yuan Tan [1,*] and Yunjie Ding [1,2,3,*]

[1] Hangzhou Institute of Advanced Studies, Zhejiang Normal University, 1108 Gengwen Road, Hangzhou 311231, China; lijie_zjnu@163.com (J.L.); wangshiyi@dicp.ac.cn (S.W.); lihuayin@zjnu.edu.cn (H.L.)

[2] Dalian National Laboratory for Clean Energy, Dalian Institute of Chemical Physics, Chinese Academy of Sciences, Dalian 116023, China

[3] The State Key Laboratory of Catalysis, Dalian Institute of Chemical Physics, Chinese Academy of Sciences, Dalian 116023, China

* Correspondence: yuantan2012@zjnu.edu.cn (Y.T.); dyj@dicp.ac.cn (Y.D.); Tel.: +86-571-82257902 (Y.T.); +86-411-84379143 (Y.D.)

Abstract: The synthesis of ester compounds is one of the most important chemical processes. In this work, Zn-Mg-Al mixed oxides with different Zn^{2+}/Mg^{2+} molar ratios were prepared via co-precipitation method and supported gold nanoclusters to study the direct oxidative esterification of aldehyde and alcohol in the presence of molecular oxygen. Various characterization techniques such as N_2-physical adsorption, X-ray diffraction (XRD), transmission electron microscopy (TEM), X-ray photoelectron spectroscopy (XPS) and CO_2 temperature programmed desorption (TPD) were utilized to analyze the structural and electronic properties. Based on the results, the presence of small amounts of Zn^{2+} ions (~5 wt.%) provoked a remarkable modification of the binary Mg-Al system, which enhanced the interaction between gold with the support and reduced the particle size of gold. For oxidative esterification reaction, the $Au_{25}/Zn_{0.05}MgAl\text{-}400$ catalyst showed the best performance, with the highest turnover frequency (TOF) of 1933 h^{-1}. The active center was believed to be located at the interface between metallic gold with the support, where basic sites contribute a lot to transformation of the substrate.

Keywords: gold catalyst; cluster; hydrotalcite; oxidative esterification; methyl isobutyrate

Citation: Li, J.; Wang, S.; Li, H.; Tan, Y.; Ding, Y. Zn Promoted Mg-Al Mixed Oxides-Supported Gold Nanoclusters for Direct Oxidative Esterification of Aldehyde to Ester. *Int. J. Mol. Sci.* **2021**, *22*, 8668. https://doi.org/10.3390/ijms22168668

Academic Editor: Raghvendra Singh Yadav

Received: 30 June 2021
Accepted: 10 August 2021
Published: 12 August 2021

1. Introduction

Ester compounds are one of the most important raw materials in chemical industry and organic synthesis and constitute numerous artificial fragrances, flavoring agents, solvent extractants and chemical intermediates [1]. The traditional synthetic method for preparation of esters involves a two-step procedure, including the oxidation of aldehyde and/or alcohol and the subsequent esterification between carboxylic acid and its activated derivatives with alcohols [2–4]. The multistep process involves the production of large amounts of toxic wastes and irremovable byproducts, which is not conducive to the development of green chemistry. Therefore, the synthesis of esters through one-step oxidative esterification from aldehydes and alcohols has attracted great interest in recent years.

Traditionally, the direct oxidative esterification requires homogeneous stoichiometric reagents, such as $KMnO_4$ [5], CrO_3 [6], H_2O_2 [7], etc., while the strong oxidants not only call for harsh conditions but also cause separation difficulties. Therefore, the green oxidants such as air and dioxygen over heterogeneous catalysts under mild conditions are highly desirable. The supported noble metal Pd-based catalysts have been reported to be efficient for one-step oxidative cross-esterification reaction between aldehyde and alcohol in the presence of molecular oxygen [8–12]. However, the selectivity of the corresponding esters is typically low and this process requires the addition of environmentally unfriendly heavy metals, such as Pb and Bi [8–10]. Furthermore, the utilization of liquid base additives is needed for Pd-based catalysts, which makes the process less green and not cost-effective [11,12].

Supported gold catalysts are a great promising candidate for this kind of reaction due to their unique selectivity under mild conditions [13–18]. Various transition-metal-oxides and mixed oxides were explored as supports for synthesis of the supported gold catalysts [15–18]. Nevertheless, their catalytic performances were largely dependent on the particle size of gold [15], the interaction between gold and the carrier [16], as well as the acidic-basic properties of the supports [15,17]. Thus, preparation of highly efficient gold catalysts for synthesis of the corresponding esters is always considered in combination of the size of gold and the properties of the supports.

Mg-Al hydrotalcite (HT) is a common solid base with unique layered structure, which plays an important role in many base-catalyzed reactions, such as alcohol oxidation [19,20], aldol condensation [21] and transesterification [22]. Their catalytic performances can be modified by incorporating various metal ions, such as Cu^{2+} [23,24], Zn^{2+} [25,26], Ni^{2+} [27], Y^{3+} [28], Zr^{4+} [29], etc. For example, Pavel et al. [28] reported yttrium-modified Mg-Al-HTs for epoxidation of styrene with hydrogen peroxide. An increment in catalytic activity was observed with the addition of Y^{3+} cations, which increased the alkalinity of catalysts accordingly. Willinton et al. [25] prepared Mg-Zn-Al HTs and their derived mixed oxides for the aldol condensation reaction. They found the presence of Zn^{2+} ions caused a clear influence on the acidic–basic properties of the catalysts when compared to the binary Mg-Al system, which influenced the activity and selectivity greatly.

Previously, our group reported that the Zn-Al HT-derived mixed oxides-supported gold catalysts were highly selective and resistant for high temperature sintering in hydrogenation reactions of functionalized nitroarenes due to the strong interaction between gold with the support [30]. Later, we found that the catalyst is also active and selective for synthesis of methyl methacrylate from methacrolein and methanol [31]. Herein, in this work, we try to incorporate Zn^{2+} ions into Mg-Al HTs for synthesis of stable gold catalysts for inorganic base-free oxidative esterification of aldehyde to ester. By adjusting the molar ratios of Mg^{2+}/Zn^{2+} ions, the textural and electronic properties of supported gold catalysts were investigated in detail, in which N_2 physical adsorption-desorption, X-ray powder diffraction (XRD), transmission electron microscopy (TEM), X-ray photoelectron spectroscopy (XPS) and CO_2 temperature programmed desorption (TPD) were successively explored. Based on these results, a clear insight into the structure-performance relationship over supported gold catalysts for direct oxidative esterification was provided.

2. Results

2.1. Compositional and Textural Analysis of the Catalysts

For synthesis of the Zn-Mg-Al mixed oxides-supported gold catalysts, Au_{25} nanoclusters were firstly prepared through the sol-gel method with cysteine as the protective ligand, defined by UV-visible spectra (Figure S1). Then, different supports with various molar ratios of Zn^{2+}/Mg^{2+} were impregnated with the solution of Au_{25}. Before each test, the samples were dried overnight and pretreated in muffle at 400 °C for 2 h to remove most of the protective ligands. After that, various characterizations of the structural and electronic properties of the catalysts were carried out under specific conditions. The actual metal loadings of gold were measured by inductively coupled plasma atomic emission spectroscopy (ICP-AES), from which all catalysts showed similar values of ~1.5 wt.% (Table 1), which agrees well with the nominal values.

Table 1. Textural properties of Zn-Mg-Al mixed oxides-supported gold catalysts.

Entry	Catalysts	Loadings of Au (%) [a]	S_{BET} (m^2g^{-1}) [b]	Volume (cm^3g^{-1}) [b]	D_{pore} (nm) [b]	Particle Size (nm) [c]	Total Basicity $(\mu mol\ g^{-1})$ [d]
1	$Au_{25}/MgAl$-400	1.52	221.1	0.51	0.57	2.6	158
2	$Au_{25}/Zn_{0.05}MgAl$-400	1.52	242.0	0.51	0.57	1.8	111
3	$Au_{25}/Zn_{0.33}MgAl$-400	1.52	132.4	0.37	0.56	1.9	81
4	Au_{25}/Zn_3MgAl-400	1.47	113.9	0.35	0.56	1.9	23
5	$Au_{25}/ZnAl$-400	1.50	91.1	0.24	0.56	1.9	29

[a]: The actual loadings of gold were evaluated by ICP-AES measurement. [b]: Specific surface area (S_{BET}), total pore volume and average pore size were calculated from N_2 physical adsorption–desorption. S_{BET} were calculated from BET equation at P/P_0 range of 0.05-0.3. Total pore volumes were calculated at $P/P_0 = 0.98$. Pore sizes were calculated from the adsorption branch of the isotherm-BJH. [c]: The mean particle sizes were calculated from TEM measurement. [d]: Total basicity was calculated from CO_2-TPD experiments.

The specific surface area, pore width and pore volume were measured by a nitrogen adsorption–desorption test. The isothermal curves and contrastive results are shown in Figure 1 and Table 1. In Figure 1a, Zn-Mg-Al mixed oxides-supported gold catalysts present typically type IV isotherms according to the IUPAC classification [32]. Furthermore, the hysteresis loops are H3 type and are located at higher relative pressures, indicating the formation of slit-shaped holes and the presence of mesopores, which were derived from the collapse of hydrotalcite structure after calcination [33]. To be noted, with the increase of Zn^{2+} ions into the Mg-Al mixed oxides, the pore volumes and pore sizes of catalysts declined, while the surface area of catalysts increased initially from 221.1 to 242.0 m^2/g and then decreased drastically to about 100 m^2/g. This is likely derived from the variation of phase structure of catalysts and it implies that the structural property of catalysts could be adjusted by the addition of Zn^{2+} ions. Later, the structural features were further analyzed by the following XRD results.

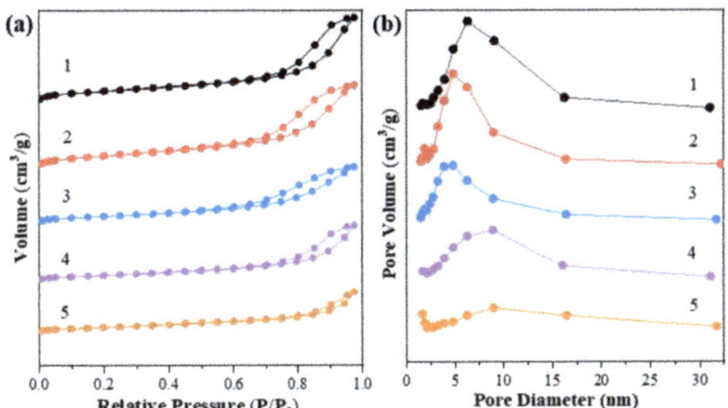

Figure 1. (**a**) N_2 physical adsorption–desorption isotherms and (**b**) BJH-pore size distributions of Zn-Mg-Al mixed oxides-supported gold catalysts: (1) $Au_{25}/MgAl$-400; (2) $Au_{25}/Zn_{0.05}MgAl$-400; (3) $Au_{25}/Zn_{0.33}MgAl$-400; (4) Au_{25}/Zn_3MgAl-400; (5) $Au_{25}/ZnAl$-400.

2.2. Structural Analysis of the Catalysts

The phase structure of Zn-modified Mg-Al-HT supported gold catalysts before and after thermal treatment was analyzed by X-ray diffraction (XRD). Figures 2 and 3 display the sample diffraction peaks. Before calcination, all samples exhibited layered characteristics of hydrotalcites, with sharp and intense lines located at 11.3°, 22.8°, 34.7°, 39.1°, 46.4°, 60.6° and 61.9° (Figure 2), corresponding to the (003), (006), (012), (015), (018), (110) and (113) planes (standard magnesium aluminum hydrotalcite JCPD NO. 00-035-0965, standard zinc aluminum hydrotalcite JCPD NO. 00-048-1023), respectively. The XRD patterns of the samples with fewer Zn^{2+} ions have broader reflections, which are close to that of the $Au_{25}/MgAl$-HT (Figure 2a), while the samples with more Zn^{2+} content present sharper reflections that are similar to the $Au_{25}/ZnAl$-HT (Figure 2e).

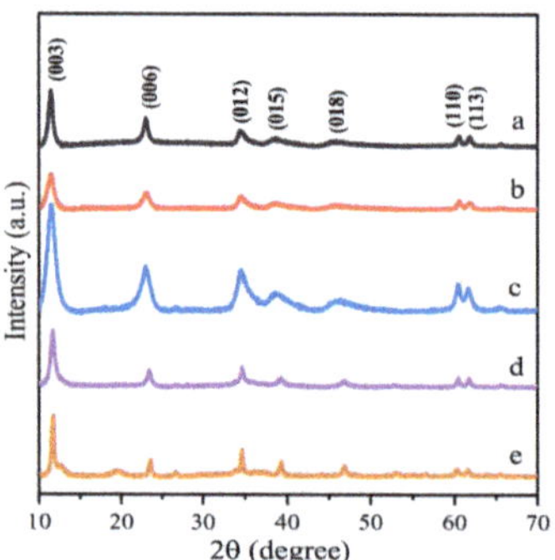

Figure 2. XRD spectra of Zn-Mg-Al HT supported gold nanoclusters: (**a**) Au_{25}/MgAl-HT; (**b**) Au_{25}/$Zn_{0.05}$MgAl-HT; (**c**) Au_{25}/$Zn_{0.33}$MgAl-HT; (**d**) Au_{25}/Zn_3MgAl-HT; (**e**) Au_{25}/ZnAl-HT.

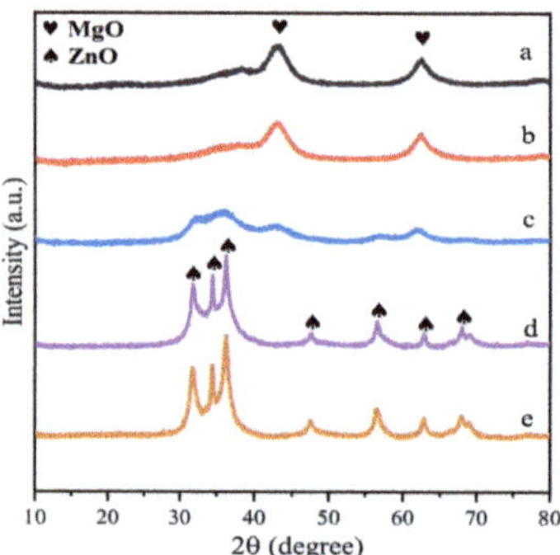

Figure 3. XRD spectra of Zn-Mg-Al mixed oxides-supported gold catalysts: (**a**) Au_{25}/MgAl-400; (**b**) Au_{25}/$Zn_{0.05}$MgAl-400; (**c**) Au_{25}/$Zn_{0.33}$MgAl-400; (**d**) Au_{25}/Zn_3MgAl-400; (**e**) Au_{25}/ZnAl-400.

According to the literature, characteristics of (003) and (110) reflections correspond to basal and in-plane spacing [33,34]. Here, the lattice parameters (a and c) in Table 2 were calculated on the basis of a rhombohedral R3m space group and hexagonal cell by the formula of a = 2 d_{110} and c = 3 d_{003} [33,34]. The parameter a represents the average distance between metal and metal in each layer of the hydrotalcite, whereas the parameter c reveals the distance between layer and layer [33]. From the results, the value of a remains constant upon zinc addition, while the value of c gradually decreases with increasing of Zn^{2+}. This may be due to the similar ionic radius of octahedral coordinated Mg^{2+}(0.72 Å) with Zn^{2+}(0.74 Å) [35]. Hence, the effect of Zn substitution on the value of a was negligible. However, the larger electronegativity of Zn (1.66) compared to Mg (1.29) would lead to an increase in layer charge density, which is directly related to the decrease of interlayer distance [36]. Table 2 presents the crystal sizes of different samples that were calculated using Scherrer's equation. The increased crystallite sizes were consistent with the textural analysis. That is, a higher degree of crystallinity suggests a loss in surface area [36].

Furthermore, the XRD patterns of calcined samples after heat treatment at 400 °C are displayed in Figure 3. It can be clearly seen that the crystal structure of the samples exhibits typical periclase phase of MgO and/or zincite phase of ZnO, with diffraction peaks at 37.0°, 43.0°, 62.3° (JCPD 01-077-2364) and 31.8°, 34.4°, 36.3°, 47.6°, 56.6°, 62.9°, 68.0° (JCPD 01-089-0510), respectively. When 5 wt.% of Mg^{2+} cations were substituted by Zn^{2+} ions (Au_{25}/$Zn_{0.05}$MgAl-400, Figure 3b), the structure of the catalyst was quite similar to the binary Au_{25}/MgAl-400. When the replacement increased to 35.6 wt.% (Au_{25}/$Zn_{0.33}$MgAl-400, Figure 3c), both MgO and ZnO phase appears in the patterns. Further increasing the replacement of Zn^{2+} to 70.9 wt.% (Au_{25}/Zn_3MgAl-400, Figure 3d), the phase of ZnO dominates the structure. It implies that the addition of Zn^{2+} ions strongly affected the ordered structure of catalysts. In addition, no diffraction peaks attributed to face-centered cubic gold were observed in all heat-treated samples, indicating that the gold particles were highly dispersed and not aggregated under high temperature calcination.

Table 2. Crystallographic parameters calculated for Zn-Mg-Al HTs supported gold nanoclusters.

Entry	Catalysts	$d_{(003)}$ (nm)	$d_{(110)}$ (nm)	a (nm) [a]	c (nm) [b]	Crystallite Size (nm) [c]	
						(003)	(110)
1	Au_{25}/MgAl-400	0.772	0.153	0.306	2.316	18	26
2	Au_{25}/$Zn_{0.05}$MgAl-400	0.769	0.153	0.306	2.307	11	22
3	Au_{25}/$Zn_{0.33}$MgAl-400	0.763	0.153	0.306	2.289	9	22
4	Au_{25}/Zn_3MgAl-400	0.760	0.153	0.306	2.280	17	31
5	Au_{25}/ZnAl-400	0.755	0.154	0.308	2.265	33	30

[a]: Lattice parameter: $a = 2d_{110}$. [b]: Lattice parameter: $c = 3d_{003}$. [c]: Average crystallite size: calculated using the Scherrer equation.

2.3. Particle Size of Gold

To further observe the dispersion of gold particles and examine the average particle sizes, transmission electron microscopy (TEM) was performed over the above catalysts. The TEM images and size distribution histograms are displayed in Figure 4. From the photos, the gold particles in all samples are almost highly dispersed, with average particle sizes of 2.6 ± 0.9, 1.8 ± 0.7, 1.9 ± 0.7, 1.9 ± 0.6, 1.9 ± 0.8 nm for the Au_{25}/MgAl-400 (Figure 4a), Au_{25}/$Zn_{0.05}$MgAl-400 (Figure 4b), Au_{25}/$Zn_{0.33}$MgAl-400 (Figure 4c), Au_{25}/Zn_3MgAl-400 (Figure 4d) and Au_{25}/ZnAl-400 (Figure 4e) catalysts, respectively. Obviously, the addition of Zn^{2+} ions greatly reduced the particle sizes of gold from 2.6 nm to 1.8 nm, and facilitated the dispersion, which probably derived from the strong interaction between gold with Zn-contained carrier [30,37].

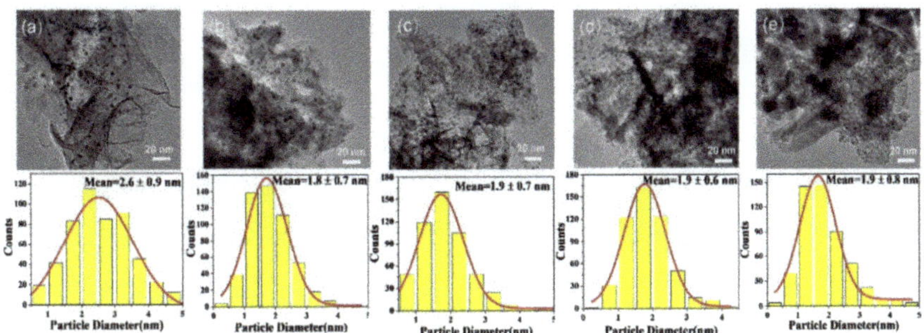

Figure 4. TEM images of Zn-Mg-Al mixed oxides-supported gold catalysts: (**a**) Au_{25}/MgAl-400; (**b**) Au_{25}/$Zn_{0.05}$MgAl-400; (**c**) Au_{25}/$Zn_{0.33}$MgAl-400; (**d**) Au_{25}/Zn_3MgAl-400; (**e**) Au_{25}/ZnAl-400.

2.4. Electronic Property of the Catalysts

To study the chemical state of elements in different samples, X-ray photoelectron spectroscopy (XPS) was performed on C1s, O1s, Au 4f, Zn 3p and Mg 2s, as shown in Figures S2 and S3, and Figure 5. The binding energy of each element was calibrated by referencing them to the energy of the C1s peak at 284.6 eV (Figure S2). All catalysts showed distinguished peaks around at 83.3 eV and 87.0 eV, respectively, which could be assigned to the Au 4f 7/2 and Au 4f 5/2 lines of metallic Au (Figure 5) [38]. The peak located at 88.2/88.3 eV is assigned to the peaks of Mg 2s and/or Zn 3p, which is identified by the purple and blue bands. Obviously, with the increase in Zn^{2+} ions, the intensity of peaks ascribed to Zn 3p increase, but the peaks of Mg 2s decrease correspondingly. Of note, a slight shift to low binding energy of Au was observed over the Au_{25}/$Zn_{0.05}$MgAl-400 (83.2 eV, Figure 5b), Au_{25}/$Zn_{0.33}$MgAl-400 (83.1 eV, Figure 5c), Au_{25}/Zn_3MgAl-400 (82.9 eV,

Figure 5d) and Au$_{25}$/ZnAl-400 (83.1 eV, Figure 5e) catalysts, indicating there might be a strong interaction between gold with Zn-contained carrier. This was consistent with the results of TEM, which shows the reduction of sizes originated from the interaction of gold with the Zn-contained carrier. The O 1s XPS spectra of different gold catalysts are presented in Figure S3, which shows two peaks. One peak located at 529.7~530.2 eV was attributed to the lattice oxygen connected with Mg^{2+} and/or Zn^{2+} [39], whereas the other peak located at 531.3~531.6 eV was assigned to hydroxide and adsorbed water [39,40], which originated from the interlaminar anions of hydrotalcites.

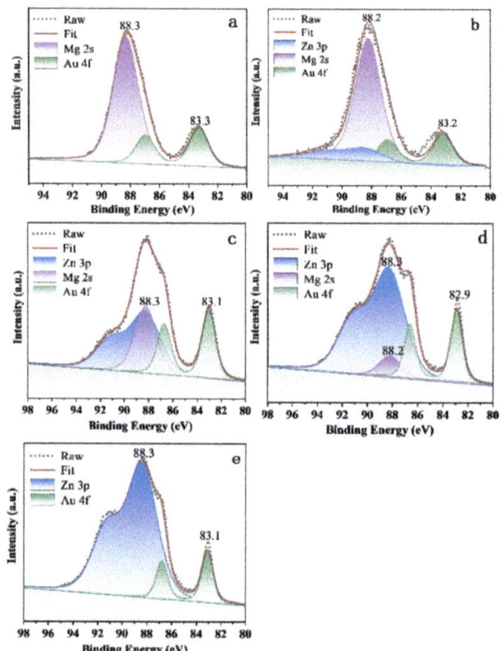

Figure 5. XPS Spectra of Au 4f-Zn 3p-Mg 2s of Zn-Mg-Al oxides-supported gold catalysts: (a) Au$_{25}$/MgAl-400; (b) Au$_{25}$/Zn$_{0.05}$MgAl-400; (c) Au$_{25}$/Zn$_{0.33}$MgAl-400; (d) Au$_{25}$/Zn$_3$MgAl-400; (e) Au$_{25}$/ZnAl-400.

2.5. Basicity of the Catalysts

Mg-Al hydrotalcite or its derived oxide is a well-known solid base, which is widely used as a catalyst or support [19–22]. In the literature, the basicity of the catalyst was reported to facilitate the oxidative esterification of aldols [41,42]. Our group also reported that the strong basic sites benefit the formation of hemiacetal intermediate, which contributes to the synthesis of the final product [31]. Hence, in this work, CO$_2$-TPD was utilized to determine the intensity of basic sites of Zn-Mg-Al mixed oxides-supported gold catalysts. The amounts of basic sites were quantitatively measured by CO$_2$-pulse-adsorption experiments. The results are summarized in Table 1. It is clear that the basicity of catalysts sharply decreased from 158 μmol·g^{-1} to 23 μmol·g^{-1} with the increase in Zn^{2+} ions. This implies the basicity of catalysts mainly derived from the MgO phase in the composite oxides [21]. Furthermore, according to the desorption temperature of carbon dioxide, the basic sites could be roughly classified into four types, namely, weak basic sites (<250 °C), medium basic sites (250–400 °C), strong basic sites (400–620 °C) and super strong basic sites (>620 °C) [41,43]. Generally, the weak basic sites were caused by hydroxide groups [44]. The medium basic sites were attributed to M^{2+}-O^{2-} pairs and the strong basic sites were assigned to lattice oxygen [44]. As displayed in Figure 6, the amounts of weak basic sites increased with the increase of Zn, but the medium basic sites and strong basic sites reduced accordingly.

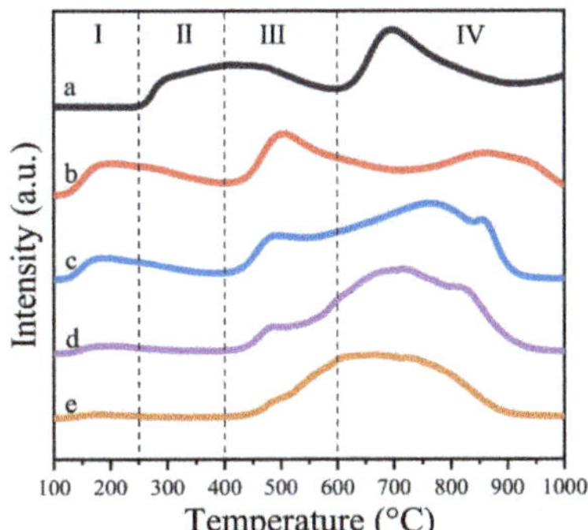

Figure 6. CO_2-TPD profiles of Zn-Mg-Al mixed oxides-supported gold catalysts: (**a**) Au_{25}/MgAl-400; (**b**) Au_{25}/$Zn_{0.05}$MgAl-400; (**c**) Au_{25}/$Zn_{0.33}$MgAl-400; (**d**) Au_{25}/Zn_3MgAl-400; (**e**) Au_{25}/ZnAl-400.

2.6. Catalytic Performances

The catalytic performances of above Zn-Mg-Al mixed oxides-supported gold catalysts were investigated in direct oxidative esterification of isobutyraldehyde with methanol for synthesis of methyl isobutyrate (Scheme 1). The results are displayed in Table 3 and Figure S4. From the results, the Au_{25}/MgAl-400 catalyst showed good performance with conversion of 78.5% and selectivity of 96.8% (Table 3, entry 1). When 5% of Zn^{2+} ions were added into the Mg-Al binary system, the conversion of isobutyraldehyde increased to 88.6%, with the selectivity of 96.6%. Meanwhile, the turnover frequency (TOF) of gold catalysts increased from 1499 to 1933 h^{-1} (Table 3, entry 2), suggesting that the catalytic performance could be well improved by the addition of Zn. However, further increase in the content of Zn^{2+} ions would not enhance the catalytic performances. On the contrary, the activity and selectivity would reduce consecutively. For example, over the Au_{25}/$Zn_{0.33}$MgAl-400 catalyst, the conversion and selectivity are 68.0% and 95.9%, respectively. Over the Au_{25}/Zn_3MgAl-400 catalyst, the conversion and selectivity decreased to 65.2% and 95.0%. On the Au_{25}/ZnAl-400 catalyst, the conversion reduced to 59.0% and the selectivity decreased to 93.0%. This meant the optimum addition amount of Zn follows the volcanic curve, and only small amount of Zn promotes the catalytic reactivity. To further affirm the result, we also performed the reaction under the same conditions three times and plotted the histogram with error bar, as displayed in Figure 7. It is clear that the Au_{25}/$Zn_{0.05}$MgAl-400 catalyst indeed shows the best performance.

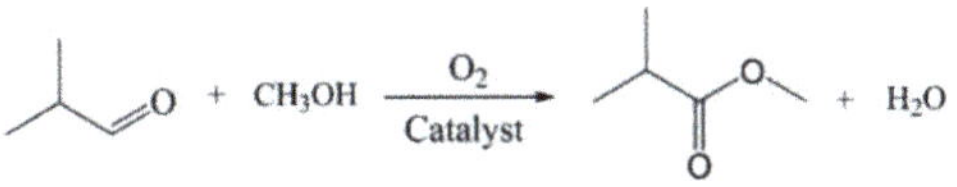

Scheme 1. Reaction pathway of isobutyraldehyde with MeOH to form methyl isobutyrate.

To further study the evolution of product distributions during the reaction process, dynamic experiments were conducted over the Au_{25}/Zn_xMgAl-400 catalysts. The yields of methyl isobutyrate and by-products of acetal and isobutyric acid with reaction time are plotted in Figure S4. From the curves, the trends on product distributions over different catalysts were somewhat different. In the process of reaction, methyl isobutyrate was formed as the sacrifice of isobutyraldehyde. At the same time, acetal, isobutyric acid and other ester compounds appeared as the by-products. Of note, the content of acetal and isobutyric acid are more with the Au_{25}/ZnAl-400 catalyst than the others, implying the transformation of isobutyraldehyde to its final target product is harder on the Au_{25}/ZnAl-400 catalyst than that of the Au_{25}/Zn_xMgAl-400 catalysts.

Table 3. Catalytic performances of Zn-Mg-Al mixed oxides-supported gold catalysts for synthesis of methyl isobutyrate from isobutyraldehyde and methanol.

Entry	Catalysts	Conversion% [a]	Selectivity % [a]				TOF (h^{-1}) [c]
			Methyl Isobutyrate	Isobutyric Acid	Acetal	Others	
1	Au_{25}/MgAl-400	78.5	96.8	0.54	0.27	2.43	1499
2	Au_{25}/$Zn_{0.05}$MgAl-400	88.6	96.6	0.73	0.20	2.45	1933
3	Au_{25}/$Zn_{0.33}$MgAl-400	68.0	95.9	0.29	0.49	3.32	1790
4	Au_{25}/Zn_3MgAl-400	65.2	95.0	0.38	0.74	3.93	1713
5	Au_{25}/ZnAl-400	59.0	93.0	0.23	1.28	5.48	1252
6	Au_{25}/$Zn_{0.05}$MgAl-400 [b]	2.12	Trace	-	-	-	-
7	$Zn_{0.05}$MgAl-400	5.26	Trace	-	-	-	-

Reaction conditions: Amounts of catalyst: 40 mg; isobutyl aldehyde: 5 mmol; methanol: 5 mL; o-xylene (interior standard): 1 mmol; reaction temperature: 80 °C; pressure of O_2: 4 atm; reaction time: 2 h. [a]: Conversion and selectivity were calculated by the results of gas chromatography; [b]: O_2 was replaced by N_2 (4 atm); [c]: TOF was calculated by moles of converted aldehyde per mole of gold per hour, the reaction rate was calculated below 20% conversion.

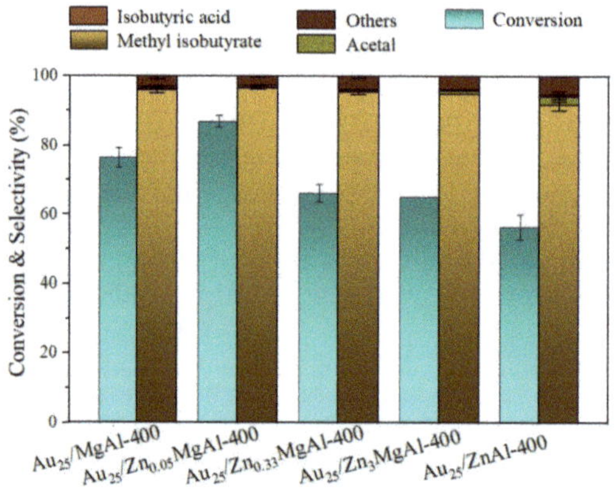

Figure 7. Catalytic performances of Zn-Mg-Al mixed oxides-supported gold catalysts for synthesis of methyl isobutyrate with error bar. Reaction conditions: catalyst: 40 mg; isobutyl aldehyde: 5 mmol; methanol: 5 mL; reaction temperature: 80 °C; pressure of O_2: 4 atm; reaction time: 2 h.

3. Discussion

According to our previous characterizations, the small amounts of Zn^{2+} ions (~5 wt.%) would lead to the reduction of gold particle size to below 2 nm (Figure 4) and enhance the interaction between gold with the support (Figure 5). Hence, we have a reason to believe that the improved activity is connected with the decreased sizes ascribed to the addition of Zn. In fact, many investigations have demonstrated that the activity of gold catalyst is markedly dependent on the particle size [45,46], since small gold particles possess more coordinated-unsaturated gold species, which supply more adsorption sites for reactants [46]. However, based on the results of catalytic performances, further increasing the content of Zn^{2+} ions would decrease the activity, even though the particle size of gold is still below 2 nm. Thus, this strongly suggests the size of gold was not the only factor that influences the catalytic performance.

Previously, we have proposed that the basicity of catalysts is critical for oxidative esterification [31]. Various works also suggest that the basic sites can accelerate the formation of hemiacetal intermediate, thus contributing to the formation of target esters [47,48]. To examine the basic sites of our catalysts, CO_2-TPD experiments were then carried out on the above catalysts. From the results, it is clear to see the total basicity of catalysts decreased gradually with the increase of Zn^{2+} ions (Table 1). Meanwhile, the intensity of strong basic sites decreased correspondingly (Figure 6). Based on the previous literature, the strong basic sites are proposed to be the key sites for enhancement of the intermediate [47,49]. Thus, with the decrease of intensity assigned to strong basic sites, the catalytic performances would be decreased. This is consistent with our result. Therefore, the activity of Zn-Mg-Al mixed oxides-supported gold catalysts were supposed to be sectionally influenced by the particle size of gold and the basicity of supports.

Furthermore, to clarify the effect of loading gold particles, comparison experiments were conducted on $Zn_{0.05}MgAl$-400 catalyst or with N_2 as the source gas (Table 3, entry 7, 6). Both situations give no products after reaction under the same conditions, suggesting that the gold particles were the critical components for oxidative esterification. Our previous work demonstrated that a calcination temperature above 300 °C was sufficient to remove most of the protective ligands [30], so that bare gold particles could be exposed on the surface of catalyst. Besides, it has been reported that the gold clusters below 2 nm could be active for dissociation of the molecular oxygen [50,51]. Hence, with respect to this work, we speculate that the role of gold is probably activating and dissociating oxygen so the oxidative esterification could be progressed successfully.

Considering the effect of basic sites for transformation of isobutyl aldehyde and the large role of dissociation of oxygen on the gold particles, the active center is believed to be located at the interface between metallic gold with the support. Therefore, the reaction mechanism over the above catalysts could be proposed based on the results and the literature [31,49,52]. That is, oxygen was dissociated into the adsorbed atoms on gold particles and reacts with methanol to form methoxy groups [46,50]. Then, the isobutyl aldehyde molecules were oxidized by atomic oxygen to form isobutyric acid or underwent nucleophilic attack by methoxy groups to form the surface hemiacetal intermediate. Later, during the oxidation of intermediate hemiacetal, the hemiacetal transforms to the target ester via β-H elimination with the assistant of atomic oxygen or forms the acetal by-product by reacting with methoxy groups.

4. Materials and Methods

4.1. Materials

Sodium hydroxide (NaOH, AR), sodium carbonate (Na_2CO_3, AR), hydrogen tetrachloroaurate hydrate ($HAuCl_4 \cdot 3H_2O$), methanol (AR) and aluminum nitrate hydrate ($Al(NO_3)_3 \cdot 9H_2O$, 99%) were purchased from Sinopharm. Sodium borohydride ($NaBH_4$, 97%) and magnesium nitrate hexahydrate ($Mg(NO_3)_3 \cdot 6H_2O$, 99%) were purchased from Shanghai Lingfeng Chemical Reagent Company. Isobutyl aldehyde (98%), cysteine (99%), methyl isobutyrate (99%), ortho-xylene (CP) and zinc nitrate hydrate ($Zn(NO_3)_2 \cdot 6H_2O$, 99%) were purchased from Aladdin Industrial Corporation. All chemicals were used directly without any purification. All glassware was washed with Aqua Regia and rinsed with ethanol and ultrapure water (18.2 MΩ).

4.2. Catalyst Preparation

The Zn-Mg-Al hydrotalcites (HTs) with different Zn^{2+}/Mg^{2+} molar ratios were prepared by a co-precipitation method. Typically, specific amounts of $Zn(NO_3)_2 \cdot 6H_2O$ (0, 0.03, 0.2, 0.42, 0.63 mol), $Mg(NO_3)_2 \cdot 6H_2O$ (0.63, 0.6, 0.42, 0.2, 0 mol) and $Al(NO_3)_3 \cdot 9H_2O$ (0.21 mol) were mixed with 200 mL of ultrapure water to obtain a solution (denoted as solution A). Na_2CO_3 (0.113 mol) and NaOH (0.438 mol) were mixed with 200 mL of ultrapure water in a 500 mL round-bottom flask to obtain solution B. Under vigorous stirring in a water bath at 70 °C, solution A was pumped into solution B very slowly. Then, the obtained mixture was aged at 70 °C for about 24 h with constant stirring. Following filtering, washing and drying overnight, the Zn-Mg-Al HTs were obtained, which were denoted as MgAl-HT; $Zn_{0.05}MgAl$-HT; $Zn_{0.33}MgAl$-HT; Zn_3MgAl-HT and ZnAl-HT (0.05, 0.33 and 3 denote the molar ratio of Zn^{2+}/Mg^{2+}).

Zn-Mg-Al mixed oxides-supported gold catalysts were prepared as follows. Firstly, gold clusters protected by thiolate-ligands (cysteine) were prepared by a $NaBH_4$ reduction method according to our previous work [30]. Typically, 1.485 mL of $HAuCl_4$ (19.12 g_{Au}/L) was dissolved in 25 mL of ultrapure water under vigorous stirring for 5 min. Then, 37.5 mL of cysteine solution (5.61 mM) and 7.5 mL of 1 M NaOH solution were successively added into the solution. After about 1 h, excessive and fresh prepared $NaBH_4$ solution was poured into the solution quickly and we continued to stir for 3 h. UV-Vis spectra were analyzed to define the products as Au_{25} nanoclusters. Then, about 2.5 g of powder of support was added into the aqueous solution of Au_{25} clusters. Then, 2 h later, the resulting mixture was filtered, washed and dried to obtain the precursor of the catalyst. Before catalytic testing, the precursors of the catalysts were calcined at 400 °C for 120 min, which were denoted as Au_{25}/MgAl-400; Au_{25}/$Zn_{0.05}$MgAl-400; Au_{25}/$Zn_{0.33}$MgAl-400; Au_{25}/Zn_3MgAl-400 and Au_{25}/ZnAl-400.

4.3. Catalyst Characterization

The actual loadings of gold in all of the catalysts were determined by inductively coupled plasma atomic emission spectroscopy (ICP-AES) on an IRIS Intrepid II XSP instrument (Thermo Electron Corporation, Waltham, Massachusetts, USA). UV-Vis absorption spectroscopy (EVO300, Waltham, Massachusetts, USA) was utilized to precisely define the atomic Au_{25} nanoclusters, with water as the reference. The X-ray diffraction (XRD) analysis was conducted on a PW3040/60 X'Pert PRO (PANalytical, Almelo, The Netherlands) diffractometer, with a Cu Kα radiation source (λ = 0.15432 nm) in a scanning angle (2θ) range of 10–90°. The N_2-physical adsorption–desorption was measured at 77 K on a Quantachrome NT3LX-2 instrument after degassing the samples at 300 °C for 120 min. The pore size distribution and the specific surface area were calculated by the BJH and BET methods. The high-resolution transmission electron microscopy (HRTEM) and TEM were recorded on a JEM-2100F microscope at 200 kV. The X-ray photoelectron spectra (XPS) were conducted on an ESCLALAB 250Xi X-ray photoelectron spectrometer equipped with a monochromatic Al and double anode Al/Mg target. The binding energy was calibrated using the C 1s peak (284.6 eV) as the reference. The temperature-programmed desorption of carbon dioxide (CO_2-TPD) experiments was studied on a Micromeritics Autochem II 2920 chemisorber equipped with a thermal conductivity detector and mass spectrometry. Firstly, about 100 mg of the catalysts was pretreated in a U-type quartz tube reactor with a He gas stream (50 cm^3/min) at 400 °C for 30 min. After the temperature decreased to 100 °C, the catalysts were saturated with 10%CO_2/He flow (30 cm^3/min) for 15 min. Then, the samples were purged in helium at 100 °C for 30 min and waited for baseline balance. Finally, the CO_2-TPD profile of the sample was recorded by increasing the temperature from 100 °C to 1000 °C at a heating rate of 10 °C/min under He flow.

4.4. Catalytic Test

Catalytic evaluation of one step oxidative esterification for synthesis of methyl isobutyrate from isobutyraldehyde and methanol was carried out in stainless steel autoclave equipped with magnetic stirring and pressure gauge. Typically, certain amounts of catalysts and the mixture of isobutyraldehyde (1 M) and methanol (5 mL) were introduced into the reactor. After sealing, the air in the autoclave was replaced with oxygen six times and then pressurized to 4 atm. Before the reaction, the reactants were gradually heated to 353 K in a water bath and initiated to stir. When the reaction finished, the reactor was placed in an ice bath immediately to terminate the reaction. The products were identified by GC−MS (Agilent 5977B MSD GC/MS) and analyzed quantitatively by GC (Agilent 7890 A) with a flame ionization detector (FID) and a HP-5 column (30 m, 0.25 mm inner diameter).

5. Conclusions

A series of Zn-Mg-Al hydrotalcites and derived mixed oxides-supported gold catalysts with different Zn^{2+}/Mg^{2+} molar ratios were prepared for one-step oxidative esterification of isobutyraldehyde with methanol to form methyl isobutyrate. Adding a small amount of Zn^{2+} ions (~5 wt.%) provoked a remarkable modification of catalytic performance for the binary Mg-Al system, which enhanced the activity of catalysts from 78.5% to 88.6% and TOF values from 1499 h^{-1} to 1933 h^{-1}. Characterizations of structure and electronic properties demonstrated the addition of Zn not only reduced the particle size of gold but also enhanced the interaction between gold with the support. Furthermore, the catalytic performance was also highly dependent on the basicity of the catalyst. Based on a comparison experiment, the active center was believed to be located at the interface between metallic gold with the support, in which basicity plays a big role in transformation of the intermediates. This work provides deep insights into one-step oxidation esterification of aldehydes and alcohols over ternary system and supported gold catalysts.

Supplementary Materials: The following are available online at https://www.mdpi.com/article/10.3390/ijms22168668/s1.

Author Contributions: Y.T. and Y.D. conceived the study; J.L. performed most of the experiments; S.W. and H.L. contributed to the characterizations. All authors contributed to the writing of the manuscript. All authors have read and agreed to the published version of the manuscript.

Funding: This research was funded by Zhejiang Provincial Natural Science Foundation of China under Grant No. LQ20B030005.

Institutional Review Board Statement: Not applicable.

Informed Consent Statement: Not applicable.

Data Availability Statement: Data is available upon request from the corresponding authors.

Acknowledgments: The authors would like to thank the public testing platform of Zhejiang Normal University and Dalian Institute of Chemical Physics, Chinese Academy of Sciences.

Conflicts of Interest: The authors declare no conflict of interest.

References

1. Ray, R.; Jana, R.D.; Bhadra, M.; Maiti, D.; Lahiri, G.K. Efficient and Simple Approaches Towards Direct Oxidative Esterification of Alcohols. *Chem. Eur. J.* **2014**, *20*, 15618–15624. [CrossRef]
2. Hou, F.; Wang, X.-C.; Quan, Z.-J. Efficient synthesis of esters through oxone-catalyzed dehydrogenation of carboxylic acids and alcohols. *Org. Biomol. Chem.* **2018**, *16*, 9472–9476. [CrossRef] [PubMed]
3. Jayachitra, G.; Yasmeen, N.; Rao, K.S.; Ralte, S.L.; Srinivasan, R.; Singh, A.K. Synthetic Communications: An International Journal for Rapid Communication of Synthetic Organic Chemistry. *Synth. Commun.* **2003**, *33*, 3461–3466.
4. Gaspa, S.; Porcheddu, A.; Luca, L.D. Metal-Free Oxidative Cross Esterification of Alcohols via Acyl Chloride Formation. *Adv. Synth. Catal.* **2016**, *358*, 154–158. [CrossRef]
5. Abiko, A.; Roberts, J.C.; Takemasa, T.; Masamune, S. $KMnO_4$ revisited: Oxidation of aldehydes to carboxylic acids in the-butyl alcohol-aqueous NaH_2PO_4 system. *Tetrahedron Lett.* **1986**, *27*, 4537–4540. [CrossRef]
6. O'Connor, B.; Just, G. ChemInform Abstract: A New Method for the Conversion of Aldehydes to Methyl Esters Using Pyridinium Dichromate and Methanol in Dimethylformamide. *Tetrahedron Lett.* **1987**, *28*, 3235–3236. [CrossRef]
7. Wu, X.F. Zinc-catalyzed oxidative esterification of aromatic aldehydes. *Tetrahedron Lett.* **2012**, *53*, 3397–3399. [CrossRef]
8. Powell, A.B.; Stahl, S.S. Aerobic Oxidation of Diverse Primary Alcohols to Methyl Esters with a Readily Accessible Heterogeneous Pd/Bi/Te Catalyst. *Org. Lett.* **2013**, *15*, 5072–5075. [CrossRef]
9. Liu, C.; Tang, S.; Zheng, L.; Liu, D.; Zhang, H.; Lei, A. Covalently Bound Benzyl Ligand Promotes Selective Palladium Catalyzed Oxidative Esterification of Aldehydes with Alcohols. *Angew. Chem. Int. Ed.* **2012**, *51*, 5662–5666. [CrossRef]
10. Biajoli, A.F.P.; Peringer, F.; Monteiro, A.L. $Pd(OAc)_2$/dppp, an efficient catalytic system for the oxidative esterification of benzaldehyde using organic halides as oxidants. *Catal. Commun.* **2017**, *89*, 48–51. [CrossRef]
11. Mannel, D.S.; King, J.; Preger, Y.; Ahmed, M.S.; Root, T.W.; Stah, S.S. Mechanistic Insights into Aerobic Oxidative Methyl Esterification of Primary Alcohols with Heterogeneous PdBiTe Catalysts. *ACS Catal.* **2018**, *8*, 1038–1047. [CrossRef]
12. Li, F.; Li, X.L.; Li, C.; Shi, J.; Fu, Y. Aerobic oxidative esterification of 5-hydroxymethylfurfural to dimethyl furan-2, 5-dicarboxylate by using homogeneous and heterogeneous PdCoBi/C catalysts under atmospheric oxygen. *Green Chem.* **2018**, *20*, 3050–3058. [CrossRef]
13. Hui, Y.; Zhang, S.; Wang, W. Recent Progress in Catalytic Oxidative Transformations of Alcohols by Supported Gold Nanoparticles. *Adv. Synth. Catal.* **2019**, *361*, 2215–2235. [CrossRef]

14. Du, J.; Fang, H.; Qu, H.; Zhang, J.; Duan, X.; Yuan, Y. Fabrication of supported Au-CuO$_x$ nanohybrids by reduction-oxidation strategy for efficient oxidative esterification of 5-hydroxymethyl-2-furfural into dimethyl furan-2,5-dicarboxylate. *Appl. Catal. A* **2018**, *567*, 80–89. [CrossRef]

15. Taketoshi, A.; Ishida, T.; Murayamaa, T.; Honmac, T.; Haruta, M. Oxidative esterification of aliphatic aldehydes and alcohols with ethanol over gold nanoparticle catalysts in batch and continuous flow reactors. *Appl. Catal.* **2019**, *585*, 117169. [CrossRef]

16. Zuo, C.; Tian, Y.; Zheng, Y.; Wang, L.; Fu, Z.; Jiao, T.; Wang, M.; Huang, H.; Li, Y. A sustainable oxidative esterification of thiols with alcohols by a cobalt nanocatalyst supported on doped carbon. *Catal. Commun.* **2019**, *124*, 51–55. [CrossRef]

17. Mishra, D.K.; Choa, J.K.; Yi, Y.; Lee, H.J.; Kim, Y.J. Hydroxyapatite supported gold nanocatalyst for base-free oxidative esterification of 5-hydroxymethyl-2-furfural to 2,5-furan dimethylcarboxylate with air as oxidant. *J. Ind. Eng. Chem.* **2019**, *70*, 338–345. [CrossRef]

18. Shahin, Z.; Rataboul, F.; Demessence, A. Study of the oxidative esterification of furfural catalyzed by Au$_{25}$(glutathione)$_{18}$ nanocluster deposited on zirconia. *Mol. Catal.* **2021**, *499*, 111265. [CrossRef]

19. Liu, L.; Li, H.; Tan, Y.; Chen, X.; Lin, R.; Yang, W.; Huang, C.; Wang, S.; Wang, X.; Liu, X.; et al. Metal-support synergy of supported gold nanoclusters in selective oxidation of alcohols. *Catalysts* **2020**, *10*, 107. [CrossRef]

20. Hajavazzade, R.; Kargarrazi, M.; Mahjoub, A.R. Phosphotungstic Acid Supported on MgAl$_2$O$_4$ Nanoparticles as an Efficient and Reusable Nanocatalyst for Benzylic Alcohols Oxidation with Hydrogen Peroxide. *J. Inorg. Organomet. Polym. Mater.* **2019**, *29*, 1523–1532. [CrossRef]

21. Kikhtyanin, O.; Tišler, Z.; Velvarská, R.; Kubicka, D. Reconstructed Mg-Al hydrotalcites prepared by using different rehydration and drying time: Physico-chemical properties and catalytic performance in aldol condensation. *Appl. Catal. A* **2017**, *536*, 85–96. [CrossRef]

22. Zheng, L.; Xia, S.; Hou, Z.; Zhang, M.; Hou, Z. Transesterification of glycerol with dimethyl carbonate over Mg-Al hydrotalcites. *Chin. J. Catal.* **2014**, *35*, 310–318. [CrossRef]

23. Benito, P.; Vaccari, A.; Antonetti, C.; Licursi, D.; Schiarioli, N.; Rodriguez-Castellón, E.; Galletti, A.M.R. Tunable copper-hydrotalcite derived mixed oxides for sustainable ethanol condensation to n-butanol in liquid phase. *J. Clean. Prod.* **2019**, *209*, 1614–1623. [CrossRef]

24. Barrett, J.A.; Jones, Z.R.; Stickelmaier, C.; Schopp, N.; Ford, P.C. A Pinch of Salt Improves n-Butanol Selectivity in the Guerbet Condensation of Ethanol over Cu-Doped Mg/Al Oxides. *ACS Sustain. Chem. Eng.* **2018**, *6*, 15119–15126. [CrossRef]

25. Hernández, W.Y.; Aliç, F.; Verberckmoes, A.; Van Der Voort, P. Tuning the acidic–basic properties by Zn substitution in Mg–Al hydrotalcites as optimal catalysts for the aldol condensation reaction. *J. Mater. Sci.* **2016**, *52*, 628–642. [CrossRef]

26. Cheng, W.; Wang, W.; Zhao, Y.; Liu, L.; Yang, J.; He, M. Influence of acid-base properties of ZnMgAl-mixed oxides for the synthesis of 1-methoxy-2-propanol. *Appl. Clay Sci.* **2008**, *42*, 111–115. [CrossRef]

27. Wu, Y.J.; Li, P.; Yu, J.G.; Cunha, A.F.; Rodrigues, A.E. Sorption-enhanced steam reforming of ethanol on NiMgAl multifunctional materials: Experimental and numerical investigation. *Chem. Eur. J.* **2013**, *231*, 36–48. [CrossRef]

28. Pavel, O.D.; Cojocaru, B.; Angelescu, E.; Pârvulescu, V.I. The activity of yttrium-modified Mg, Al hydrotalcites in the epoxidation of styrene with hydrogen peroxide. *Appl. Catal. A* **2011**, *403*, 83–90. [CrossRef]

29. Nowicki, J.; Lach, J.; Organek, M.; Sabura, E. Transesterification of rapeseed oil to biodiesel over Zr-dopped MgAl hydrotalcites. *Appl. Catal. A* **2016**, *524*, 17–24. [CrossRef]

30. Tan, Y.; Liu, X.Y.; Zhang, L.; Wang, A.; Li, L.; Pan, X.; Miao, S.; Haruta, M.; Wei, H.; Wang, H.; et al. ZnAl-Hydrotalcite-Supported Au$_{25}$ Nanoclusters as Precatalysts for Chemoselective Hydrogenation of 3-Nitrostyrene. *Angew. Chem. Int. Ed.* **2017**, *56*, 2709–2713. [CrossRef] [PubMed]

31. Li, H.; Tan, Y.; Chen, X.; Yang, W.; Huang, C.; Li, J.; Ding, Y. Efficient Synthesis of Methyl Methacrylate by One Step Oxidative Esterification over Zn-Al-Mixed Oxides Supported Gold Nanocatalysts. *Catalysts* **2021**, *11*, 162. [CrossRef]

32. Jiang, X.; Zhang, S.; Yue, M. Copper functionalized solid base prepared by potassium modification of N-doped carbons to promote the conversion of ethanol to butanol. *Microporous Mesoporous Mater.* **2021**, *311*, 110731. [CrossRef]

33. Pavel, O.D.; Tichit, D.; Tichit, D.; Marcu, I.C. Acido-basic and catalytic properties of transition-metal containing Mg–Al hydrotalcites and their corresponding mixed oxides. *Appl. Clay Sci.* **2012**, *61*, 52–58. [CrossRef]

34. Zheng, Y.; Li, N.; Zhang, W. Preparation of nanostructured microspheres of Zn–Mg–Al layered double hydroxides with high adsorption property. *Colloids Surf. A Physicochem. Eng. Asp.* **2012**, *415*, 195–201. [CrossRef]

35. Shannon, R.D. Revised effective ionic radii and systematic studies of interatomic distances in halides and chalcogenides. *Acta Cryst.* **1976**, *32*, 751–767. [CrossRef]

36. Valente, J.S.; Tzompantzi, F.; Prince, J.; Cortez, J.G.H.; Gomez, R. Adsorption and photocatalytic degradation of phenol and 2,4 dichlorophenoxiacetic acid by Mg–Zn–Al layered double hydroxides. *Appl. Catal. B* **2009**, *90*, 330–338. [CrossRef]

37. Tan, Y.; Liu, X.; Zhang, L.; Liu, F.; Wang, A.; Zhang, T. Producing of cinnamyl alcohol from cinnamaldehyde over supported gold nanocatalyst. *Chin. J. Catal.* **2021**, *42*, 470–481. [CrossRef]

38. Zuo, C.; Tian, Y.; Zheng, Y.; Wang, L.; Fu, Z.; Jiao, T.; Wang, M.; Huang, H.; Li, Y. One step oxidative esterification of methacrolein with methanol over AuCeO$_2$/γ-Al$_2$O$_3$ catalysts. *Catal. Commun.* **2019**, *124*, 51–55. [CrossRef]

39. Bhuvaneswari, K.; Palanisamy, G.; Pazhanivel, T.; Bharathi, G.; Nataraj, D. Photocatalytic Performance on Visible Light Induced ZnS QDs-MgAl Layered Double Hydroxides Hybrids for Methylene Blue Dye Degradation. *ChemistrySelect* **2018**, *3*, 13419–13426. [CrossRef]

40. Das, S.; Patnaik, S.; Parida, K. Dynamic charge transfer through Fermi level equilibration in the p-CuFe$_2$O$_4$/n-NiAl LDH interface towards photocatalytic application. *Catal. Sci. Technol.* **2020**, *10*, 6285–6298. [CrossRef]
41. Gao, J.; Fan, G.; Yang, L.; Cao, X.; Zhang, P.; Li, F. Oxidative Esterification of Methacrolein to Methyl Methacrylate over Gold Nanoparticles on Hydroxyapatite. *ChemCatChem* **2017**, *9*, 1230–1241. [CrossRef]
42. Tian, Y.; Li, Y.; Zheng, Y.; Wang, M.; Zuo, C.; Huang, H.; Yin, D.; Fu, Z.; Tan, J.; Zhou, Z. Nano-Au/MCeO$_x$ Catalysts for the Direct Oxidative Esterification of Methylacrolein to Methyl Esters. *Ind. Eng. Chem. Res.* **2019**, *58*, 19397–19405. [CrossRef]
43. Ren, J.; Mebrahtu, C.; Palkovits, R. Ni-based catalysts supported on Mg-Al hydrotalcites with different morphologies for CO$_2$ methanation: Exploring the effect of metal-support interaction. *Catal. Sci. Technol.* **2020**, *10*, 1902–1913. [CrossRef]
44. Dębek, R.; Radlik, M.; Motak, M.; Galvez, M.E.; Turek, W.; Costa, P.D.; Grzybek, T. Ni-containing Ce-promoted hydrotalcite derived materials as catalysts for methane reforming with carbon dioxide at low temperature-On the effect of basicity. *Catal. Today* **2015**, *257*, 59–65. [CrossRef]
45. Fang, J.; Zhang, B.; Yao, Q.; Yang, Y.; Xie, J.; Yan, N. Recent advances in the synthesis and catalytic applications of ligand-protected, atomically precise metal nanoclusters. *Coord. Chem. Rev.* **2016**, *322*, 1–29. [CrossRef]
46. Taketoshi, A.; Haruta, M. Size- and Structure-specificity in Catalysis by Gold Clusters. *Chem. Lett.* **2014**, *43*, 380–387. [CrossRef]
47. Wan, X.; Deng, W.; Zhang, Q.; Wang, Y. Magnesia-supported gold nanoparticles as efficient catalysts for oxidative esterification of aldehydes or alcohols with methanol to methyl esters. *Catal. Today* **2014**, *233*, 147–154. [CrossRef]
48. Trimpalis, A.; Giannakakis, G.; Cao, S.; Flytzani-Stephanopoulos, M. NiAu single atom alloys for the selective oxidation of methacrolein with methanol to methyl methacrylate. *Catal. Today* **2020**, *355*, 804–814. [CrossRef]
49. Li, Y.; Zheng, Y.; Wang, L.; Fu, Z. Oxidative Esterification of Methacrolein to Methyl Methacrylate over Supported Gold Catalysts Prepared by Colloid Deposition. *ChemCatChem* **2017**, *9*, 1960–1968. [CrossRef]
50. Yoon, B.; Häkkinen, H.; Landman, U. Interaction of O$_2$ with Gold Clusters: Molecular and Dissociative Adsorption. *J. Phys. Chem. A* **2003**, *107*, 4066–4071. [CrossRef]
51. Turner, M.; Golovko, V.B.; Vaughan, O.; Abdulkin, P.; Berenguer-Murcia, A.; Tikhov, M.S.; Johnson, B.F.G.; Lambert, R.M. Selective oxidation with dioxygen by gold nanoparticle catalysts derived from 55-atom clusters. *Nature* **2008**, *454*, 981–983. [CrossRef] [PubMed]
52. Xu, B.; Liu, X.; Haubrich, J.; Friend, C.M. Vapour-phase gold-surface-mediated coupling of aldehydes with methanol. *Nat. Chem.* **2010**, *2*, 61–65. [CrossRef] [PubMed]

International Journal of
Molecular Sciences

Article

Preparation and Size Control of Efficient and Safe Nanopesticides by Anodic Aluminum Oxide Templates-Assisted Method

Chunxin Wang, Bo Cui, Yan Wang, Mengjie Wang, Zhanghua Zeng, Fei Gao, Changjiao Sun, Liang Guo, Xiang Zhao * and Haixin Cui *

Institute of Environment and Sustainable Development in Agriculture, Chinese Academy of Agricultural Sciences, Beijing 100081, China; wangchunxin@caas.cn (C.W.); cuibo@caas.cn (B.C.); wangyan03@caas.cn (Y.W.); wangmengjie@caas.cn (M.W.); zengzhanghua@caas.cn (Z.Z.); gaofei@caas.cn (F.G.); sunchangjiao@caas.cn (C.S.); guoliang01@caas.cn (L.G.)
* Correspondence: zhaoxiang@caas.cn (X.Z.); cuihaixin@caas.cn (H.C.)

Abstract: Efficient and safe nanopesticides play an important role in pest control due to enhancing target efficiency and reducing undesirable side effects, which has become a hot spot in pesticide formulation research. However, the preparation methods of nanopesticides are facing critical challenges including low productivity, uneven particle size and batch differences. Here, we successfully developed a novel, versatile and tunable strategy for preparing buprofezin nanoparticles with tunable size via anodic aluminum oxide (AAO) template-assisted method, which exhibited better reproducibility and homogeneity comparing with the traditional method. The storage stability of nanoparticles at different temperatures was evaluated, and the release properties were also determined to evaluate the performance of nanoparticles. Moreover, the present method is further demonstrated to be easily applicable for insoluble drugs and be extended for the study of the physicochemical properties of drug particles with different sizes.

Keywords: buprofezin; nanoparticles; AAO template; tunable particle size

check for
updates

Citation: Wang, C.; Cui, B.; Wang, Y.; Wang, M.; Zeng, Z.; Gao, F.; Sun, C.; Guo, L.; Zhao, X.; Cui, H. Preparation and Size Control of Efficient and Safe Nanopesticides by Anodic Aluminum Oxide Templates-Assisted Method. *Int. J. Mol. Sci.* **2021**, *22*, 8348. https://doi.org/10.3390/ijms22158348

Academic Editor: Raghvendra Singh Yadav

Received: 2 July 2021
Accepted: 16 July 2021
Published: 3 August 2021

Publisher's Note: MDPI stays neutral with regard to jurisdictional claims in published maps and institutional affiliations.

1. Introduction

Pesticides are widely used in protecting crops and improving crop yield to meet agricultural production [1,2]. However, the low effective utilization rate of pesticides is a serious problem. Therefore, improving the utilization rate of pesticides has become the focus of pesticide formulation research [3]. According to Noyes–Whitney equation, the dissolution rate of drugs is directly proportional to the surface area of drug particles, and therefore the most effective method is to reduce the particle size of drug particles [4]. Nanopesticide formulations have been developed to maximize pesticide utilization while minimizing the side effects [5–8].

Currently, the research of the nanopesticide falls into two categories: one is to use nanotechnology including emulsification and homogenization to prepare nanoparticles and the other is to use nanomaterials to construct nanoparticles [9–11]. The former is mainly affected by external factors including equipment and process parameters, it faces critical challenges including relatively large particle size difference, batch differences and low production efficiency [12–15]. Functionalized abamectin poly (lactic acid) nanoparticles were prepared by the ultrasonic emulsification to explore the adhesion of nanoparticles to the leaf surface of crops. Antagonistic effect of azoxystrobin poly (lactic acid) microspheres with controllable particle size was studied. These studies showed a large influence of the preparation conditions on the drug particles [16,17]. The latter can accurately control the particle size and improve the production efficiency [18–22]. Pyrimethanil-loaded mesoporous silica nanoparticles was synthesis to explore the distribution and dissipation in cucumber plants. Abamectin using porous silica nanoparticles as carriers were constructed to evaluate controlled-release performance. The nanoparticles prepared by nanomaterials can ensure the uniformity of particle size [23,24].

As known to us, anodized aluminum oxide (AAO) templates have been widely used for forming nanotubes and nanowires. The tiny holes on the surface of AAO are arranged in an orderly manner. In addition, the pore channels of the AAO template were parallel to each other, and the pore size uniformity of the template was good. Based on the above principle, the AAO template has been applied for forming nanoparticles by making full use of the characteristics of AAO template. However, AAO templates have never been used for preparing pesticide nanoparticles. In this work, we demonstrated for the first time that the AAO template method can in fact be extended for preparing uniform pesticide nanoparticles comparing to the conventional reprecipitation approach [25–28].

Buprofezin as an insect growth regulator has the characteristics of high activity, high selectivity and long residual period, which can inhibit chitin synthesis and interferes with metabolism in insect pests [29]. However, insecticidal application of poorly soluble drugs, such as buprofezin, in farmland systems is limited. In addition, it will directly pollute the environment after large-scale application of buprofezin [30]. Therefore, developing nanopesticide formulations and clarifying the effects of particle size on physicochemical properties and biological activity of pesticides is the key to solve the problem of poor dispersibility of insoluble pesticides [31–35].

In this study, the buprofezin nanoparticles (BNPs) with different sizes were developed via AAO template-assisted process. The particles with similar surface characteristics and extremely narrow size distribution were obtained using AAO templates with different pore sizes, which could also exclude the impact of shape and surface charge compared with the traditional method. The effect of drug concentration and number cycles on the size of particle was investigated. At the same time, the stability and release properties of nanoparticles were evaluated to characterize the effects of particle size and dispersion on their physicochemical properties. In conclusion, this research not only established an innovative and simple method for the potential application of nanopesticides with controllable particle size, excellent monodispersity and uniform morphology in plant protection, but also laid a foundation for the study of the physicochemical properties of drug particles with different sizes.

2. Results and Discussion

2.1. Characterization of the BNPs

In the BNPs fabrication process, BNPs were prepared using 100 nm AAO template in Figure 1. The mean particle size of BNPs based on a scanning electron microscope (SEM) image (Figure 1b) and a transmission electron microscope (TEM) image (Figure 1c) were 100 nm and 103 nm, respectively, which was smaller than the hydrated particle size (Figure 1a). SEM and TEM shows the real particle size in a dried state, whereas dynamic light scattering (DLS) provides the micelle core and swollen corona [36,37].

These results indicated that BNPs exhibited almost spherical morphology and comparatively monodisperse distribution. As known to us, the size of each aperture of the template is consistent, and the template is arranged in order, so the prepared drug particles are uniform in size. During the whole preparation process, drug molecules were gradually deposited in the pore of AAO template, and the particle size was limited by the pore size finally [25,27]. Therefore, the template with different pore size can be used to prepare nanoparticles with different sizes, which laid a foundation for the study of the physicochemical properties of drug particles with different sizes.

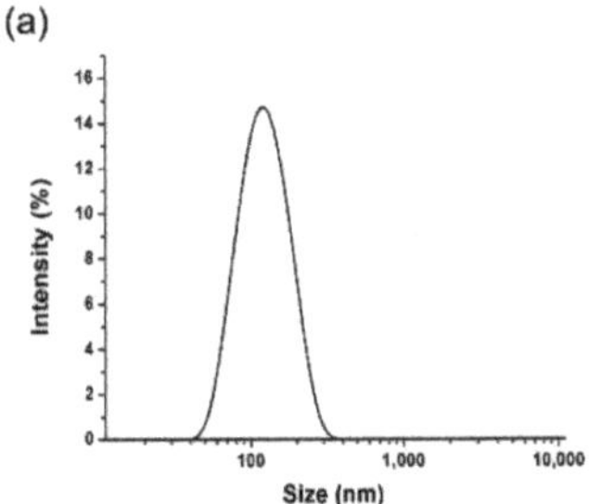 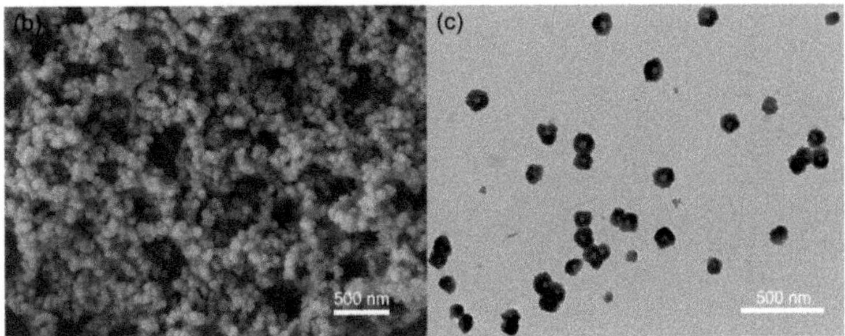

Figure 1. Characterization of particle size and morphology of the buprofezin nanoparticles. (**a**) DLS image of BNPs, (**b**) SEM image of BNPs, (**c**) TEM image of BNPs.

The SEM images of AAO templates after immersing in methanol, tetrahydrofuran and acetone were shown in supporting Information Figure S1. The pore size of the template immersed in tetrahydrofuran and acetone increased to 132 nm, which changed the structure of the template and affected the preparation of drug particles with different sizes. The main reason is that the structure of the template was destroyed in tetrahydrofuran and acetone solution, but this did not happen in methanol solution. The pore size of the template immersed in methanol has not changed. Therefore, methanol was selected as the solvent of the drug.

To confirm the difference between the properties of BNPs prepared with the present method and free drug molecules. We also developed buprofezin particles (BPs) using reprecipitation. The shape of drug particles was irregular and the size distribution of the particles was not uniform in Figure S2.

2.2. Optimization of Preparation Parameters

The optimization of the preparation process is very important for the size and distribution of the drug particles. As shown in Figure 2a, the AAO templates with pore size of 100 nm were immersed in the lower concentration of the drug solution, which can form smaller nanoparticles. The particle size of the BNPs with 20 mg/mL was larger than that of the BNPs with 5 and 10 mg/mL. The active components entering into the pore size of the template in unit time increased with the increase of drug concentration, thus enlarging the particle size. The particle size of the particles varied little at the concentration of 20 and 40 mg/mL. In a certain concentration range, the size of drug particles became larger with the increase of the concentration of solution, but it was always controlled in the range of template pore size. It was also observed that the particle size increased with the number of cycles (2, 5, 10 cycles). The particle size changed little at 5 and 10 cycles. The drug particles will grow along with the increase of the number of cycles, but the size of the particles is always affected by the pore size of the template. Overall, the amount of drug nanoparticles is less in the case of low concentration and low cycle times, which is not suitable for mass production. In conclusion, the optimal preparation conditions for BNPs were as follows: drug concentration (20 mg/mL), cycle number (5 cycle) and actuation duration (5 min).

2.3. Release Behavior

BNPs with 100 nm size were selected to confirm the better physicochemical properties comparing to the free drug molecules. The drug release profile is of importance in applying the proposed system for practical drug delivery. Figure 3 indicated that the prepared BNPs had characteristics of rapid release compared with free buprofezin particles prepared by the reprecipitation method, which was due to the dissolution rate. The cumulative release rate of the BPs was only 40.2% after 216 h because of the poor water-solubility. In contrast, the cumulative release rate of the BNPs reached 92.5% after 216 h due to the decrease of the particle size and the increase of specific surface area.

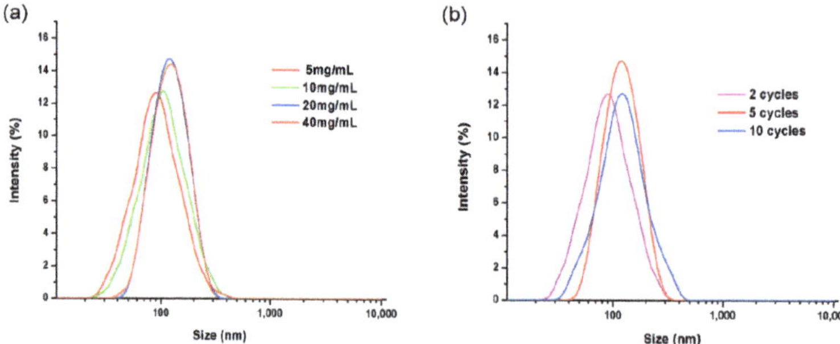

Figure 2. BNPs fabricated with AAO templates with 100 nm pore sizes in different drug concentration (5, 10, 20, 40 mg/mL) with different cycles (2, 5, 10 cycles). (**a**) DLS image of BNPs in different drug concentration (5, 10, 20, 40 mg/mL), (**b**) DLS image of BNPs with different cycles (2, 5, 10 cycles).

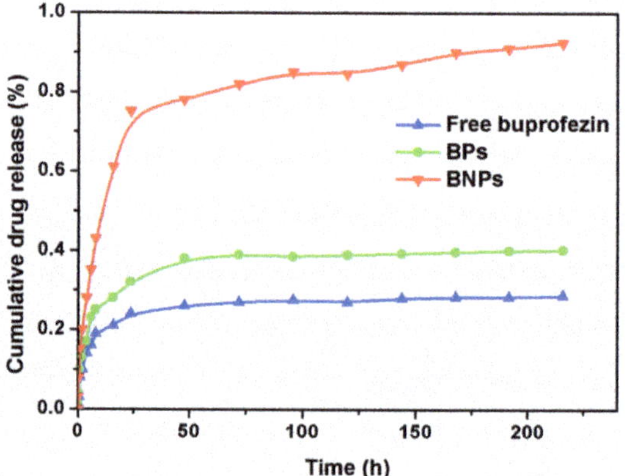

Figure 3. Cumulative drug release from free buprofezin, BPs and BNPs in PBS medium.

As far as we know, drug release performance is related to drug solubility. The particle size of the drug prepared by precipitation method was large and uneven, which did not improve the water solubility of the drug, so the release of the drug was relatively slow [38–40]. After the nanoparticles were prepared using the template method, the particle size of the drug became smaller, resulting in the increase of specific surface area and enhancement of the solubility, so the drug release was faster and more durable [41,42]. These results indicated that BNPs showed good release properties, raising the dissolution rate of drugs and increasing the utilization efficiency of pesticides.

2.4. Stability

The mean particle size and PDI were measured to assess the storage stability of the drug nanoparticles. In Figure 4, the mean particle sizes of BNPs increased to 110 nm and the PDI maintained below 0.3 after storage at 0 °C for 7 days, which means that particles of similar size are more concentrated. The mean particle sizes of BNPs increased to 125 nm and the PDI maintained about 0.3 after storage at 54 °C for 14 days. The particle size of DLS is consistent with that of SEM. As shown in Figure 5, the nanoparticles at different storage temperature exhibited almost spherical morphology and relatively monodisperse distribution.

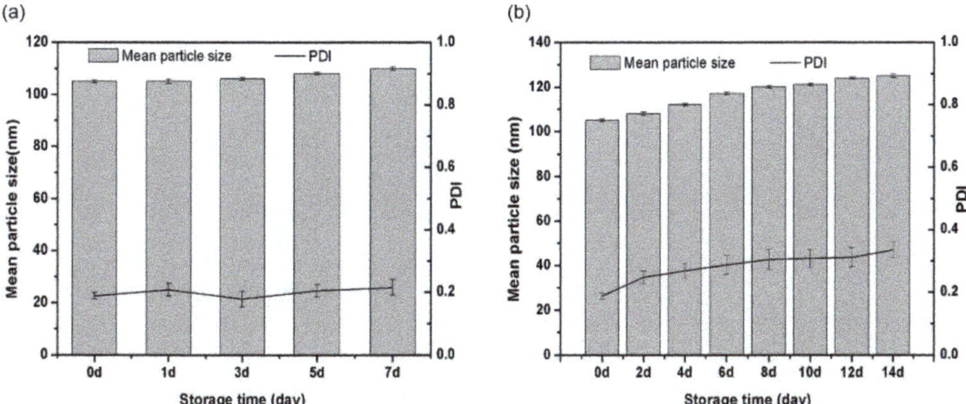

Figure 4. The mean particle size and PDI of the BNPs at different storage temperature. (**a**) BNPs at 0 °C for 7 days, (**b**) BNPs at 54 °C for 14 days.

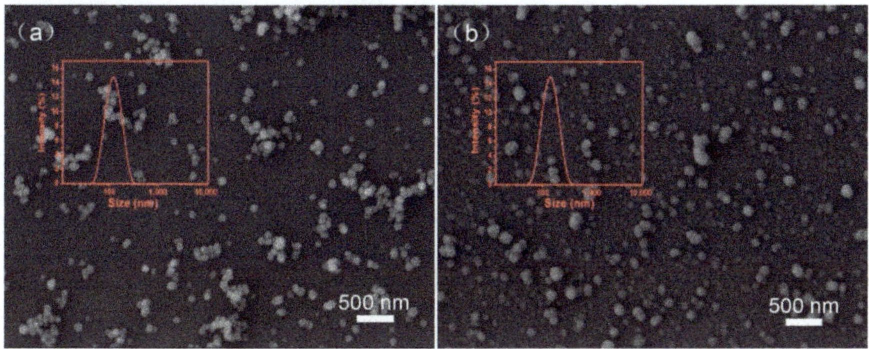

Figure 5. The DLS and SEM image of the BNPs at different storage temperature. (**a**) BNPs at 0 °C for 7 days, (**b**) BNPs at 54 °C for 14 days.

As mentioned in the literature, the stability of drug particles is related to the uniformity of drug particle size. Small particles tend to grow to large particles, which lead to agglomeration of particles. The higher the disorder degree of particles, the more obvious the aggregation degree of particles. The BNPs has the better uniformity in size, which reduces the aggregation between particles to maintain the stability of particles [43–46]. In addition, there is no external energy in the whole preparation process, and the surface energy of particles is low, which reduces the aggregation between particles [47]. These results indicated that BNPs showed good storage stability at 0 °C for 7 days and at 54 °C for 14 days.

2.5. Promotion and Application

To demonstrate the universality of the current method for pesticides. BNPs with different sizes were prepared using AAO templates with pore size of 20 nm and 200 nm. As shown in Figure 6a, the statistical mean particle size and PDI of nanoparticles using 20 nm AAO template were 22 nm and 0.182, respectively. The mean particle size and PDI of nanoparticles using 200 nm AAO template were 196 nm and 0.201, respectively. The SEM image showed that the particles have spherical shape, uniform size and uniform dispersion.

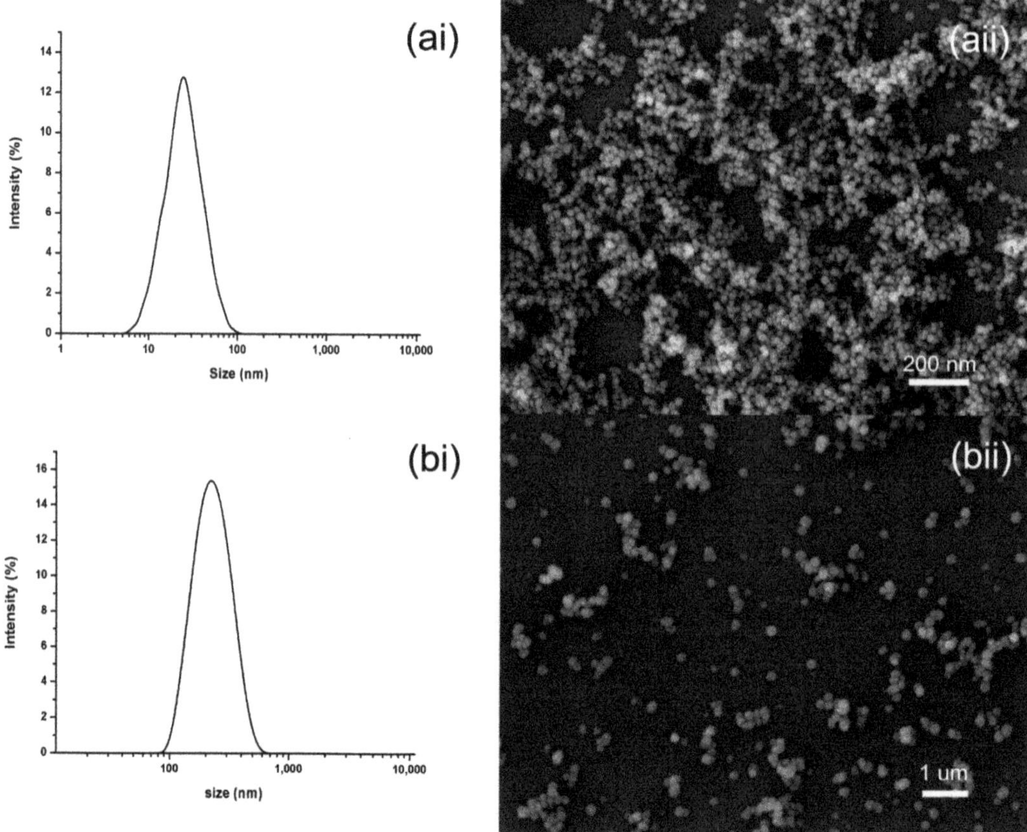

Figure 6. The DLS and SEM characterization of BNPs using AAO templates with pore size of 20 nm and 200 nm. (**ai**) DLS image of BNPs using AAO templates with pore size of 20 nm, (**aii**) SEM image of BNPs using AAO templates with pore size of 20 nm, (**bi**) DLS image of BNPs using AAO templates with pore size of 200 nm, (**bii**) SEM image of BNPs using AAO templates with pore size of 200 nm.

These results indicated that the particle size of BNPs was limited by the pore size of the template and would not exceed the pore size. The channels in the template are not connected with each other, which prevents the adhesion and growth of particles, so the dispersion and uniformity of particles are very good. In conclusion, the template pore dimension would restrict continuous growth of the nanoparticles, and then determines the size of nanoparticles, which used to explore the properties of drug particles with different sizes.

As depicted in Figure 7, the nanodrugs can be released not only in 0.1 mol/L NaOH solution but also in 1 mol/L diluted hydrochloric acid solution by dissolution of AAO.

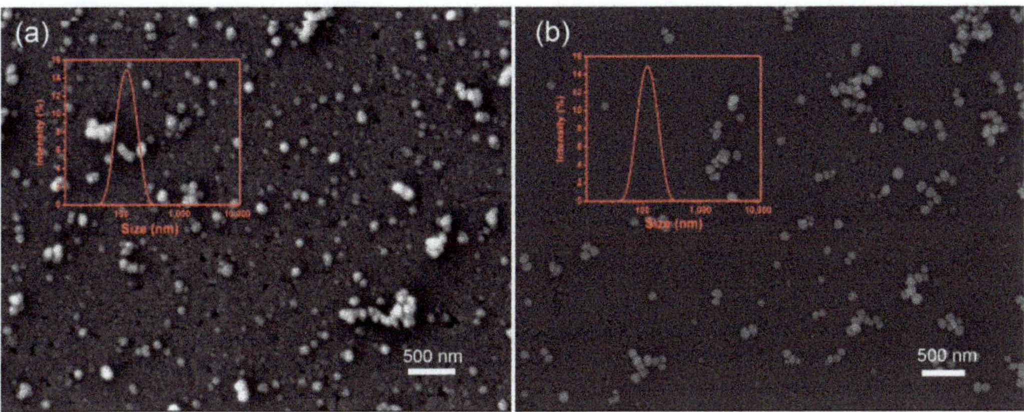

Figure 7. The DLS and SEM characterization of BNPs released from dilute acid and ultrasonic system. (**a**) BNPs released from dilute acid, (**b**) BNPs released from ultrasonic systems.

The drugs can be collected until the template was completely dissolved in the above alkaline solution for 2 h. In addition, the template can be completely dissolved in 10 h under the above acidic solution. Moreover, it was observed that nanoparticles can be released from AAO template by ultrasonication without dissolving the AAO template. The DLS results showed the particle size was 105 nm in acid solution, and maintained 104 nm in ultrasonic condition.

As known to us, the dissolution of the template did not affect the performance of the drug particles, nor did the ultrasonication. The SEM images showed that these nanoparticles released from the AAO template presented almost spherical and exhibit good dispersion. The AAO templates were completely dissolved in the above NaOH solution and hydrochloric acid solution in supporting Information Figure S3, which avoided the interference of the template to the DLS and SEM results. These results demonstrated that the present method was versatile for hydrophobic drugs and can be easily applied for preparing nanoparticles even if the drugs are sensitive to diluted acid or base.

To demonstrate the versatility of the AAO template-assisted method, we further fabricated nanoparticles of another hydrophobic drugs (avermectin and pyraclostrobin) using the present methods. Avermectin was the representative of insecticide and pyrazoxystrobin was the representative of fungicide. Considering the sensitivity of drugs to acid and base, the ultrasonic method was used to obtain nanoparticles in Figure 8. The DLS results showed the particle size of the avermectin was 108 nm, and the particle size of pyraclostrobin 102 nm.

As known to us, the size of nanoparticles is not affected by different hydrophobic pesticides, which is related to the pore size of the template. The SEM images showed that the nanoparticles were spherical and had good dispersion. In conclusion, the nanoparticles containing multiple drugs can be prepared, which demonstrated that the strategy was versatile and convenient for hydrophobic drugs.

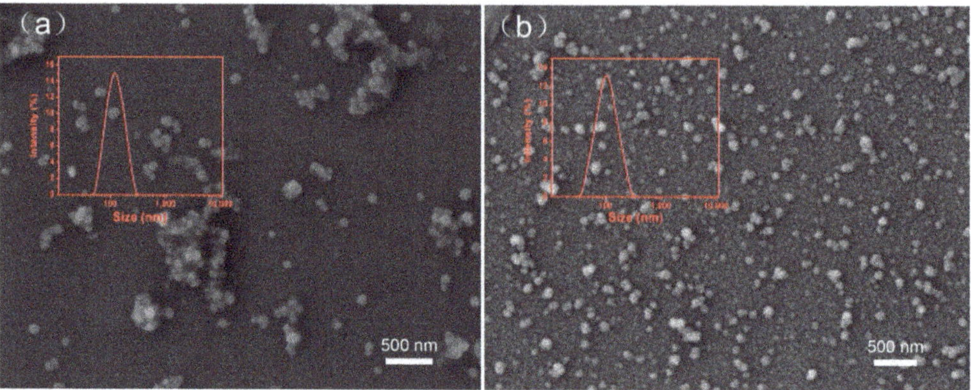

Figure 8. The DLS and SEM characterization of different drug nanoparticles. (**a**) Avermectin, (**b**) pyraclostrobin.

3. Materials and Methods

3.1. Materials

Buprofezin (97%) was purchased from Hubei Jiufenglong Chemical Co., Ltd (Beiing, China). Anodized aluminum oxide (AAO) membranes were purchased from Beijing Zhongjingkeyi Technology Co., Ltd (Beiing, China). Methanol, Tetrahydrofuran, acetone, hydrochloric acid (HCl) and sodium hydroxide (NaOH) were purchased from Sinopharm Chemical Reagent Co., Ltd (Beiing, China). Milli-Q water (15.0 MΩ cm^{-1}, total organic carbon $\leq$ 4 ppb) was used in all analytical experiments.

3.2. Methods

3.2.1. Selection of Organic Solvent

The AAO templates with 100 nm pore size were immersed in methanol, tetrahydrofuran and acetone for several minutes, respectively. The pore size of the template immersed in tetrahydrofuran and acetone increased relative to the theoretical value, which was not suitable for the preparation of nanoparticles. The pore size of the template in methanol did not change, so methanol as a stable solvent was selected for the preparation of buprofezin nanoparticles.

3.2.2. Preparation of Buprofezin Nanoparticles (BNPs) by AAO Method

The BNPs were prepared using AAO template with 100 nm pore size by the following steps. The buprfezin was dissolved in methanol. The AAO templates were first washed sequentially with methanol followed by immersion in buprfezin solutions for 5 min. Then, the soaked templates were still dried at room temperature to remove the organic solvent. During the whole soaking process, the drug was dispersed in the pores of the membrane. The whole process was repeated for several times. The dried templates were dissolved in NaOH solution to dissolve template materials. The discrete nanoparticles were collected by multiple centrifugations and redispersion. The extracted pure nanodrug was finally freeze-dried.

3.2.3. Analysis of BNPs Collection Methods

The BNPs with 100 nm size was collected using 0.1 mol/L NaOH solution to dissolve template materials by multiple centrifugations and redispersion. In order to demonstrate the universality of the current method for drugs. We also explored whether drugs can be released by dissolution of AAO using diluted hydrochloric acid solution (1 mol/L). In addition, we also demonstrated whether the nanoparticles can also be obtained by ultrasonication of the drug-loaded AAO template in water without dissolving the AAO template.

3.2.4. Optimization of Preparation Process of BNPs

Optimization of the preparation process is very important for the size and distribution of drug particles. In this study, particle size and polydispersity index (PDI) were used as the evaluation index to optimize the drug concentration and reaction times. The effect of drug concentration on particle size was investigated in the concentration gradient range of 5, 10, 20 and 40 mg/mL. The effect of soaking and drying cycles on the particle size in drug solutions with a concentration of 20 mg/mL was analyzed by setting 2, 5 and 10 cycles.

3.2.5. Characterization of BNPs

The particle sizes and morphologies of the samples were characterized using SEM (JSM-7401F, JEOL Ltd., Tokyo, Japan). Of the sample solution, 3.5 uL was dripped onto the silicon slice, then dried and sprayed with gold for 40 s. The SEM images were captured at 3 kV voltage and 10 mA current. The morphology of the nanosuspension was characterized by TEM (HT7700, Hitachi Ltd., Tokyo, Japan) with 80 kV accelerating voltage. Two microliters of diluted solution were dripped onto a carbon-coated copper grid and were dried at room temperature for TEM measurement. The hydrated mean particle sizes of the BNPs were examined using dynamic light scattering (DLS, Zetasizer Nano ZS90, Malvern Instruments Ltd., Malvern, UK). The polydispersity index (PDI) was used to characterize particle size distribution. The PDI value less than 0.3 indicated good dispersion. The measurement was carried out in triplicate for each sample.

3.2.6. In Vitro Drug-Release

The drug-release behaviors of BNPs were investigated by ultraviolet visible spectrophotometer (UV, TU1901, Shimadzu, Tokyo, Japan). Five microliters BNPs and free buprofezin were suspended in PBS solution (Ph 7.4, 5 mL) and the solution was transferred to dialysis bags (2000 MWCO), respectively. Finally, the treated dialysis bags were placed in a brown bottle with PBS solution (95 mL) for 216 h. Two milliliters of solution was taken to measure the absorbance by UV spectrophotometer at 246 nm at a specific time. During the dialysis, the solution volume was maintained constant by supply 2 mL of buffer after each sampling. The cumulative release of buprofezin in the solution at different time was calculated according to the standard curve. The assay was performed three times for each sample.

3.2.7. Stability

The BNPs with 100 nm size was selected as an example for stability test. The freeze-dried solid samples were stored at different temperatures (0 °C for 7 days and at 54 °C for 14 days) to explore physicochemical stability. The sample was taken out and diluted 0.5% (w/w) to determine the hydrated particle size and morphology at a specific time. The stability was evaluated by observing the changes of particle size and PDI during the whole storage process.

3.2.8. Statistical Analysis

The statistical data was presented as mean ± standard deviation (SD). Least significant difference (LSD) was used to analyze data. A probability (p) of less than 0.05 means significant differences.

4. Conclusions

In summary, we successfully developed a novel, versatile and controllable approach for solving the problem of dissolution and dispersion of hydrophobic pesticides. The AAO template-assisted method produced nanoparticles of different sizes with excellent monodispersion and uniform morphology, which had better reproducibility and homogeneity comparing with the reprecipitation method. In addition, the pore size of the template restricts the growth of the nanoparticles and then controls the size of the drug particles. The BNPs with the mean size of 100 nm using AAO template-assisted method exhibited better release performance due to the larger specific surface area and faster dissolution rate. The good dispersion and homogeneity of nanoparticles play an important role in maintaining the stability of particles.

In addition, the pore size of the template restricts the continuous growth of the nanoparticles and then controls the size of the drug particles. Furthermore, the nanoparticles could release from the AAO template by dissolving template or ultrasonication even if the drugs are sensitive to diluted acid or base. Moreover, the AAO template method can be applied to other hydrophobic pesticides and extended for fabricating nanoparticles with different functional agents, and used to prepare the nanoparticles containing multiple drugs. In conclusion, the novel AAO template method was not only a versatile and convenient method for the preparing hydrophobic nanodrugs, but also laid a foundation for exploring properties of drug particles with different sizes.

Supplementary Materials: The supplementary material are available online at https://www.mdpi.com/article/10.3390/ijms22158348/s1.

Author Contributions: Conceptualization and methodology, H.C., X.Z. and C.W.; data analysis, C.W., B.C. and Z.Z.; writing—original draft, C.W.; writing—review and editing, M.W., F.G., C.S. and L.G.; funding acquisition, C.W., B.C., Y.W. and X.Z. All authors have read and agreed to the published version of the manuscript.

Funding: This work was supported by the Agricultural Science and Technology Innovation Program (CAAS-ZDRW202008), the National Key Research and Development Program of China (2016YFD0200500), the National Natural Science Foundation of China (31701825, 31901912, 5193248), the Basic Scientific Research Foundation of National Non-Profit Scientific Institute of China (BSRF201907, BSRF202006).

Institutional Review Board Statement: Not applicable.

Informed Consent Statement: Not applicable.

Data Availability Statement: Not applicable.

Conflicts of Interest: The authors declare that they have no competing interest.

References

1. Cooper, J.; Dobson, H. The benefits of pesticides to mankind and the environment. *Crop. Prot.* **2007**, *26*, 1337–1348. [CrossRef]
2. Campos, E.V.R.; De Oliveira, J.L.; Fraceto, L.F. Applications of Controlled Release Systems for Fungicides, Herbicides, Acaricides, Nutrients, and Plant Growth Hormones: A Review. *Adv. Sci. Eng. Med.* **2014**, *6*, 373–387. [CrossRef]
3. Rossall, S. Fungicide Resistance in Crop Protection: Risk and Management. *Plant Pathol.* **2012**, *61*, 820. [CrossRef]
4. Murdande, S.B.; Shah, D.A.; Dave, R.H. Impact of Nanosizing on Solubility and Dissolution Rate of Poorly Soluble Pharmaceuticals. *J. Pharm. Sci.* **2015**, *104*, 2094–2102. [CrossRef]
5. Usman, M.; Farooq, M.; Wakeel, A.; Nawaz, A.; Cheema, S.A.; Rehman, H.U.; Ashraf, I.; Sanaullah, M. Nanotechnology in agriculture: Current status, challenges and future opportunities. *Sci. Total Environ.* **2020**, *721*, 137778. [CrossRef]
6. Raj, S.N.; Anooj, E.; Rajendran, K.; Vallinayagam, S. A comprehensive review on regulatory invention of nano pesticides in Agricultural nano formulation and food system. *J. Mol. Struct.* **2021**, *1239*, 130517. [CrossRef]
7. Benelli, G.; Pavela, R.; Maggi, F.; Petrelli, R.; Nicoletti, M. Commentary: Making Green Pesticides Greener? The Potential of Plant Products for Nanosynthesis and Pest Control. *J. Clust. Sci.* **2017**, *28*, 3–10. [CrossRef]
8. Rai, M.; Ingle, A. Role of nanotechnology in agriculture with special reference to management of insect pests. *Appl. Microbiol. Biotechnol.* **2012**, *94*, 287–293. [CrossRef]
9. Zhao, X.; Cui, H.X.; Wang, Y.; Cui, B.; Zeng, Z.H. Development strategies and prospects of nano-based smart pesticide formulation. *J. Agric. Food. Chem.* **2017**, *66*, 6504–6512. [CrossRef]
10. Mishra, P.; Tyagi, B.K.; Chandrasekaran, N.; Mukherjee, A. Biological nanopesticides: A greener approach towards the mos-quito vector control. *Environ. Sci. Pollut. Res. Int.* **2017**, *25*, 1–13.
11. Shang, Y.; Hasan, K.; Ahammed, G.J.; Li, M.; Yin, H.; Zhou, J. Applications of Nanotechnology in Plant Growth and Crop Protection: A Review. *Molecules* **2019**, *24*, 2558. [CrossRef] [PubMed]
12. Wang, C.X.; Guo, L.; Yao, J.W.; Wang, A.Q.; Gao, F.; Zhao, X.; Zeng, Z.H.; Wang, Y.; Sun, C.J.; Cui, H.X. Preparation, characterization and antifungal activity of pyraclostrobin solid nanodispersion by self-mulsifying technique. *Pest. Manag. Sci.* **2019**, *75*, 2785–2793. [CrossRef] [PubMed]
13. Cui, B.; Lv, Y.; Gao, F.; Wang, C.; Zeng, Z.; Wang, Y.; Sun, C.; Zhao, X.; Shen, Y.; Liu, G.; et al. Improving abamectin bioavailability via nanosuspension constructed by wet milling technique. *Pest Manag. Sci.* **2019**, *75*, 2756–2764. [CrossRef] [PubMed]

14. Dong, J.; Liu, X.; Chen, Y.; Yang, W.; Du, X. User-safe and efficient chitosan-gated porous carbon nanopesticides and nanoherbicides. *J. Colloid Interface Sci.* **2021**, *594*, 20–34. [CrossRef]
15. Wang, C.; Cui, B.; Guo, L.; Wang, A.; Zhao, X.; Wang, Y.; Sun, C.; Zeng, Z.; Zhi, H.; Chen, H.; et al. Fabrication and Evaluation of Lambda-Cyhalothrin Nanosuspension by One-Step Melt Emulsification Technique. *Nanomaterials* **2019**, *9*, 145. [CrossRef]
16. Yu, M.; Sun, C.; Xue, Y.; Liu, C.; Qiu, D.; Cui, B.; Zhang, Y.; Cui, H.; Zeng, Z. Tannic acid-based nanopesticides coating with highly improved foliage adhesion to enhance foliar retention. *RSC Adv.* **2019**, *9*, 27096–27104. [CrossRef]
17. Yao, J.; Cui, B.; Zhao, X.; Zhi, H.; Zeng, Z.; Wang, Y.; Sun, C.; Liu, G.; Gao, J.; Cui, H. Antagonistic Effect of Azoxystrobin Poly (Lactic Acid) Microspheres with Controllable Particle Size on Colletotrichum higginsianum Sacc. *Nanomaterials* **2018**, *8*, 857. [CrossRef] [PubMed]
18. Cao, L.; Zhou, Z.; Niu, S.; Cao, C.; Li, X.; Shan, Y.; Huang, Q. Positive-Charge Functionalized Mesoporous Silica Nanoparticles as Nanocarriers for Controlled 2,4-Dichlorophenoxy Acetic Acid Sodium Salt Release. *J. Agric. Food Chem.* **2018**, *66*, 6594–6603. [CrossRef]
19. Baba, K.; Pudavar, H.E.; Roy, I.; Ohulchanskyy, T.Y.; Chen, Y.; Pandey, R.K.; Prasad, P.N. New method for delivering a hydrophobic drug for photodynamic therapy using pure nanocrystal form of the drug. *Mol. Pharm.* **2006**, *4*, 289–297. [CrossRef]
20. Priyanka, P.; Kumar, D.; Yadav, K.; Yadav, A. Nanopesticides: Synthesis, Formulation and Application in Agriculture. *Nanobiotechnol. Appl. Plant. Prot.* **2019**, *7*, 129–143.
21. Liu, F.; Wen, L.-X.; Li, Z.-Z.; Yu, W.; Sun, H.-Y.; Chen, J.-F. Porous hollow silica nanoparticles as controlled delivery system for water-soluble pesticide. *Mater. Res. Bull.* **2006**, *41*, 2268–2275. [CrossRef]
22. Zhao, P.; Cao, L.; Ma, D.; Zhou, Z.; Huang, Q.; Pan, C. Translocation, distribution and degradation of prochloraz-loaded mesoporous silica nanoparticles in cucumber plants. *Nanoscale* **2018**, *10*, 1798–1806. [CrossRef] [PubMed]
23. Zhao, P.; Cao, L.; Ma, D.; Zhou, Z.; Huang, Q.; Pan, C. Synthesis of Pyrimethanil-Loaded Mesoporous Silica Nanoparticles and Its Distribution and Dissipation in Cucumber Plants. *Molecules* **2017**, *22*, 817. [CrossRef]
24. Wang, Y.; Cui, H.; Sun, C.; Zhao, X.; Cui, B. Construction and evaluation of controlled-release delivery system of Abamectin using porous silica nanoparticles as carriers. *Nanoscale Res. Lett.* **2014**, *9*, 655. [CrossRef]
25. Tripathy, J.; Wiley, J.B. Fabrication of thick porous anodized aluminum oxide templates. *J. Solid State Electrochem.* **2015**, *19*, 1447–1452. [CrossRef]
26. Jo, H.; Haberkorn, N.; Pan, J.-A.; Vakili, M.; Nielsch, K.; Theato, P. Fabrication of Chemically Tunable, Hierarchically Branched Polymeric Nanostructures by Multi-branched Anodic Aluminum Oxide Templates. *Langmuir* **2016**, *32*, 6437–6444. [CrossRef]
27. Zeng, Z.; Zhou, Q.; Yang, Z.; Miao, Q.; Gao, X.; Zhou, G.; Zhang, Z. Size-Controlled Growth of High-Density Ordered Nanomagnet Arrays by Template-Assisted Method. *J. Nanosci. Nanotechn.* **2016**, *16*, 12231–12236. [CrossRef]
28. Zhang, J.; Li, Y.; An, F.-F.; Zhang, X.; Chen, X.; Lee, C.-S. Preparation and Size Control of Sub-100 nm Pure Nanodrugs. *Nano Lett.* **2015**, *15*, 313–318. [CrossRef]
29. Martin, N.; Workman, P. Buprofezin: A selective pesticide for greenhouse whitefly control. In Proceedings of the New Zealand Weed and Pest Control Conference, Quality Inn, Palmerston North, New Zealand, 8 January 1986; Volume 39, pp. 234–236. [CrossRef]
30. Ishaaya, I.; Degheele, D. Buprofezin: A Novel Chitin Synthesis Inhibitor Affecting Specifically Planthoppers. In *Whiteflies and Scale Insects*; Springer: Berlin/Heidelberg, Germany, 1998; Volume 5, pp. 74–91.
31. Kah, M.; Hofmann, T. Nanopesticide research: Current trends and future priorities. *Environ. Int.* **2014**, *63*, 224–235. [CrossRef] [PubMed]
32. Skrbek, K.; Bartněk, V.; Lojka, M.; Sedmidubsk, D.; Jankovsk, O. Synthesis and Characterization of the Properties of Ceria Nanoparticles with Tunable Particle Size for the Decomposition of Chlorinated Pesticides. *Appl. Sci.* **2020**, *10*, 5224. [CrossRef]
33. Kolaib, E.; Sharma, R.K. Nanodispersions Platform for Solubility Improvement. *Int. J. Res. Pharm. Biomed. Sci.* **2013**, *4*, 636–643.
34. Dinh, H.T.; Tran, P.; Duan, W.; Lee, B.-J.; Tran, T.T. Nano-sized solid dispersions based on hydrophobic-hydrophilic conjugates for dissolution enhancement of poorly water-soluble drugs. *Int. J. Pharm.* **2017**, *533*, 93–98. [CrossRef]
35. Knieke, C.; Rawtani, A.; Davé, R.N. Concentrated fenofibrate nanoparticle suspensions from melt emulsification for en-hanced drug dissolution. *Chem. Eng. Technol.* **2014**, *37*, 157–167. [CrossRef]
36. Du, X.; Xu, S.H.; Sun, Z.W.; Aa, Y. Effect of the hydrodynamic radius of colloid microspheres on the Estimation of the coag-ulation rate constant. *Acta. Phys. Chim. Sin.* **2010**, *26*, 2807–2812.
37. Zhou, H.; Sun, X.; Zhang, L.; Zhang, P.; Li, J.; Liu, Y.-N. Fabrication of Biopolymeric Complex Coacervation Core Micelles for Efficient Tea Polyphenol Delivery via a Green Process. *Langmuir* **2012**, *28*, 14553–14561. [CrossRef] [PubMed]
38. Yin, M.M.; Zhu, X.Y.; Chen, F.L. Release performance and sustained-release efficacy of emamectin benzoate-loaded polylactic acid microspheres. *J. Integr. Agric.* **2018**, *17*, 640–647. [CrossRef]
39. Zou, G.; Wang, J. The affect of suitable compatibility of auxiliary to increase the suspensility of pesticide wettable powder. Jiangxi. *Chem. Ind.* **2002**, *4*, 134–136.
40. Kwok, P.; Chan, H.-K. Nanotechnology Versus other Techniques in Improving Drug Dissolution. *Curr. Pharm. Des.* **2014**, *20*, 474–482. [CrossRef]
41. Jelvehgari, M.; Valizadeh, H.; Montazam, S.H.; Abbaszadeh, S. Experimental Design to Predict Process Variables in the Microcrystals of Celecoxib for Dissolution Rate Enhancement Using Response Surface Methodology. *Adv. Pharm. Bull.* **2015**, *5*, 237. [CrossRef]

42. Mohamed, F.; Roberts, M.; Seton, L.; Ford, J.L.; Levina, M.; Siahboomi, A. The effect of HPMC particle size on the drug re-lease rate and the percolation threshold in extended-release mini-tablets. *Drug. Dev. Ind. Pharm.* **2015**, *41*, 70–78. [CrossRef] [PubMed]
43. Wu, L.; Zhang, J.; Watanabe, W. Physical and chemical stability of drug nanoparticles. *Adv. Drug Deliv. Rev.* **2011**, *63*, 456–469. [CrossRef]
44. Luckham, P.F. Physical stability of suspension concentrates with particular reference to pharmaceutical and pesticide formulations. *Pest Manag Sci.* **2010**, *25*, 25–34. [CrossRef]
45. Yang, M.; Ma, H. Effect of polydispersity on the relative stability of hard-sphere crystals. *J. Chem. Phys.* **2008**, *128*, 134510. [CrossRef] [PubMed]
46. Song, S.; Liu, X.; Jiang, J.; Qian, Y.; Zhang, N.; Wu, Q. Stability of triazophos in self-nanoemulsifying pesticide delivery system. *Colloids Surf. A Physicochem. Eng. Asp.* **2009**, *350*, 57–62. [CrossRef]
47. Müller, E.; Vogelsberger, W.; Fritsche, H.-G. The dependence of the surface energy of regular clusters and small crystallites on the particle size. *Cryst. Res. Technol.* **1988**, *23*, 1153–1159. [CrossRef]

International Journal of
Molecular Sciences

Article

Synthesis and Characterization of Polyvinylpyrrolidone-Modified ZnO Quantum Dots and Their In Vitro Photodynamic Tumor Suppressive Action

Tianming Song [1], Yawei Qu [2], Zhe Ren [1], Shuang Yu [1], Mingjian Sun [2], Xiaoyu Yu [3,*] and Xiaoyang Yu [1,*]

[1] The Higher Educational Key Laboratory for Measuring & Control Technology and Instrumentations of Heilongjiang Province, Harbin University of Science and Technology, Harbin 150080, China; 1610600001@stu.hrbust.edu.cn (T.S.); 1910600001@stu.hrbust.edu.cn (Z.R.); yushuang@hrbust.edu.cn (S.Y.)

[2] Department of Control Science and Engineering, Harbin Institute of Technology, Harbin 150000, China; 1910600002@stu.hrbust.edu.cn (Y.Q.); sunmingjian@hit.edu.cn (M.S.)

[3] School of Atmospheric Sciences, Sun Yat-sen University, Guangzhou 519000, China

* Correspondence: yuxy69@mail.sysu.edu.cn (X.Y.); yuxiaoyang@hrbust.edu.cn (X.Y.)

Abstract: Despite the numerous available treatments for cancer, many patients succumb to side effects and reoccurrence. Zinc oxide (ZnO) quantum dots (QDs) are inexpensive inorganic nanomaterials with potential applications in photodynamic therapy. To verify the photoluminescence of ZnO QDs and determine their inhibitory effect on tumors, we synthesized and characterized ZnO QDs modified with polyvinylpyrrolidone. The photoluminescent properties and reactive oxygen species levels of these ZnO/PVP QDs were also measured. Finally, in vitro and in vivo experiments were performed to test their photodynamic therapeutic effects in SW480 cancer cells and female nude mice. Our results indicate that the ZnO QDs had good photoluminescence and exerted an obvious inhibitory effect on SW480 tumor cells. These findings illustrate the potential applications of ZnO QDs in the fields of photoluminescence and photodynamic therapy.

Keywords: zinc oxide; quantum dot; polyvinylpyrrolidone; photodynamic therapy; photosensitizers; reactive oxygen species; nanoparticle; SW480 cancer cell

check for **updates**

Citation: Song, T.; Qu, Y.; Ren, Z.; Yu, S.; Sun, M.; Yu, X.; Yu, X. Synthesis and Characterization of Polyvinylpyrrolidone-Modified ZnO Quantum Dots and Their In Vitro Photodynamic Tumor Suppressive Action. *Int. J. Mol. Sci.* **2021**, *22*, 8106. https://doi.org/10.3390/ijms22158106

Academic Editor: Raghvendra Singh Yadav

Received: 30 June 2021
Accepted: 22 July 2021
Published: 28 July 2021

Publisher's Note: MDPI stays neutral with regard to jurisdictional claims in published maps and institutional affiliations.

1. Introduction

In recent years, cancer has become one of the most fatal diseases threatening human health and the second leading cause of mortality worldwide [1,2]. The current cancer treatments include surgery, radiotherapy, and chemotherapy. However, these therapies are associated with complications, such as tissue trauma, side effects, and reoccurrence.

Photodynamic therapy (PDT) is a cancer treatment modality that can be applied to treat various tumors. In PDT, light and photosensitizers (PS) are used to destroy tumor cells by generating reactive oxygen species (ROS) [3–5]. The basic principle of photodynamic therapy is that the use of a photosensitizer exposed to a certain wavelength of light will be excited to produce ROS, which can kill cancer cells [6]. Currently, the commonly used light source is ultraviolet (UV) light, and various new photosensitizers are under development, particularly nanomaterials [7], for use in research in the PDT field. Compared with other treatment modalities, PDT is noninvasive, has a lower incidence of trauma, and causes less toxicity and fewer side effects [8]. Currently, PDT has been approved for treating superficial tumors, including skin cancer [9], superficial bladder cancer [10], lung cancer [11], cervical cancer [12], and head and neck cancers [13,14].

Zinc oxide (ZnO), an inexpensive and versatile inorganic nanomaterial with photo-electric properties, is often used in photocatalysis and on ceramic surfaces. Recently, the synthesis and biomedical application of ZnO nanoparticles have gained attention [15–25] because of their anticancer, antibacterial, antioxidant, antidiabetic, and anti-inflammatory activities, as well as for drug delivery and bioimaging applications. ZnO NPs were also reported to induce cytotoxicity in a variety of cancer cells [26–36]. Zinc (Zn^{2+}), a component of ZnO, is widely recognized as an essential micronutrient for humans and, thus, is safe for application. Furthermore, ZnO particles have been designated as Generally Recognized as Safe by the Food and Drug Administration [37].

ZnO quantum dots (QDs) are used for conducting PDT as a PS and for biological imaging. Gao et al. [38] combined ZnO with X-ray, MRI, and other imaging methods in vivo to monitor tumor growth and metastasis in real time. Fluorescence imaging, in contrast, exhibits the advantages of low cost, high sensitivity, and optimal spatial resolution among molecular imaging techniques. ZnO QDs possess excellent photodynamic properties and have potential utility in photoluminescence imaging. ZnO QDs are typically used in bioimaging [39,40] and drug delivery because of their high biocompatibility [41,42]. Many researchers aim to improve the stability and water solubility of ZnO QDs. PVP can serve as a surface stabilizer, growth modifier, nanoparticle dispersant, and reducing agent [43]. Several methods for synthesizing QDs have been reported, including coprecipitation, sol–gel, solid-state, and hydrothermal methods [44–56]. Each method has various advantages and limitations. For example, nanoparticles synthesized by the coprecipitation method are larger than those obtained by other methods, whereas the stability of the aqueous solution synthesized by the sol–sol method is lower than that observed by other methods. However, there are several limitations of ZnO QDs, including their poor water stability and easy agglomeration, preventing their application in the biological domain [57–59].

In this study, we developed an optimized method for synthesizing ZnO QDs modified with polyvinylpyrrolidone (PVP40) to improve their stability in aqueous solutions and investigated their characteristics and optical properties. Next, we assessed the cytotoxicity of ZnO/PVP QDs in SW480 cancer cells and HEK-293T human kidney cells to improve the biomedical applications of ZnO/PVP QDs. We evaluated the applications of ZnO/PVP QDs for both in vitro and in vivo PDT and analyzed the mechanisms and principles of PDT in cancer treatments. Our results indicate that ZnO QDs have considerable potential applications in the fields of photoluminescence and photodynamic tumor suppression.

2. Results and Discussion

2.1. Synthesis of ZnO/PVP QDs

ZnO/PVP QDs were synthesized using an improved sol–gel method. To verify that the solution contained ZnO QDs, the solution was subjected to irradiation under UV light at a wavelength of 365 nm. The solution turned from clear and transparent to emit yellow fluorescence. Based on the yellow fluorescence, the fluorescence wavelength was approximately 500–600 nm. The fluorescence spectrum of this solution was red-shifted (Figure 1). The Stokes shift of ZnO occurs only at the quantum level. Upon comparison with the absorption spectrum, it was found that the ZnO QDs were synthesized successfully.

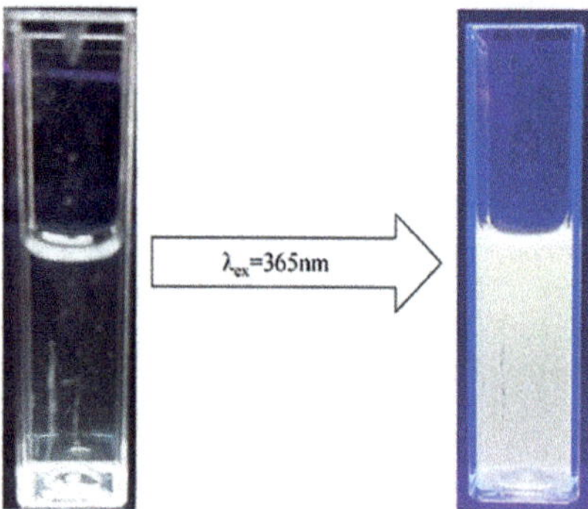

Figure 1. Fluorescence of ZnO quantum dots at λ_{ex} = 365 nm. Left: image of ZnO quantum dots under white light; right: image of ZnO quantum dots under 365-nm ultraviolet light.

2.2. Characterization of ZnO/PVP QDs

The ZnO QD particles were approximately 10 nm, as measured by dynamic light scattering (DLS), indicating that the synthesis of ultrasmall ZnO QDs was successful (Figure 2a). The zeta potential was -3.6 mV on the surfaces of the ZnO/PVP QDs (Figure S1b). Transmission electron microscopy (TEM) showed that the ZnO/PVP particles were well-dispersed in the ethanol solution, with fine, round, and granular characteristics. This also supports that ZnO QDs modified with PVP were more dispersive and less prone to agglomeration. The manual measurement of 50 units of ZnO particles showed that ZnO/PVP possessed appreciable homogeneity at sizes of ~5 nm. The particle sizes appeared larger by DLS compared to those by TEM because of hydration. Upon enlargement of the TEM image, the lattice fringes of ZnO were observed, indicating that the ZnO nanocrystals were successfully synthesized (Figure 2b). The X-ray diffraction pattern of the 2θ values of the ZnO/PVP QDs was obtained in the range of 20–80° (Figure S1a). The peaks were centered at $2\theta = 31.75°$, $34.28°$, $36.15°$, $47.58°$, $56.51°$, $62.86°$, and $66.75°$, indexing the (100), (002), (101), (102), (110), (103), and (112) diffraction planes, respectively, of ZnO/PVP. The ZnO/PVP QDs belong to the hexagonal wurtzite structure (JCPDS PDF #36-1451). The average crystallite size of the ZnO/PVP QDs (8 nm) was calculated using the Debye-Scherrer formula: $D = 0.89\lambda/\beta\cos\theta$, where λ is the X-ray wavelength, $\lambda = 0.154184$ nm, β is the peak width at half-maximum, and θ is the Bragg diffraction angle. This result is consistent with those of TEM and DLS. UV-Vis and fluorescence spectroscopy analyses were also performed using ZnO/PVP. The fluorescence spectra of ZnO/PVP showed strong absorption in the 250–360-nm UV light band. The spectral absorption of the ZnO QDs was enhanced with decreasing wavelengths, further demonstrating the excellent UV absorption of the ZnO QDs. In addition, the ZnO/PVP spectrum showed a plateau at 320–350 nm. This was one of the unique absorption peaks of the ZnO QDs as synthesized using the sol–gel method, confirming that the ZnO QDs particles were well-distributed (Figure 2c). The fluorescence spectra of ZnO/PVP displayed a broad peak at 500–650 nm. This conforms with the luminescence rule of the ZnO QDs and the yellow fluorescence of the ZnO QDs visible to the naked eye under ultraviolet light irradiation (Figure 1). The fluorescence intensity of ZnO/PVP reached 20,000 a.u. at 560 nm (Figure 2d); emissions at 560 nm can be used for fluorescent labeling. For example, when ZnO/PVP is loaded with anticancer drugs into tumor cells, the location of the drugs is determined by fluorescence, which is important when investigating treatment mechanisms.

2.3. Quantum Yield of the ZnO/PVP QDs

Quantum yield refers to the utilization of quantum principles in photochemical reactions; it is defined as the ratio of the number of photons emitted to the number of photons absorbed. This is measured by obtaining the ratio of the fluorescence intensity to the intensity of the absorption. It is typically calculated by comparing the quantum yield of a material with that of a reference material in a certain range.

The quantum yield of ZnO/PVP was calculated according to:

$$Y_u = Y_s \times \frac{F_u}{F_s} \times \frac{A_s}{A_u} \tag{1}$$

In the formula, Y_u represents the quantum yield of the unknown sample, Y_s represents the fluorescence quantum yield of the reference materials, F_u represents the integral fluorescence intensity of the dilute solution of the sample to be measured, F_s represents the integral fluorescence intensity of the dilute solution of the reference materials, and A_u and A_s represent the maximum absorbance values of the sample and reference at the excitation wavelength, respectively. The reference material used in this experiment was rhodamine B, which has a quantum yield of 67% at an absorption wavelength of 365 nm.

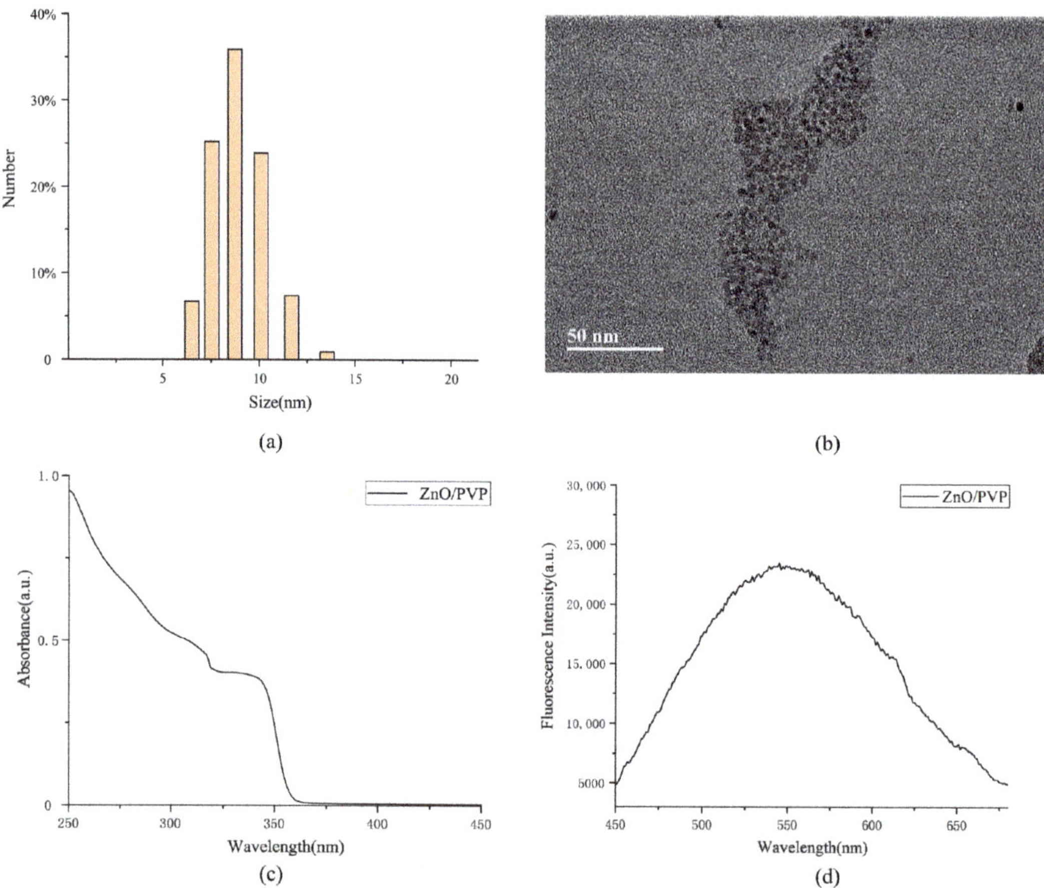

Figure 2. Characterizations of the ZnO/PVP QDs. (**a**) Particle sizes of the ZnO/PVP QDs as measured by DLS, (**b**) TEM image of the ZnO/PVP QDs, (**c**) UV-Vis absorption spectra of the ZnO/PVP QDs, and (**d**) fluorescence spectra of the ZnO/PVP QDs.

The quantum yield of the ZnO/PVP synthesized in this study was 8.7%. This value agrees with the law of quantum yield, indicating an excellent quantum yield of the ZnO QDs. van Dijken et al. [60] showed that smaller fresh ZnO NPs had higher quantum yields. Modified PVP40 on the surface of ZnO protects the acetate groups, prevents tumor growth and aggregation, and maintains optimal luminescent properties.

2.4. In Vitro Stability of ZnO/PVP QDs

The ZnO/PVP QD solution was stored in a tin-coated centrifuge tube for 2 weeks. The results of the DLS and TEM showed no significant changes in the ZnO/PVP QDs, and their sizes did not increase significantly. The UV-Vis absorption and fluorescence spectra showed decreased absorption to various degrees. This may be attributed to lower agglomeration of the ZnO QDs, partial restoration of defects on their surfaces, and reduction of the surface area, which may affect their optical properties.

Shi et al. [61] reported that unmodified ZnO QD solutions are turbid and agglomerate only after 3 days, suggesting that ZnO QDs without surface modifications are unstable. In this study, the ethanol solution of the ZnO QDs was stored for 14 days, causing the solution to become clear and transparent and allowing fluorescence to be observed under a UV light. The possibility of extensive agglomeration of the ZnO QDs was low. This showed that the polyethylene glycol modification enhanced the stability of the solution, allowing the ZnO QDs to remain stable after 14 days (Figure 3).

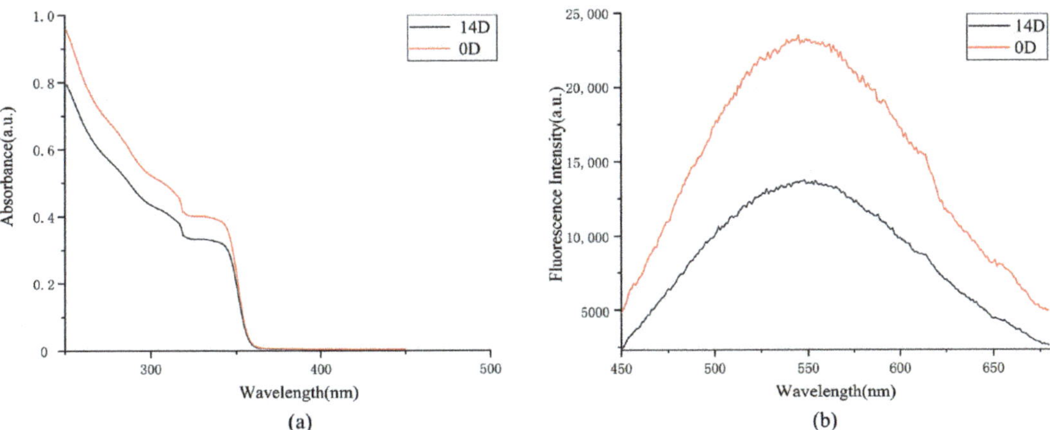

Figure 3. Absorption and fluorescence spectra of the ZnO/PVP QDs stored in an ethanol solution for two weeks. (**a**) absorption spectra of the ZnO/PVP QDs and (**b**) fluorescence spectra of the ZnO/PVP QDs.

2.5. Photoluminescent Properties of the ZnO/PVP QDs

The photoluminescent properties of the ZnO QDs at different concentrations were determined using a fluorescence microplate reader. The fluorescence intensities of the ZnO QDs increased with the increasing concentrations, indicating that the fluorescence intensities of the ZnO QDs were concentration-dependent. The fluorescence intensity of the 50-µg/mL ZnO/PVP solution was 32,000 a.u., which is consistent with the trend observed for the fluorescence spectra of the ZnO QDs. The change in fluorescence intensities of the ZnO QDs gradually decreased, which may be associated with photon-quenching caused by the increasing concentrations. This also showed that the PVP40 modification exerted no remarkable effect on the optical properties of ZnO (Figure 4a).

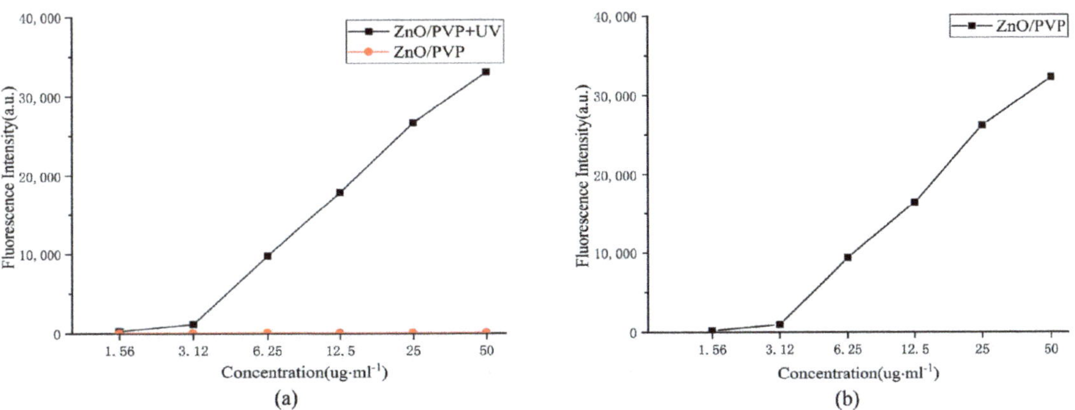

Figure 4. Fluorescence intensities of the ZnO/PVP QDs. (**a**) Fluorescence intensities of the ZnO/PVP QDs and (**b**) fluorescence intensities of the ZnO/PVP QDs at 525 nm.

2.6. In Vitro ROS Production

The ROS levels were determined using 2′,7′-dichlorodihydrofluorescein diacetate (DCFH-DA), which is oxidized by ROS to produce highly fluorescent DCF. The fluorescence of DCF was measured with a fluorescence microplate reader at excitation and emission wavelengths of 488 and 535 nm, respectively. DCHF-DA was added to the ZnO QD solutions at different concentrations, and the fluorescence intensities were measured (Figure 4b). Fluorescence at 525 nm was detected at a concentration of 6.25 μg/mL under UV irradiation, indicating that a ROS was produced at this concentration. The fluorescence intensity of the ZnO/PVP solution at a concentration of 50 μg/mL was 6900 a.u. However, fluorescence at 525 nm was not detected when no UV irradiation was applied.

2.7. Cytotoxicity of ZnO/PVP QDs

SW480 cells were incubated with different concentrations of ZnO/PVP QDs, and their survival rates were determined in a 3-(4,5-dimethylthiazol-2-yl)-2,5-diphenyltetrazolium bromide (MTT) assay. HEK293T cells were treated in a manner for comparison with SW480 cells. The MTT assay was conducted to measure the cellular metabolic activity as an indicator of cell viability, proliferation, and cytotoxicity. This colorimetric assay was based on the reduction of a yellow tetrazolium salt MTT to purple formazan crystals by metabolically active cells. Viable cells can reduce the MTT reagent, where apoptotic cells cannot. The cell survival rate gradually decreased from a concentration of 12.5 μg/mL and was 54% when the concentration reached 50 μg/mL (Figure 5a). The cytotoxicity of ZnO/PVP may be attributed to the low biocompatibility of PVP40.

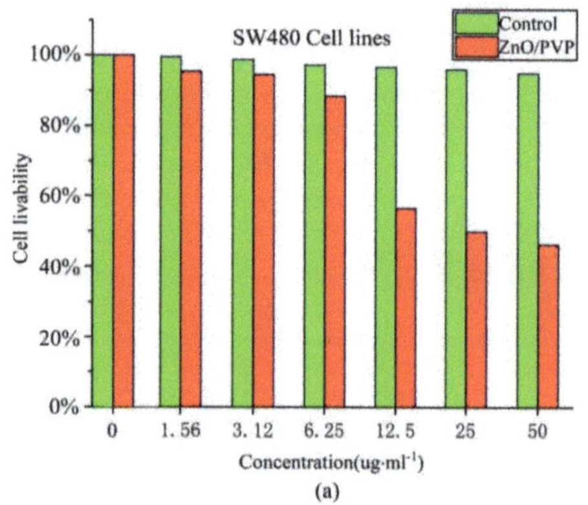

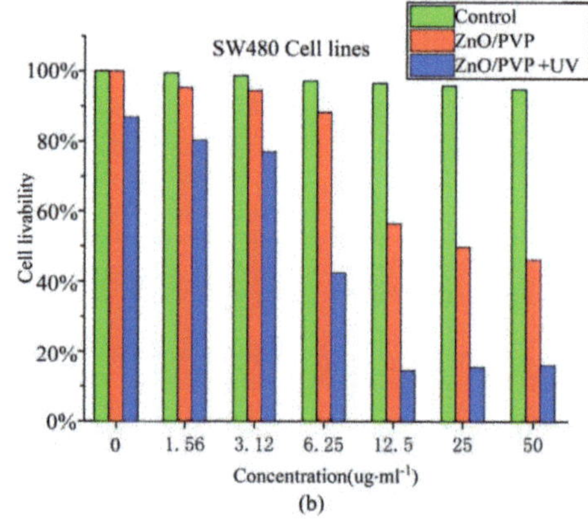

Figure 5. Cell survival rates. (**a**) Cell survival rates of the ZnO/PVP QDs at different concentrations. (**b**) Cell survival rates of the ZnO/PVP and ZnO/PVP+UV groups at different concentrations.

2.8. Photodynamic Experiment of the ZnO/PVP QDs In Vitro

To evaluate the PDT efficacy of the ZnO/PVP QDs, the viabilities of SW480 and HEK293T cells subjected to UV irradiation were evaluated using an MTT assay. ZnO/PVP showed excellent biocompatibility, with no remarkable changes in cell viabilities under suitable concentrations and without UV treatment. However, ZnO/PVP QDs subjected to UV irradiation showed evident tumor inhibition. Furthermore, the cell mortalities gradually increased with the increasing concentrations, with the cell viability decreasing to 15% at a concentration of 50 μg/mL. Near-complete apoptotic tumor cells were also observed, indicating that ZnO QDs have excellent photodynamic effects in vitro. ZnO/PVP showed evident tumor inhibitory effects at a concentration of 6.25 μg/mL when subjected to UV irradiation. Additionally, the cell survival rate of the group at this concentration was 45% lower than that of the no UV treatment group at the same concentration (Figure 5b).

The percentages of apoptotic and necrotic SW480 and HEK-293T cells were analyzed by flow cytometry (Figure S3). The number of necrotic cells decreased after treatment with 25-µg/mL ZnO QDs in both cell types.

The half-maximal inhibitory concentration of ZnO/PVP under UV irradiation was estimated as 21.688 µg/mL. This result is consistent with the trend in ROS produced by ZnO QDs in vitro. Therefore, tumor inhibition of the ZnO QDs may be caused by the excitation of ZnO/PVP by UV light. This causes the generation of electrons and holes that can be transferred to the surface, subsequently generating ROS and inducing tumor cell apoptosis. The beneficial effects observed in the photodynamic in vitro study of the ZnO/PVP QDs motivated us to study their application in PDT in vivo.

2.9. Photodynamic Experiment of ZnO/PVP QDs In Vivo

After dividing tumor-bearing nude mice into four groups, the tumors were measured and weighed every 3 days. The tumor volumes of the control, UV, ZnO/PVP, and ZnO/PVP + UV groups were 5180 ± 759, 5356 ± 795, 5562 ± 480, and 2011 ± 37 mm^3, respectively, on day 28 (Figure 6). The tumor inhibition rate of the ZnO/PVP+UV group was 61.1%, which was significantly higher than that of the control, UV, and ZnO/PVP groups ($p < 0.05$ in all instances). Moreover, the tumor inhibition effects of ZnO/PVP were observed in the early stages. The sizes of the tumors in nude mice were visible to the naked eye after 7 days (Figure 7 and Table 1).

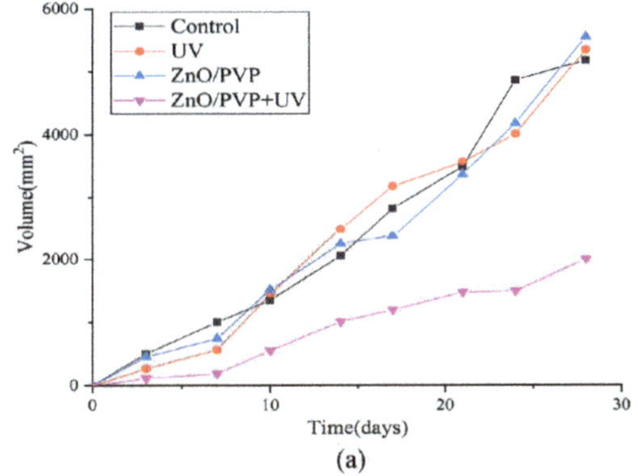

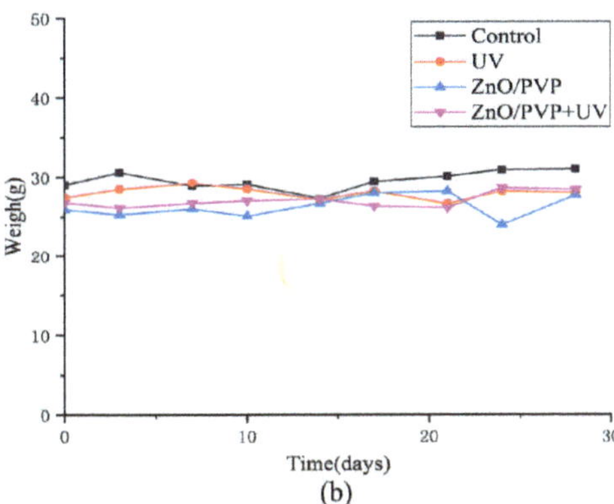

Figure 6. Influence of the treatment on the tumor volumes and weight of tumor-bearing mice treated with PDT. (**a**) Influence of the treatment on tumor volumes of groups of tumor-bearing mice treated with PDT. (**b**) Influence on the tumor weights of groups of tumor-bearing mice treated with PDT.

Table 1. Effects of UV ultraviolet irradiation and non-irradiation on the tumor volumes in mice treated with ZnO QDs.

Group	Number of Nude Mice	Tumor Volume of Nude Mice (mm^3)				Tumor Inhibition Rate (%)
		M1	M2	M3	$\chi \pm S$	
Control group	3	5292	4200	6050	5180 ± 759	–
UV group	3	4400	5320	6348	5356 ± 795	−3.3
ZnO/PVP group	3	5000	4630.5	4400	4543 ± 102	12.3
ZnO/PVP+UV group	3	2025	2048	1960	2011 ± 37	61.1

M1–3: mice 1–3.

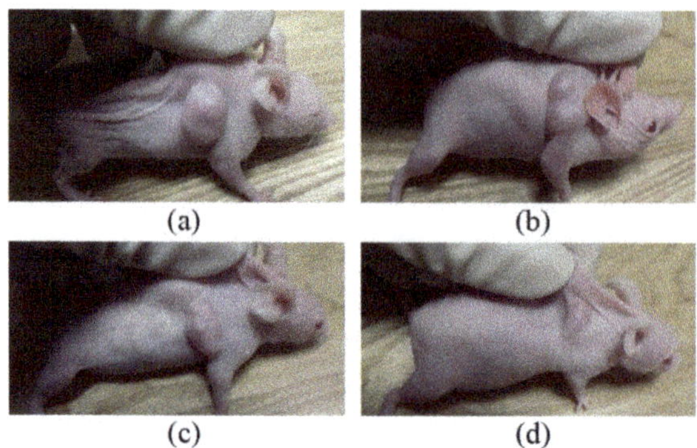

Figure 7. Changes in the tumors of tumor-bearing mice on day 7. (**a**) Control group, (**b**) UV group, (**c**) ZnO/PVP group, and (**d**) ZnO/PVP+UV group.

The tumors were dissected and weighed after the mice were euthanized. The weights of the tumors from the control, UV, ZnO/PVP, and ZnO/PVP+UV groups were 2.17 ± 0.07, 2.17 ± 0.12, 2.08 ± 0.08, and 1.78 ± 0.10 g, respectively (Table 2). The tumor weights in the ZnO/PVP+UV group were significantly lower than those in the control, UV, and ZnO/PVP groups ($p < 0.05$ in all instances).

Table 2. Effects of UV irradiation and non-irradiation on the tumor weights in mice treated with ZnO QDs.

Group	Number of Nude Mice	Tumor Volume of Nude Mice (mm^3)			
		M1	M2	M3	$\chi \pm S$
Control group	3	2.27	2.16	2.08	2.17 ± 0.07
UV group	3	2.3	2.22	2.01	2.17 ± 0.12
ZnO/PVP group	3	2.14	2.13	2.24	2.17 ± 0.04
ZnO/PVP+UV group	3	1.5	1.43	1.52	1.48 ± 0.03

M1–3: mice 1–3.

We observed no significant weight changes between each group throughout the treatment, and the weight fluctuations in each group were less than 1 g (Figure 6b). These results indicate that the injection of the ZnO QDs had no evident toxic side effects on the nude mice, suggesting the good biocompatibility of the ZnO QDs. The basic principle of photodynamic therapy is that the use of a photosensitizer exposed to a certain wavelength of light will be excited to produce ROS, which can kill cancer cells. The tumor inhibition of ZnO QDs may be caused by the excitation of ZnO/PVP by UV light. The results of the in vivo experiments also proved the photodynamic effect on SW480 tumor cells.

Our results demonstrated that ZnO has strong photodynamic therapeutic effects. As the main mechanism of PDT in cancer treatment, ZnO QDs irradiated with ultraviolet light can generate ROS, such as superoxide anion, hydroxyl radical, hydrogen peroxide, singlet oxygen, and peroxyl radicals. These species disrupt the cell membrane integrity and damage lysosomes, mitochondria, proteins, intranuclear macromolecules, and other important physiological functions to finally cause cell apoptosis, thus effectively killing tumor cells in tumor treatments (Figure S4).

Despite the promising therapeutic potential applications of photodynamic tumor suppression, some limitations remain to be addressed. For example, UV exposure at a higher level may cause skin burns. Although the ZnO QDs showed therapeutic potential, further studies are warranted prior to clinical application.

3. Materials and Methods

3.1. Reagents and Cell Lines

Zinc acetate dihydrate [Zn(OAc)$_2$·2H$_2$O] and PVP40 [(C$_6$H$_9$NO)$_n$] were purchased from Sigma-Aldrich (St. Louis, MO, USA). Lithium hydroxide (LiOH), anhydrous ethanol, and cyclohexane were purchased from Sinopharm Chemical Reagent (Shanghai, China). All reagents used in this study were of analytical grade. The human colon cancer cell line SW480 was acquired from the cell bank of the Chinese Academy of Sciences (Shanghai, China). The cell lines were cultured in 90% DMEM/L-15 (Gibco, Grand Island, NY, USA) and supplemented with 10% heat-inactivated fetal bovine serum (Gibco). All cultures were maintained in an incubator at 37°C with 5% CO$_2$ in a humidified atmosphere.

3.2. Synthesis of the ZnO/PVP QDs

First, 220 mg of zinc acetate dihydrate and 66 mg of PVP40 were dissolved in 10 mL of ethanol. The solutions were heated under reflux at 70 °C with magnetic stirring for 1.5 h to allow the formation of a colorless transparent ZnO precursor solution. Meanwhile, 15 mg of LiOH was added to 6 mL of ethanol and incubated at 50 °C for 20 min with magnetic stirring. Subsequently, 10 mL of ZnO precursor solution was added to the LiOH solution and incubated at 50 °C for 1 h with magnetic stirring. Next, 15 mL of ZnO/PVP QD ethanol solution was added to 30 mL of *n*-hexane, and the solution was left overnight at 25 °C. The obtained solution was then centrifuged (Allegra® X-30 Centrifuges; Beckman Coulter, Brea, CA, USA) at 2000 rpm for 10 min, and the resulting supernatant was discarded. Finally, 15 mL of anhydrous ethanol was added to dissolve the precipitate, yielding a clear and transparent solution. ZnO/PVP QDs were obtained.

3.3. Characterizations of the ZnO/PVP QDs

The morphology and particle size of the ZnO/PVP QDs were characterized by performing TEM (JEM-2100; JEOL, Tokyo, Japan) and DLS (Zetasizer Nano Z; Malvern Panalytical, Malvern, UK), respectively. X-ray diffraction ((Ultima IV; Rigaku, Tokyo, Japan) of the ZnO/PVP QDs was recorded at room temperature at a scan rate of 2.4°/min from 20° to 80°. The optical properties of the ZnO/PVP QDs were determined using both a UV-Vis spectrophotometer (LAMBDA 365; PerkinElmer, Waltham, MA, USA) and fluorescence spectrophotometer (FL8500; PerkinElmer).

3.4. Quantum Yield of ZnO/PVP QDs

To measure the quantum yield of the obtained ZnO/PVP QDs, we used a standard fluorescent dye (rhodamine B) with a known quantum yield of 67% at 365 nm. ZnO/PVP and rhodamine B were each diluted with ethanol. The fluorescence spectrometer was used at excitation and emission wavelengths of 365 and 450–680 nm, respectively. The slits, including the excitation and emission, were set to 3 nm. The solutions of ZnO/PVP and rhodamine B were measured at the same excitation wavelengths and slit widths. The absorption spectra were recorded at this wavelength.

3.5. In Vitro Stability Study of the ZnO/PVP QDs

The color, DLS, and wavelength of the ZnO/PVP QD ethanol solution were observed after storage in a tin-coated centrifuge tube for 2 weeks.

3.6. Fluorescence Intensities of the ZnO/PVP QDs

Specific volumes (100 µL) of different concentrations of ZnO/PVP solution (1.56, 3.125, 6.25, 12.5, 25, and 50 µg/mL) were pipetted into the wells of a 96-well plate. The fluorescence intensities were measured using a fluorescence microplate reader (Fluoroskan FL; Thermo Scientific, Waltham, MA, USA) (excitation wavelength: 365 nm, emission wavelength: 570 nm) to obtain the fluorescence intensities at each concentration of the ZnO QDs.

3.7. Measurement of ROS Levels

The ROS levels were measured using dissolved DCFH-DA, a fluorescent dye that enables the visualization of ROS. The DCFH-DA was dissolved in dimethyl sulfoxide to obtain a 48-μg/mL DCFH-DA solution. Specific volumes (100 μL) of different concentrations of ZnO/PVP solution (1.56, 3.125, 6.25, 12.5, 25, and 50 μg/mL) were pipetted into the wells of a 96-well plate, and 10 μL of DCFH-DA (48 μg/mL) was added to each well. UV light at a wavelength of 365 nm (the source was situated 5 cm above the plate) was used to irradiate the 96-well plate for 5 and 10 min. ROS detection was performed via a fluorescence quantitative analysis using a fluorescence microplate reader (Fluoroskan, FL, USA).

3.8. Cytotoxicity Analysis

SW480 cells in the logarithmic growth phase were digested with 0.25% pancreatin, evenly seeded into a 96-well plate (1×10^4 cells/mL), and cultured for 24 h. The prepared ZnO/PVP solutions were divided into six groups and diluted to concentrations of 1.56, 3.125, 6.25, 12.5, 25, and 50 μg/mL using the cell culture medium. The solutions in each group were pipetted into five wells, and a blank control group consisting of only the culture medium was prepared. HEK293T cells were treated in the same manner. The MTT method was used to determine the cell viability.

3.9. In Vitro Photodynamic Study

SW480 cells in the logarithmic growth phase were digested with 0.25% pancreatin, evenly seeded into the wells of a 96-well plate (1×10^4 cells/mL), and cultured for 12 h. The prepared ZnO/PVP solutions were divided into six groups and diluted to concentrations of 1.56, 3.125, 6.25, 12.5, 25, and 50 μg/mL using a cell culture medium. The solutions in each group were pipetted into five wells, and a blank control group consisting of only the culture medium was prepared. HEK293T cells were treated in the same manner. The cells were irradiated for 10 min using UV light at a wavelength of 365 nm and then incubated for 48 h at 25 °C. Cell viability was determined using the MTT method.

3.10. Experimental Animals

All animal experiments were performed in accordance with the guidelines of the Institutional Animal Care and Use Committee of General Hospital of Chinese People's Armed Police Forces, which also approved the research procedures (Permit Number 2011-0039). The surgeries were performed using isoflurane gas anesthesia (3% isoflurane–air mixture), and all efforts were made to minimize suffering.

Female nude mice (aged 7 to 8 weeks old) were purchased from the Laboratory Animal Center of the Chinese Academy of Medical Sciences (Beijing, China) and housed under specific pathogen-free conditions at 25 °C with a relative humidity of 50%. All mice were fed sterilized pellets and allowed access to water ad libitum under a 12-h light and dark cycle. The whole-body weights and tumor volumes of the mice were monitored daily. Mice that lost >20% of their initial body weights were euthanized by carbon dioxide asphyxiation, and the experiment was terminated. No mice died prior to application of the humane endpoint.

Approximately 2×10^6 SW480 cells in 50 μL were implanted subcutaneously in the left underarm of each mouse to enable the formation of solid tumors. The mice were used for the experiments once the tumors reached a diameter of 0.5 cm.

3.11. In Vivo Photodynamic Study

The tumor-bearing mice were divided into the following four groups: control, UV, ZnO/PVP, and ZnO/PVP+UV groups. After 48 h, each group was again treated with UV, ZnO/PVP, and ZnO/PVP+UV separately. The body weight and tumor volume of each nude mouse were measured daily.

The tumor volume was calculated according to:

$$V_t = W_t{}^2 \times L_t/2 \tag{2}$$

V_t represents the volume of the tumor, W_t represents the width of the tumor, and L_t represents the length of the tumor. The tumor inhibition rate was defined according to:

$$R_{ti} = 1 - V_e/V_c \times 100\% \tag{3}$$

R_{ti} represents the tumor inhibition rate, V_e represents the tumor volume of the experimental group, and V_c represents the tumor volume of the control group.

The mice were euthanized using carbon dioxide, and their tumors were dissected and weighed at the end of the experiment.

3.12. Statistical Analysis

Statistical significance was determined by one-way analysis of variance, followed by the Newman–Keuls test. All statistical analyses were performed using SPSS 16.0 software (SPSS, Inc., Chicago, IL, USA). $p < 0.05$ was considered to indicate statistically significant results.

4. Conclusions

We synthesized ZnO/PVP QDs via an improved sol–gel method and demonstrated their stability and optical properties. Furthermore, we investigated the photoluminescent properties of the ZnO/PVP QDs. PVP40 accumulated at the defect sites on the ZnO surfaces, a mechanism that may be related to photon-quenching at the increasing concentrations. Our in vitro and in vivo studies demonstrated that ZnO/PVP considerably enhanced the PDT efficacy. Therefore, further research of ZnO/PVP bioimaging in live animals is warranted. Based on these results, the ZnO/PVP QDs have potential in the field of photodynamic tumor suppression. We analyzed the mechanism of the photocatalytic killing of cancer, which provided a theoretical foundation for applying ZnO in tumor PDT treatments and a basis for the further synthesis, selection, and modification of new nanomaterials.

Supplementary Materials: The following are available online at https://www.mdpi.com/article/10.3390/ijms22158106/s1.

Author Contributions: Conceptualization, Y.Q. and T.S.; methodology, Y.Q. and M.S.; validation, T.S. and Z.R.; formal analysis, T.S. and Z.R.; investigation, T.S. and X.Y. (Xiaoyang Yu).; resources, Y.Q.; data curation, Y.Q. and T.S.; writing—original draft preparation, T.S.; writing—review and editing, T.S. and X.Y. (Xiaoyu Yu).; supervision, X.Y. (Xiaoyang Yu).; project administration, Y.Q.; and funding acquisition, X.Y. (Xiaoyu Yu). and S.Y. All authors have read and agreed to the published version of the manuscript.

Funding: This research was funded by the National Key Research and Development Project, grant number 2016YFA0602701, and National Key Scientific Instrument and Equipment Development Project, grant number 42027804.

Institutional Review Board Statement: All animal experiments were performed in accordance with the guidelines of the Institutional Animal Care and Use Committee (IACUC) at Peking University (Permit Number 2011-0039), which also approved the research procedures.

Data Availability Statement: The data presented in this study are available in this article.

Acknowledgments: The authors would like to thank the research staff for their contributions to this project.

Conflicts of Interest: The authors declare no conflict of interest.

References

1. Jemal, A.; Bray, F.; Center, M.M.; Ferlay, J.; Ward, E.; Forman, D. Global cancer statistics. *CA Cancer J. Clin.* **2011**, *61*, 69–90. [CrossRef]
2. Bray, F.; Ferlay, J.; Soerjomataram, I.; Siegel, R.L.; Torre, L.A.; Jemal, A. Global cancer statistics 2018: GLOBOCAN estimates of incidence and mortality worldwide for 36 cancers in 185 countries. *CA Cancer J. Clin.* **2018**, *68*, 394–424. [CrossRef] [PubMed]
3. Choudhary, S.; Nouri, K.; Elsaie, M.L. Photodynamic therapy in dermatology: A review. *Lasers Med. Sci.* **2009**, *24*, 971–980. [CrossRef] [PubMed]
4. Moor, A.C.E. Signaling pathways in cell death and survival after photodynamic therapy. *J. Photochem. Photobiol. B Biol.* **2000**, *57*, 1–13. [CrossRef]
5. Barra, F.; Roscetto, E.; Soriano, A.; Vollaro, A.; Postiglione, I.; Pierantoni, G.; Palumbo, G.; Catania, M. Photodynamic and antibiotic therapy in combination to fight biofilms and resistant surface bacterial infections. *Int. J. Mol. Sci.* **2015**, *16*, 20417–20430. [CrossRef] [PubMed]
6. Robertson, C.A.; Evans, D.H.; Abrahamse, H. Photodynamic therapy (PDT): A short review on cellular mechanisms and cancer research applications for PDT. *J. Photochem. Photobiol. B Biol.* **2009**, *96*, 1–8. [CrossRef]
7. Kim, M.M.; Darafsheh, A. Light sources and dosimetry techniques for photodynamic therapy. *Photochem. Photobiol.* **2020**, *96*, 280–294. [CrossRef]
8. Rong, P.; Yang, K.; Srivastan, A.; Kiesewetter, D.O.; Yue, X.; Wang, F.; Nie, L.; Bhirde, A.; Wang, Z.; Liu, Z.; et al. Photosensitizer loaded nano-graphene for multimodality imaging guided tumor photodynamic therapy. *Theranostics* **2014**, *4*, 229–239. [CrossRef]
9. Allison, R.R.; Sibata, C.H.; Downie, G.H.; Cuenca, R.E. A clinical review of PDT for cutaneous malignancies. *Photodiagn. Photodyn. Ther.* **2006**, *3*, 214–226. [CrossRef]
10. Nseyo, U.O.; De Haven, J.; Dougherty, T.J.; Potter, W.R.; Merrill, D.L.; Lundahl, S.L.; Lamm, D.L. Photodynamic therapy (PDT) in the treatment of patients with resistant superficial bladder cancer: A long term experience. *J. Clin. Laser Med. Surg.* **1998**, *16*, 61–68. [CrossRef] [PubMed]
11. Moghissi, K.; Dixon, K.; Thorpe, J.A.C.; Stringer, M.; Oxtoby, C. Photodynamic therapy (PDT) in early central lung cancer: A treatment option for patients ineligible for surgical resection. *Thorax* **2007**, *62*, 391–395. [CrossRef]
12. Muroya, T.; Suehiro, Y.; Umayahara, K.; Akiya, T.; Iwabuchi, H.; Sakunaga, H.; Sakamoto, M.; Sugishita, T.; Tenjin, Y. Photodynamic therapy (PDT) for early cervical cancer. *Gan Kagaku Ryoho* **1996**, *23*, 47–56.
13. Biel, M.A. Photodynamic therapy of head and neck cancers. In *Methods in Molecular Biology*; Humana Press: Tortowa, NJ, USA, 2010; pp. 281–293.
14. Biel, M. Advances in photodynamic therapy for the treatment of head and neck cancers. *Lasers Surg. Med.* **2006**, *38*, 349–355. [CrossRef]
15. Han, Z.; Wang, X.; Heng, C.; Han, Q.; Cai, S.; Li, J.; Qi, C.; Liang, W.; Yang, R.; Wang, C. Synergistically enhanced photocatalytic and chemotherapeutic effects of aptamer-functionalized ZnO nanoparticles towards cancer cells. *Phys. Chem. Chem. Phys.* **2015**, *17*, 21576–21582. [CrossRef]
16. Xiong, H.-M.; Liu, D.-P.; Xia, Y.-Y.; Chen, J.-S. Polyether-grafted ZnO nanoparticles with tunable and stable photoluminescence at room temperature. *Chem. Mater.* **2005**, *17*, 3062–3064. [CrossRef]
17. Jamieson, T.; Bakhshi, R.; Petrova, D.; Pocock, R.; Imani, M.; Seifalian, A.M. Biological applications of quantum dots. *Biomaterials* **2007**, *28*, 4717–4732. [CrossRef]
18. Wu, W.; Shen, J.; Banerjee, P.; Zhou, S. A Multifuntional nanoplatform based on responsive fluorescent plasmonic ZnO-Au@PEG hybrid nanogels. *Adv. Funct. Mater.* **2011**, *21*, 2830–2839. [CrossRef]
19. Pan, Z.-Y.; Liang, J.; Zheng, Z.-Z.; Wang, H.-H.; Xiong, H.-M. The application of ZnO luminescent nanoparticles in labeling mice. *Contrast Media Mol. Imaging* **2011**, *6*, 328–330. [CrossRef] [PubMed]
20. Xiong, H.-M. ZnO Nanoparticles applied to bioimaging and drug delivery. *Adv. Mater.* **2013**, *25*, 5329–5335. [CrossRef] [PubMed]
21. Urban, B.E.; Neogi, P.; Senthilkumar, K.; Rajpurohit, S.K.; Jagadeeshwaran, P.; Kim, S.; Fujita, Y.; Neogi, A. Bioimaging using the optimized nonlinear optical properties of ZnO nanoparticles. *IEEE J. Sel. Top. Quantum Electron.* **2012**, *18*, 1451–1456. [CrossRef]
22. Zhang, H.-J.; Xiong, H.-M. Biological applications of ZnO nanoparticles. *Curr. Mol. Imaging* **2013**, *2*, 177–192. [CrossRef]
23. Staedler, D.; Magouroux, T.; Rachid, H.; Joulaud, C.; Extermann, J.; Schwung, S.; Passemard, S.; Kasparian, C.; Clarke, G.; Gerrmann, M.; et al. Harmonic nanocrystals for biolabeling: A survey of optical properties and biocompatibility. *ACS Nano* **2012**, *6*, 2542–2549. [CrossRef]
24. Premanathan, M.; Karthikeyan, K.; Jeyasubramanian, K.; Manivannan, G. Selective toxicity of ZnO nanoparticles toward Gram-positive bacteria and cancer cells by apoptosis through lipid peroxidation. *Nanomed. Nanotechnol. Biol. Med.* **2011**, *7*, 184–192. [CrossRef]
25. Wang, A.; Qi, W.; Wang, N.; Zhao, J.; Muhammad, F.; Cai, K.; Ren, H.; Sun, F.; Chen, L.; Guo, Y.; et al. A smart nanoporous theranostic platform for simultaneous enhanced MRI and drug delivery. *Microporous Mesoporous Mater.* **2013**, *180*, 1–7. [CrossRef]
26. Martinez-Carmona, M.; Gun'ko, Y.; Vallet-Regi, M. ZnO nanostructures for drug delivery and theranostic applications. *Nanomaterials* **2018**, *8*, 268. [CrossRef] [PubMed]
27. Fang, X.; Jiang, L.; Gong, Y.; Li, J.; Liu, L.; Cao, Y. The presence of oleate stabilized ZnO nanoparticles (NPs) and reduced the toxicity of aged NPs to Caco-2 and HepG2 cells. *Chem. Biol. Interact.* **2017**, *278*, 40–47. [CrossRef] [PubMed]

28. Aswathanarayan, J.B.; Vittal, R.R.; Muddegowda, U. Anticancer activity of metal nanoparticles and their peptide conjugates against human colon adenorectal carcinoma cells. *Artif. Cells Nanomed. Biotechnol.* **2018**, *46*, 1444–1451. [CrossRef] [PubMed]
29. Lu, J.; Tang, M.; Zhang, T. Review of toxicological effect of quantum dots on the liver. *J. Appl. Toxicol.* **2019**, *39*, 72–86. [CrossRef]
30. Marfavi, Z.H.; Farhadi, M.; Jameie, S.B.; Zahmatkeshan, M.; Pirhajati, V.; Jameie, M. Glioblastoma U-87MG tumour cells suppressed by ZnO folic acid-conjugated nanoparticles: An in vitro study. *Artif. Cells Nanomed. Biotechnol.* **2019**, *47*, 2783–2790. [CrossRef]
31. Liu, Z.-Y.; Shen, C.-L.; Lou, Q.; Zhao, W.B.; Wei, J.Y.; Liu, K.K.; Zang, J.H.; Dong, L.; Shan, C.-X. Efficient chemiluminescent ZnO nanoparticles for cellular imaging. *J. Lumin.* **2020**, *221*, 117111. [CrossRef]
32. Wang, Y.; He, L.; Yu, B.; Chen, Y.; Shen, Y.; Cong, H. ZnO quantum dots modified by pH-activated charge-reversal polymer for tumor targeted drug delivery. *Polymers* **2018**, *10*, 1272. [CrossRef]
33. Li, Y.; Zhang, C.; Liu, L.; Gong, Y.; Xie, Y.; Cao, Y. The effects of baicalein or baicalin on the colloidal stability of ZnO nanoparticles (NPs) and toxicity of NPs to Caco-2 cells. *Toxicol. Mech. Methods* **2018**, *28*, 167–176. [CrossRef]
34. Jiang, J.; Pi, J.; Cai, J. The advancing of zinc oxide nanoparticles for biomedical applications. *Bioinorg. Chem. Appl.* **2018**, *2018*, 1062562. [CrossRef]
35. Yang, Y.; Song, Z.; Wu, W.; Xu, A.; Lv, S.; Ji, S. ZnO quantum dots induced oxidative stress and apoptosis in HeLa and HEK-293T cell lines. *Front. Pharmacol.* **2020**, *11*, 131. [CrossRef] [PubMed]
36. Sowik, J.; Miodyńska, M.; Bajorowicz, B.; Mikolajczyk, A.; Lisowski, W.; Klimczuk, T.; Kaczor, D.; Medynska, A.Z.; Malankowska, A. Optical and photocatalytic properties of rare earth metal-modified ZnO quantum dots. *Appl. Surf. Sci.* **2019**, *464*, 651–663. [CrossRef]
37. Hong, H.; Wang, F.; Zhang, Y.; Graves, S.A.; Eddine, S.B.Z.; Yang, Y.; Theuer, C.P.; Nickles, R.J.; Wang, X.; Cai, W. Red fluorescent zinc oxide nanoparticle: A novel platform for cancer targeting. *ACS Appl. Mater. Interfaces* **2015**, *7*, 3373–3381. [CrossRef] [PubMed]
38. Liu, Y.; Ai, K.; Yuan, Q.; Lu, L. Fluorescence-enhanced gadolinium-doped zinc oxide quantum dots for magnetic resonance and fluorescence imaging. *Biomaterials* **2011**, *32*, 1185–1192. [CrossRef] [PubMed]
39. Tang, X.; Choo, E.S.G.; Li, L.; Ding, J.; Xue, J. Synthesis of ZnO nanoparticles with tunable emission colors and their cell labeling applications. *Chem. Mater.* **2010**, *22*, 3383–3388. [CrossRef]
40. Wang, S.-L.; Liu, K.-K.; Shan, C.-X.; Liu, E.-S.; Shen, D.-Z. Oleylamine-assisted and temperature-controlled synthesis of ZnO nanoparticles and their application in encryption. *Nanotechnology* **2019**, *30*, 015702. [CrossRef]
41. Zhang, J.; Zhang, R.; Zhao, L.-H.; Sun, S.-Q. Synthesis of water-soluble gamma-aminopropyl triethoxysilane-capped ZnO:MgO nanocrystals with biocompatibility. *CrystEngComm* **2012**, *14*, 613–619. [CrossRef]
42. Yang, W.; Zhang, B.; Ding, N.; Ding, W.; Wang, L.; Yu, M.; Zhang, Q. Fast synthesize ZnO quantum dots via ultrasonic method. *Ultrason. Sonochem.* **2016**, *30*, 103–112. [CrossRef] [PubMed]
43. Koczkur, K.M.; Mourdikoudis, S.; Polavarapu, L.; Skrabalak, S.E. Polyvinylpyrrolidone (PVP) in nanoparticle synthesis. *Dalton Trans.* **2015**, *44*, 17883–17905. [CrossRef] [PubMed]
44. Jacobsson, T.J.; Viarbitskaya, S.; Mukhtar, E.; Edvinsson, T. A size dependent discontinuous decay rate for the exciton emission in ZnO quantum dots. *Phys. Chem. Chem. Phys.* **2014**, *16*, 13849–13857. [CrossRef]
45. Chen, B.W.; Bi, Y.; Luo, X.; Zhang, L. Photoluminescence of monolithic zinc oxide aerogel synthesised by dispersed inorganic sol–gel method. *Mater. Technol.* **2014**, *30*, 65–69. [CrossRef]
46. Senthilkumar, K.; Senthilkumar, O.; Yamauchi, K.; Sato, M.; Morito, S.; Ohba, T.; Nakamura, M.; Fujita, Y. Preparation of ZnO nanoparticles for bio-imaging applications. *Phys. Status Solidi* **2009**, *246*, 885–888. [CrossRef]
47. Djurišić, A.B.; Leung, Y.H. Optical properties of ZnO nanostructures. *Small* **2006**, *2*, 944–961. [CrossRef]
48. Jones, M.; Nedeljkovic, J.; Ellingson, R.J.; Nozik, A.J.; Rumbles, G. Photoenhancement of luminescence in colloidal CdSe quantum dot solutions. *J. Phys. Chem. B* **2003**, *107*, 11346–11352. [CrossRef]
49. Moussodia, R.-O.; Balan, L.; Merlin, C.; Mustin, C.; Schneider, R. Biocompatible and stable ZnO quantum dots generated by functionalization with siloxane-core PAMAM dendrons. *J. Mater. Chem.* **2010**, *20*, 1147–1155. [CrossRef]
50. Moussodia, R.-O.; Balan, L.; Schneider, R. Synthesis and characterization of water-soluble ZnO quantum dots prepared through PEG-siloxane coating. *New J. Chem.* **2008**, *32*, 1388. [CrossRef]
51. Saliba, S.; Serrano, C.V.; Keilitz, J.; Kahn, M.L.; Mingotaud, C.; Haag, R.; Marty, J.-D. Hyperbranched polymers for the formation and stabilization of ZnO nanoparticles. *Chem. Mater.* **2010**, *22*, 6301–6309. [CrossRef]
52. Hancock, J.M.; Rankin, W.M.; Hammad, T.M.; Salem, J.S.; Chesnel, K.; Harrison, R.G. Optical and magnetic properties of ZnO nanoparticles doped with Co, Ni and Mn and synthesized at low temperature. *J. Nanosci. Nanotechnol.* **2015**, *15*, 3809–3815. [CrossRef]
53. Ostrovsky, S.; Kazimirsky, G.; Gedanken, A.; Brodie, C. Selective cytotoxic effect of ZnO nanoparticles on glioma cells. *Nano Res.* **2009**, *2*, 882–890. [CrossRef]
54. Guo, D.; Wu, C.; Jiang, H.; Li, Q.; Wang, X.; Chen, B. Synergistic cytotoxic effect of different sized ZnO nanoparticles and daunorubicin against leukemia cancer cells under UV irradiation. *J. Photochem. Photobiol. B Biol.* **2008**, *93*, 119–126. [CrossRef]
55. Ng, S.M.; Wong, D.S.N.; Phung, J.H.C.; Chin, S.F.; Chua, H.S. Integrated miniature fluorescent probe to leverage the sensing potential of ZnO quantum dots for the detection of copper (II) ions. *Talanta* **2014**, *119*, 639. [CrossRef]

56. Zhang, Y.; Chen, W.; Wang, S.; Liu, Y.; Pope, C. Phototoxicity of zinc oxide nanoparticle conjugates in human ovarian cancer NIH: OVCAR-3 cells. *J. Biomed. Nanotechnol.* **2008**, *4*, 432–438. [CrossRef]
57. Depan, D.; Misra, R.D. Structural and physicochemical aspects of silica encapsulated ZnO quantum dots with high quantum yield and their natural uptake in HeLa cells. *J. Biomed. Mater. Res. Part A* **2014**, *102*, 2934–2941. [CrossRef]
58. Sandmann, A.; Kompch, A.; Mackert, V.; Liebscher, C.H.; Winterer, M. Interaction of L-cysteine with ZnO: Structure, surface chemistry and optical properties. *Langmuir* **2015**, *31*, 5701–5711. [CrossRef] [PubMed]
59. Mirzaei, H.; Darroudi, M. Zinc oxide nanoparticles: Biological synthesis and biomedical applications. *Ceram. Int.* **2017**, *43*, 907–914. [CrossRef]
60. Van Dijken, A.; Meulenkamp, E.A.; Vanmaekelbergh, D.; Meijerink, A. Identification of the transition responsible for the visible emission in ZnO using quantum size effects. *J. Lumin.* **2000**, *90*, 123–128. [CrossRef]
61. Shi, H.Q.; Li, W.N.; Sun, L.W.; Liu, Y.; Xiao, H.M.; Fu, S.Y. Synthesis of silane surface modified ZnO quantum dots with ultrastable, strong and tunable luminescence. *Chem. Commun.* **2011**, *47*, 11921–11923. [CrossRef] [PubMed]

International Journal of
Molecular Sciences

Article

Effect of ZnSO$_4$, MnSO$_4$ and FeSO$_4$ on the Partial Hydrogenation of Benzene over Nano Ru-Based Catalysts

Haijie Sun [1], Yiru Fan [1], Xiangrong Sun [1], Zhihao Chen [2,*], Huiji Li [1], Zhikun Peng [3,*] and Zhongyi Liu [3]

[1] School of Chemistry and Chemical Engineering, Zhengzhou Normal University, Zhengzhou 450044, China; sunhaijie406@zznu.edu.cn (H.S.); FyR158905@sohu.com (Y.F.); sun1398041920@sohu.com (X.S.); huijili@zznu.edu.cn (H.L.)

[2] Zhengzhou Tobacco Research Institute of CNTC, Zhengzhou 450001, China

[3] College of chemistry and molecular engineering, Zhengzhou University, Zhengzhou 450001, China; liuzhongyi406@sohu.com

* Correspondence: chenzh@ztri.com.cn (Z.C.); pengzhikun@zzu.edu.cn (Z.P.); Tel.: +86-371-6767-2762 (Z.C.)

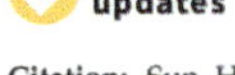
check for
updates

Citation: Sun, H.; Fan, Y.; Sun, X.; Chen, Z.; Li, H.; Peng, Z.; Liu, Z. Effect of ZnSO$_4$, MnSO$_4$ and FeSO$_4$ on the Partial Hydrogenation of Benzene over Nano Ru-Based Catalysts. *Int. J. Mol. Sci.* **2021**, *22*, 7756. https://doi.org/10.3390/ijms22147756

Academic Editor: Raghvendra Singh Yadav

Received: 30 June 2021
Accepted: 16 July 2021
Published: 20 July 2021

Publisher's Note: MDPI stays neutral with regard to jurisdictional claims in published maps and institutional affiliations.

Abstract: Nano Ru-based catalysts, including monometallic Ru and Ru-Zn nanoparticles, were synthesized via a precipitation method. The prepared catalysts were evaluated on partial hydrogenation of benzene towards cyclohexene generation, during which the effect of reaction modifiers, i.e., ZnSO$_4$, MnSO$_4$, and FeSO$_4$, was investigated. The fresh and the spent catalysts were thoroughly characterized by XRD, TEM, SEM, XPS, XRF, and DFT studies. It was found that Zn^{2+} or Fe^{2+} could be adsorbed on the surface of a monometallic Ru catalyst, where a stabilized complex could be formed between the cations and the cyclohexene. This led to an enhancement of catalytic selectivity towards cyclohexene. Furthermore, electron transfer was observed from Zn^{2+} or Fe^{2+} to Ru, hindering the catalytic activity towards benzene hydrogenation. In comparison, very few Mn^{2+} cations were adsorbed on the Ru surface, for which no cyclohexene could be detected. On the other hand, for Ru-Zn catalyst, Zn existed as rodlike ZnO. The added ZnSO4 and FeSO$_4$ could react with ZnO to generate (Zn(OH)$_2$)$_5$(ZnSO$_4$)(H$_2$O) and basic Fe sulfate, respectively. This further benefited the adsorption of Zn^{2+} or Fe^{2+}, leading to the decrease of catalytic activity towards benzene conversion and the increase of selectivity towards cyclohexene synthesis. When 0.57 mol·L^{-1} of ZnSO$_4$ was applied, the highest cyclohexene yield of 62.6% was achieved. When MnSO$_4$ was used as a reaction modifier, H$_2$SO$_4$ could be generated in the slurry via its hydrolysis, which reacted with ZnO to form ZnSO$_4$. The selectivity towards cyclohexene formation was then improved by the adsorbed Zn^{2+}.

Keywords: benzene; partial hydrogenation; cyclohexene; reaction modifier; ZnSO$_4$; MnSO$_4$; FeSO$_4$

1. Introduction

Cyclohexene, a compound with an unstable double bond, can be utilized to synthesize adipic acid, cyclohexanol, and cyclohexanone [1], which are important monomers for the production of nylon 6 and nylon 66. Dehydration of cyclohexanol, dehydrochlorination of halogenated cyclohexane, and Birch reduction methods were traditionally applied for the production of cyclohexene. However, high cost of crude material, side products, and environmental issues are generally addressed, limiting the industrialization of cyclohexene production [2]. In comparison, because of safety, energy conservation, environmental friendliness, and high carbon atom economy, great attention is drawn to the production of cyclohexene via partial hydrogenation of benzene, of which the reaction mechanism is given in Scheme 1. As can been seen in Scheme 1, cyclohexane formation could be done either by direct hydrogenation of benzene or via cyclohexene as an intermediate. It was also reported by Rochin et al. [3] that 1,3-cyclohexadiene could not be detected since diene is extremely unstable due to the further hydrogenation. On the other hand, unfortunately, it is thermodynamically difficult to obtain cyclohexene, since the standard free energy change for cyclohexane formation from benzene hydrogenation is −98 kJ mol^{-1}, which is much lower than that for selective hydrogenation to cyclohexene, i.e., -23 kJ mol^{-1} [4]. Therefore, it is of great necessity to develop a suitable catalytic system with high selectivity towards cyclohexene generation.

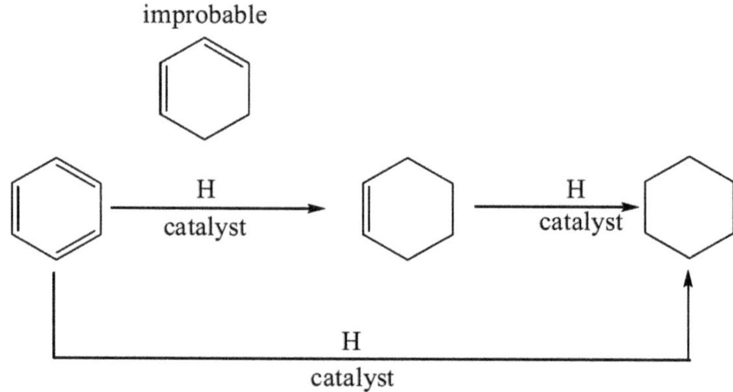

Scheme 1. Reaction mechanism for hydrogenation of benzene.

A reaction modifier is one of the most effective approaches to improve the selectivity towards cyclohexene [5–7] by adding one or several compounds into the reaction slurry to modify the surface of catalysts. In general, cyclohexene can hardly be obtained without applying any reaction modifiers over Ru catalysts [8]. $ZnSO_4$, $FeSO_4$, $CdSO_4$, $CoSO_4$, $NiSO_4$, $CuSO_4$, and $MgSO_4$ were reported as reaction modifiers for partial hydrogenation of benzene towards cyclohexene formation over Ru-based catalysts, among which $ZnSO_4$ and $FeSO_4$ were considered the most effective ones. Notably, it is still controversial in regard to the mechanism for the effect of the reaction modifier. For instance, on one hand, Struijk et al. proposed that the more difficult it is to reduce the cations, the higher are selectivity and yield towards cyclohexene generation [9]. Furthermore, Costa et al. proposed that the oxidation of $ZnNb_2O_6$ contributed to catalytic selectivity over Ru catalysts [10], while Spod et al. stated $Zn(OH)_3^-$ played a significant role in improving the selectivity over Ru catalysts [11]. On the other hand, it was reported by Wang et al. that the binding energy of $Zn2p_{3/2}$ over Ru-Zn/m-ZrO_2 is close to that observed for metallic Zn, declaring that Zn^{2+} was reduced into metallic Zn by the spilled H on the Ru surface [12]. Interestingly, both metallic Zn and Zn^{2+} were observed over Ru-Zn/HAP by Zhang et al. [13]. In general, the effect of Zn can be addressed as follows: (1) the bonding status of hydrogen on the catalyst surface was modified by Zn [14]. The total amount of bonded hydrogen was declined, while the amount of weak bonded hydrogen was enhanced, leading to lower activity towards benzene conversion and higher selectivity towards cyclohexene, respectively; (2) Zn could cover part of the active sites, which is favorable for complete hydrogenation of benzene towards cyclohexane generation [12]; (3) some electrons could be transferred from Zn to Ru, and $Ru^{\delta-}$ might benefit the formation of cyclohexene [12,13]; (4) the adsorption of cyclohexene on the Ru surface could be retarded by the presence of Zn, hindering the further hydrogenation of cyclohexene to cyclohexane [15]. Generally, these controversies are mainly attributed to the development of technologies for the catalyst characterization. Therefore, for the better design of the Ru-based catalytic system on partial hydrogenation of benzene towards cyclohexene production, it is of great significance to reveal the state of Zn and the effect of reaction additives.

Based on the aforementioned situation, $ZnSO_4$, $FeSO_4$, and $MnSO_4$ were applied for the partial hydrogenation of benzene towards cyclohexene formation over monometallic Ru catalyst and Ru-Zn catalyst. The mechanism in which the reaction modifier affects the catalytic activity of nano Ru-based catalysts was thoroughly investigated by XRD, XRF, TEM, and XPS.

2. Results and Discussions

2.1. Ru Catalyst

Figure 1 illustrates the XRD patterns of the monometallic Ru catalyst after the catalytic experiments applying different amount of $ZnSO_4$, $FeSO_4$, and $MnSO_4$ as reaction modifiers. As can be observed, characteristic diffractions corresponding to metallic Ru were shown over all measured samples, demonstrating that Ru existed mainly as the metallic state during the hydrogenation reaction. Furthermore, the particle size of the monometallic Ru catalyst before and after catalytic experiments was calculated via the Scherrer equation and is listed in Table 1. In comparison to the fresh sample, no obvious change was observed for Ru particle size after the hydrogenation reaction with different amounts of $ZnSO_4$, $FeSO_4$, and $MnSO_4$ as reaction modifiers. The particle size of all samples ranged from 3.5 nm to 3.7 nm, which was extremely comparable to that of fresh Ru nanoparticle. This suggests that the particle size of Ru was not affected by the utilization of a reaction modifier.

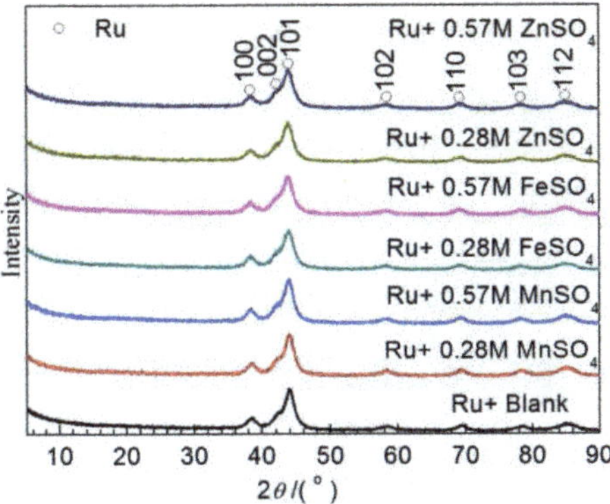

Figure 1. XRD patterns of the monometallic Ru catalyst after catalytic experiments applying different amounts of $ZnSO_4$, $FeSO_4$, and $MnSO_4$ as reaction modifiers.

Table 1. Composition and particle size of the monometallic Ru before and after catalytic experiments with different amounts of reaction modifiers as well as the pH value of slurry.

Conditions	$n(Zn)/n(Ru)$ [1] (mol/mol)	$n(Mn)/n(Ru)$ [1] (mol/mol)	$n(Fe)/n(Ru)$ [1] (mol/mol)	Particle Size [2] (nm)	pH Value [3]
Ru catalyst	-	-	-	3.6	-
Ru + Blank	-	-	-	3.6	6.95
Ru + 0.28 M MnSO₄	-	0.0019	-	3.7	3.23
Ru + 0.57 M MnSO₄	-	0.0023	-	3.6	3.11
Ru + 0.28 M FeSO₄	-	-	0.0358	3.5	4.66
Ru + 0.57 M FeSO₄	-	-	0.0295	3.7	3.83
Ru + 0.28 M ZnSO₄	0.0307	-	-	3.6	4.35
Ru + 0.57 M ZnSO₄	0.0258	-	-	3.7	3.53

[1] Determined by XRF instrument; [2] determined by XRD instrument; [3] pH value of the slurry after catalytic experiments at room temperature.

Table 1 gives the composition of the Ru catalyst before and after catalytic experiments as well as the pH value of the slurry. As observed, only Ru was detected over the fresh Ru catalyst and Ru after hydrogenation reaction in pure distilled water. In comparison, Zn, Fe, and Mn were observed when 0.28 mol L^{-1} of $ZnSO_4$, $FeSO_4$, and $MnSO_4$ were applied as reaction modifiers, respectively. Moreover, the molar ratio of added metal to Ru ranged from Fe > Zn > Mn, implying that the ability of adsorption on the Ru surface ranged from $FeSO_4$ > $ZnSO_4$ > $MnSO_4$. Additionally, the pH value of slurry after the catalytic experiments ranged from the same order, demonstrating that $MnSO_4$ provided the most acidic condition through its hydrolysis. This suggests that the more easily the salt was hydrolyzed, the more difficult it was to adsorb on the Ru surface. Interestingly, when the concentration of applied reaction modifier was increased to 0.57 mol L^{-1}, the amount of adsorbed metal was decreased. This can be rationalized considering that, by enhancing the concentration of the added modifier, the slurry became more acidic due to the hydrolysis of the metal salts, for which the adsorption of the corresponding metal was retarded. This was also confirmed by the pH value of the slurry after the hydrogenation reaction.

Figure 2 demonstrates the TEM images of the fresh Ru nanoparticles (Figure 2a) and the SEM image as well as the element mapping image of the Ru catalyst after catalytic experiments adding 0.57 mol L^{-1} of $ZnSO_4$, $FeSO_4$, and $MnSO_4$ (Figure 2b–d). As can be seen from Figure 2a, the fresh monometallic Ru catalyst displayed a circular or an elliptical shape. The particle size of Ru was around 3.5 nm, which was in good agreement with XRD results. In addition, it was shown from Figure 2b,c that plenty of Zn and Fe were adsorbed on the Ru surface. On the contrary, only a few Mn were observed from Figure 2c. This was consistent with XRF results (i.e., n(Mn)/n(Ru) was only 0.0023).

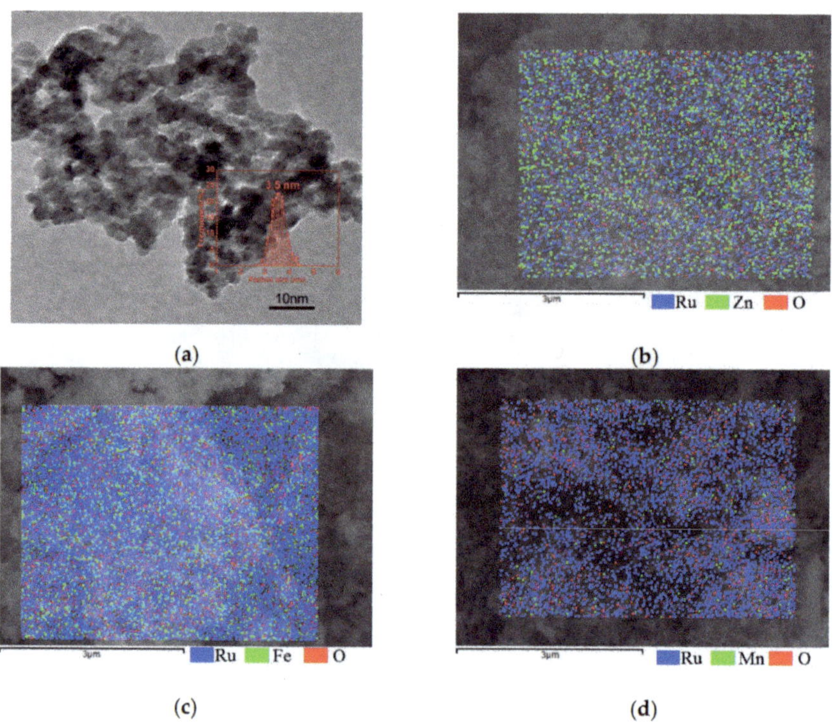

Figure 2. TEM images (**a**) as well as particle size distribution of the fresh Ru nanoparticles (insert image) and the SEM image as well as the element mapping image of Ru catalyst after catalytic experiments adding 0.57 mol L^{-1} of $ZnSO_4$ (**b**), $FeSO_4$ (**c**), and $MnSO_4$ (**d**).

Figure 3 shows the XPS spectra of the monometallic Ru catalyst after hydrogenation reaction applying 0.57 mol L^{-1} of $ZnSO_4$, $FeSO_4$, and $MnSO_4$ as the reaction modifiers, respectively. As can be observed from Figure 3a, two peaks related to binding energies (BE) of 709.8 eV and 712.2 eV were attributed to Fe^{2+} and Fe^{3+}, respectively. The peak at 719.3 eV was related to Fe^{3+} [16]. Furthermore, the peak area of Fe^{2+} was significantly larger than that of Fe^{3+}, indicating that the adsorbed Fe mainly existed as Fe^{2+}. From Figure 3b,c, the BE of Zn $2p_{3/2}$ and Zn LMM was shown at 1021.5 eV and 988.9 eV, respectively. The BE was the same as Zn^{2+}, which was reported by Sun et al. [17–19], suggesting that the adsorbed Zn mainly existed as Zn^{2+}. Unfortunately, due to the extremely low amount of adsorbed Mn, no reflection attributed to Mn^{2+} was observed. As illustrated in Figure 3d, the BE of

Ru 3d$_{5/2}$ was 280.1 eV when no reaction modifier was added, which belonged to metallic Ru. This further proved Ru existed mainly as the metallic state during the hydrogenation reaction. Applying MnSO$_4$ as a reaction modifier, the BE of Ru 3d$_{5/2}$ was still observed at 280.1 eV, implying that no electron transfer occurred between Ru and Mn. On the other hand, when ZnSO$_4$ and FeSO$_4$ were used, the BE of Ru 3d$_{5/2}$ increased from 280.1 eV to 280.6 eV and 280.9 eV, respectively. This can be rationalized by the fact that some electrons were transferred from Ru to Zn and Fe, generating Ru$^{\delta+}$. In addition, it was deemed that more electrons were transferred from Ru to Fe than from Ru to Zn.

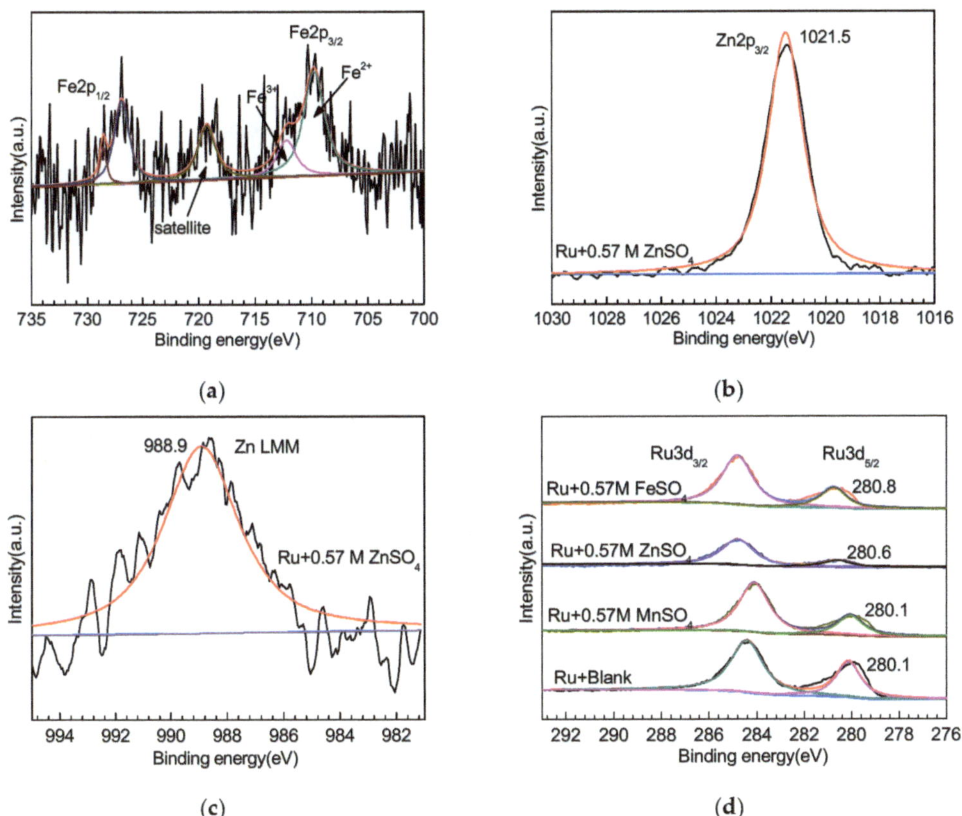

Figure 3. XPS profiles of the monometallic Ru catalyst after hydrogenation reaction applying 0.57 mol L^{-1} of ZnSO$_4$, FeSO$_4$, and MnSO$_4$ as the reaction modifiers. (**a**) Fe 2p over the monometallic Ru catalyst applying 0.57 mol L^{-1} FeSO$_4$; (**b**) Zn 2p$_{3/2}$ over the monometallic Ru catalyst applying 0.57 mol L^{-1} ZnSO$_4$; (**c**) Zn LMM over the monometallic Ru catalyst applying 0.57 mol L^{-1} ZnSO$_4$; and (**d**) Ru 3d applying 0.57 mol L^{-1} of ZnSO$_4$, FeSO$_4$, and MnSO$_4$ as the reaction modifiers.

The privileged structure of the formed complex between single/double cyclohexene molecules and cations is shown in Figure 4, and the corresponding Gibbs free energy ($\Delta_f G$) is given in Table 2. As can be seen, the $\Delta_f G$ of the formed complex between Mn^{2+} and a single cyclohexene molecule was -1723 kJ mol^{-1}, and lower $\Delta_f G$ of -1794 kJ mol^{-1} was observed over the complex formed by Fe^{2+} and a single cyclohexene molecule. The most negative $\Delta_f G$ of -1866 kJ mol^{-1} was obtained over the complex formed between Zn^{2+} and a single cyclohexene molecule. Furthermore, the same trend of $\Delta_f G$ was noticed over the formed complex between two cyclohexene molecules and the cations. This showed that Zn^{2+} was the most suitable cation to stabilize cyclohexene by forming a complex.

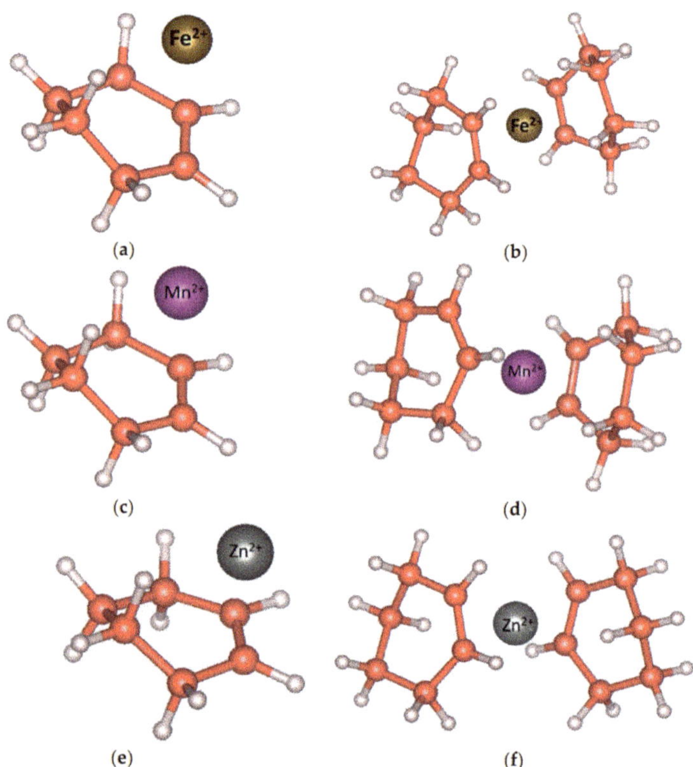

Figure 4. The privileged structure of the formed complex between single/double cyclohexene molecule and Fe^{2+}, Mn^{2+}, as well as Zn^{2+}. (**a**) Fe^{2+} and single cyclohexene molecule; (**b**) Fe^{2+} and double cyclohexene molecules; (**c**) Mn^{2+} and single cyclohexene molecule; (**d**) Mn^{2+} and double cyclohexene molecules; (**e**) Zn^{2+} and single cyclohexene molecule; and (**f**) Zn^{2+} and double cyclohexene molecules.

Table 2. The Gibbs free energy of the formed complex between single/double cyclohexene molecules and cations.

Complex	$\Delta_f G/(\text{kJ}\cdot\text{mol}^{-1})$
Cyclohexene Mn(II)	−1723
(Cyclohexene)$_2$Mn(II)	−1598
Cyclohexene Fe(II)	−1794
(Cyclohexene)$_2$Fe(II)	−1661
Cyclohexene Zn(II)	−1866
(Cyclohexene)$_2$Zn(II)	−1815

Catalytic activity towards benzene conversion and cyclohexene selectivity over monometallic Ru catalyst with different reaction modifiers are illustrated in Figure 5. Notably, complete conversion of benzene was observed within 5 min when $MnSO_4$ (both 0.28 and 0.57 mol L^{-1}) was applied, and no cyclohexene was detected. Similar results were obtained when no reaction modifier was added, where 100% of benzene was converted to cyclohexane within 5 min of reaction time. In comparison, when $ZnSO_4$ or $FeSO_4$ were used, catalytic activity towards benzene conversion over monometallic Ru was retarded, while cyclohexene selectivity was increased. With increasing the concentration of used reaction modifier (e.g., $ZnSO_4$ or $FeSO_4$), catalytic activity towards benzene conversion was enhanced and selectivity to cyclohexene dropped. In addition, higher benzene conversion and lower cyclohexene selectivity was observed with applying $ZnSO_4$ than that obtained with utilizing $FeSO_4$ as the reaction modifier. Based on the characterization results, it is deemed that the effect of reaction modifier is mainly attributed to the adsorption of cation. When $MnSO_4$ is added, Mn^{2+} was barely adsorbed on the Ru surface, thus the catalytic activity and selectivity towards cyclohexene formation was not influenced. Additionally, the amount of adsorbed Zn^{2+} or Fe^{2+} was declined with raising the concentration of applied $ZnSO_4$ or $FeSO_4$, respectively. This leads to the higher catalytic activity towards benzene conversion and lower selectivity towards cyclohexene generation. How the adsorbed Zn^{2+} and Fe^{2+} affecting the partial hydrogenation of benzene over Ru catalysts can be addressed as following: 1. some electrons could be transferred from Ru to Zn^{2+}/Fe^{2+} to generate $Ru^{\delta+}$, which is less active for activation the π orbital of benzene. Only two of the π orbital of benzene was activated, resulting in the synthesis of cyclohexene via a partial hydrogenation procedure [20]; 2. more stabilized complex could be formed between Zn^{2+}/Fe^{2+} and cyclohexene, improving the selectivity towards cyclohexene production [21,22]; 3. the adsorbed Zn^{2+} and Fe^{2+} could cover parts of the active sites of Ru for H_2 dissociation, hence benefiting the cyclohexene formation [9]. Thus, with increasing the concentration of $ZnSO_4/FeSO_4$, the amount of adsorbed Zn^{2+}/Fe^{2+} was declined, hence the selectivity towards cyclohexene was decreased. Furthermore, more Fe^{2+} was adsorbed on the Ru surface than that observed for Zn^{2+}, leading to more electron transfer and more active sites coverage for Fe than that for Zn. It is therefore lower catalytic activity towards benzene conversion and higher cyclohexene selectivity was achieved over monometallic Ru catalyst with applying $FeSO_4$ than using $ZnSO_4$.

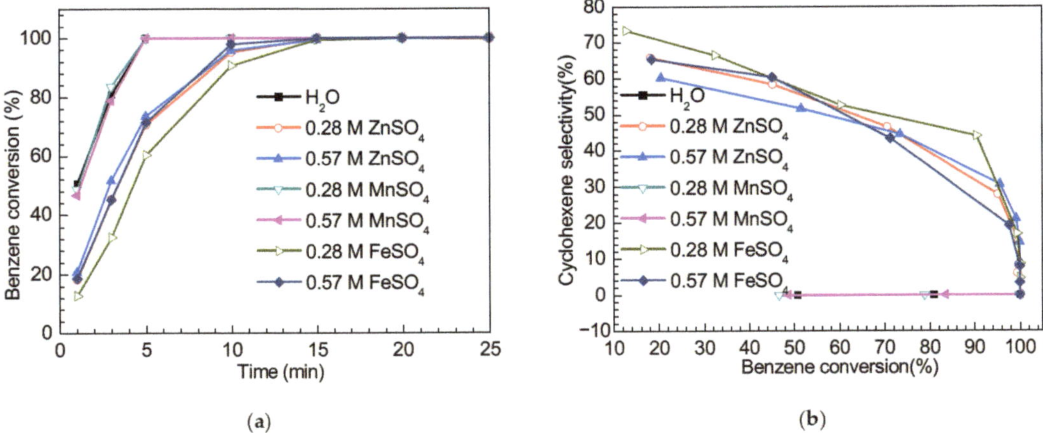

Figure 5. Catalytic activity towards benzene conversion and cyclohexene selectivity over the monometallic Ru catalyst with different reaction modifiers (m_{cat} = 1.8 g, V_{H2O} = 280 mL, $V_{benzene}$ = 140 mL, T = 423 K, p H_2 = 5.0 MPa) (**a**): benzene conversion; (**b**): cyclohexene selectivity.

2.2. Ru-Zn Catalyst

Figure 6 exhibits the XRD patterns of the Ru-Zn catalyst after catalytic experiments applying different amounts of $ZnSO_4$, $FeSO_4$, and $MnSO_4$ as reaction modifiers. As can be observed, characteristic diffractions corresponding to metallic Ru with the hexagonal phase and ZnO with the hexagonal phase were observed over Ru-Zn after hydrogenation without adding any reaction modifier, demonstrating that Ru and Zn existed mainly as metallic Ru and ZnO, respectively. Applying $MnSO_4$ or $FeSO_4$ (both 0.28 and 0.57 mol L^{-1}), the reflection to ZnO could no longer be observed. This can be attributed to the fact that $MnSO_4$ or $FeSO_4$ could react with ZnO to form new compounds, which were dissolvable under the reaction conditions. When $ZnSO_4$ was added as a reaction modifier, a new diffraction related to $(Zn(OH)_2)_5(ZnSO_4)(H_2O)$ was observed, indicating that ZnO could react with $ZnSO_4$ to generate $(Zn(OH)_2)_5(ZnSO_4)(H_2O)$. The same results were reported by Sun et al. [23] and Yan et al. [24].

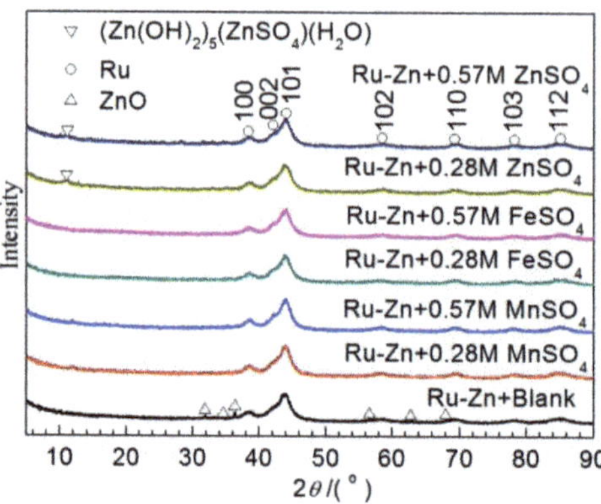

Figure 6. XRD patterns of the Ru-Zn catalyst after catalytic experiments applying different amounts of $ZnSO_4$, $FeSO_4$, and $MnSO_4$ as reaction modifiers.

Table 3 lists the composition and the particle size of the Ru-Zn catalyst after catalytic experiments as well as the pH value of the slurry. Without using any reaction modifier, the molar ratio of Zn to Ru was close to that of the fresh Ru-Zn catalyst (e.g., n(Zn)/n(Ru) = 0.2645), suggesting that Zn was barely lost during the catalytic experiment. The slurry was close to neutral (pH = 6.95). When 0.28 mol L^{-1} of $MnSO_4$ was used, n(Zn)/n(Ru) was decreased from 0.2645 to 0.2228, and n(Mn)/n(Ru) was also raised in comparison to that observed over the monometallic Ru catalyst (e.g., 0.0247 vs. 0.0019). This was mainly because the acidic condition was neutralized by ZnO. Hence, the pH value of the slurry increased, resulting in more Mn^{2+} and SO_4^{2-} being adsorbed on the Ru surface. While the concentration of used $MnSO_4$ was increased to 0.57 mol L^{-1}, the slurry became more acidic, leading to less adsorption of Zn^{2+}, Mn^{2+}, and SO_4^{2-} than that obtained with 0.28 mol L^{-1} of $MnSO_4$. Furthermore, with addition of 0.28 mol L^{-1} of $ZnSO_4$, the n(Zn)/n(Ru) slightly dropped in comparison to that detected without any reaction modifier (e.g., 0.2360 vs. 0.2645). This can be rationalized considering that part of ZnO was dissolved under such acidic conditions. However, a significant increase of S was observed, i.e., n(S)/n(Ru) = 0.0216, and the molar ratio of Zn to Ru was obviously higher than that observed over the monometallic Ru catalyst under the same conditions (e.g., 0.2360 vs. 0.0307). This was mainly due to the formation of $(Zn(OH)_2)_5(ZnSO_4)(H_2O)$ salt from the reaction between ZnO and $ZnSO_4$, which benefited the adsorption of Zn^{2+}. By enhancing the concentration of $ZnSO_4$ from 0.28 mol L^{-1} to 0.57 mol L^{-1}, the pH value decreased from 6.08 to 5.61. This resulted in the lower content of formed $(Zn(OH)_2)_5(ZnSO_4)(H_2O)$, hence the adsorbed Zn^{2+}. On the other hand, when $FeSO_4$ was applied, a severe decline of adsorbed Zn was noticed. It was deemed that $FeSO_4$ could react with ZnO to form salt-like basic Fe sulfate salt, which depressed the adsorption of Zn. Moreover, the particle size of all Ru-Zn catalysts ranged from 3.7 nm to 3.9 nm after the experiments, suggesting that particle size of Ru-Zn was not affected by the utilization of reaction modifiers.

The TEM and the SEM images as well as the element mapping image of the spent Ru-Zn catalyst without using any reaction modifier are illustrated in Figure 7a,b, respectively. It can be seen from Figure 7a that the spent Ru-Zn catalyst displayed a circular or an elliptical shape, of which particle size was around 3.4 nm. Moreover, rodlike ZnO was observed, indicating that ZnO could not be uniformly dispersed on the surface of the catalyst. As observed from Figure 7b, the rodlike shape of the Zn element was also observed, which was consistent with the TEM results. Figure 7c displays the TEM image of the spent Ru-Zn catalyst applying $ZnSO_4$ (0.57 mol L^{-1}) as the reaction modifier. No rodlike ZnO was detected, suggesting that ZnO was consumed through the reaction with $ZnSO_4$. Furthermore, the particle size of Ru-Zn was mainly distributed at 3.3 nm, which was comparable to that obtained over spent Ru-Zn without using any reaction modifier. This implied that the particle size of Ru-Zn could not be affected by utilization of the reaction modifier. Figure 7d demonstrates the SEM image as well as the element mapping image of the spent Ru-Zn catalyst using $ZnSO_4$ as a reaction modifier. No rodlike ZnO could be observed, and Zn species was uniformly dispersed on the surface of the Ru-Zn catalyst. This can be rationalized considering that $(Zn(OH)_2)_5(ZnSO_4)(H_2O)$ could be generated and uniformly dispersed on the surface of the Ru-Zn catalyst. When $FeSO_4$ was added (0.57 mol L^{-1}), as shown in Figure 7e, both Zn and Fe species were uniformly dispersed on the surface of the Ru-Zn catalyst. Besides, rodlike ZnO disappeared. A similar phenomenon was observed when $MnSO_4$ (0.57 mol L^{-1}) was applied, where Zn and Mn species were also uniformly dispersed on the surface of the Ru-Zn catalyst. Interestingly, more Mn was observed in comparison to that detected over the monometallic Ru catalyst, indicating an enhancement of adsorbed Mn on the Ru-Zn surface.

Table 3. Composition and particle size of monometallic Ru before and after catalytic experiments with different amounts of reaction modifiers as well as the pH value of slurry.

Conditions	n(Zn)/n(Ru) [1] (mol/mol)	n(Mn)/n(Ru) [1] (mol/mol)	n(Fe)/n(Ru) [1] (mol/mol)	n(S)/n(Ru) (mol/mol)	Particle [2] Size(nm)	pH [3] Value
Ru-Zn(0.27) + Blank	0.2645	-	-	-	3.7	6.95
Ru-Zn(0.27) + 0.28 M $MnSO_4$	0.1528	0.0247	-	0.0078	3.8	6.98
Ru-Zn(0.27) + 0.57 M $MnSO_4$	0.1413	0.0225	-	0.0053	3.9	6.85
Ru-Zn(0.27) + 0.28 M $FeSO_4$	0.0624	-	0.2422	0.0032	3.7	6.92
Ru-Zn(0.27) + 0.57 M $FeSO_4$	0.0542	-	0.2782	0.0048	3.8	6.51
Ru-Zn(0.27) + 0.28 M $ZnSO_4$	0.2360	-	-	0.0216	3.8	6.08
Ru-Zn(0.27) + 0.57 M $ZnSO_4$	0.2135	-	-	0.0183	3.9	5.61

[1] Determined by XRF instrument; [2] determined by XRD instrument; [3] pH value of the slurry after catalytic experiments at room temperature.

Figure 8 shows the XPS profiles of the Ru-Zn catalyst after catalytic experiments applying $ZnSO_4$, $FeSO_4$, and $MnSO_4$ (all 0.57 mol L^{-1}) as the reaction modifier, respectively. The BE of Zn $2p_{3/2}$ and Zn LMM over Ru-Zn catalyst using $ZnSO_4$ or $MnSO_4$ as reaction modifier was demonstrated in Figure 8a,b. With addition of $ZnSO_4$, the BEs of Zn $2p_{3/2}$ and Zn LMM over Ru-Zn catalyst were observed at 1021.0 eV and 988.5 eV, respectively. In comparison, when $MnSO_4$ was applied, the BEs of Zn $2p_{3/2}$ and Zn LMM over the Ru-Zn catalyst were observed at 1021.7 eV and 987.1 eV, respectively. It was deemed that Zn mainly existed as Zn^{2+}, as the BE of Zn LMM for metallic Zn was 990.1 eV [25]. More importantly, more electrons were transferred from Zn to Ru using $ZnSO_4$ than those obtained using $MnSO_4$ as the reaction modifier. As can be observed from Figure 8c, two peaks related to binding energies of 710.7 eV and 713.4 eV were attributed to Fe^{2+} and Fe^{3+}, respectively. Besides, the satellite diffraction at 718.6 eV belonged to Fe^{2+} [26]. Furthermore, the peak area of Fe^{2+} was obviously larger than that of Fe^{3+}, suggesting that the adsorbed Fe mainly existed as Fe^{2+}. As observed in Figure 8d, the BE of Mn $2p_{3/2}$ was shown at 644.2 eV, which was attributed to Mn^{2+} [27]. This implied that the adsorbed Mn was not reduced. Furthermore, as illustrated in Figure 8e, the BE of Ru $3d_{5/2}$ was 278.9 eV when no reaction modifier was added, which belonged to metallic Ru. This revealed that undispersed ZnO of Ru-Zn could not affect the electronic structure of Ru. Applying $MnSO_4$ as a reaction modifier, the BE of Ru $3d_{5/2}$ was slightly increased from 278.9 eV to 279.0 eV. Based on the aforementioned discussion, it was already concluded that adsorbed Mn could not affect the electronic structure of Ru. Therefore, combined with the shift for BE of Zn $2p_{3/2}$ and Zn LMM, it was deemed that the slight raise for the BE of Ru $3d_{5/2}$ was attributed to the adsorbed Zn. In addition, an obvious increase of the BE of Ru $3d_{5/2}$ was noticed when $ZnSO_4$ was applied, i.e., from 278.9 eV to 280.2 eV. This can be rationalized in terms of the formation of $(Zn(OH)_2)_5(ZnSO_4)(H_2O)$, which contained more Zn species. Higher amounts of adsorbed Zn resulted in the higher BE of Ru $3d_{5/2}$, implying that more $Ru^{\delta+}$ was generated. Moreover, the highest BE of Ru $3d_{5/2}$ was observed with addition of FeSO$_4$ (e.g., 280.8 eV), demonstrating that the most electrons were transferred from Ru to Fe.

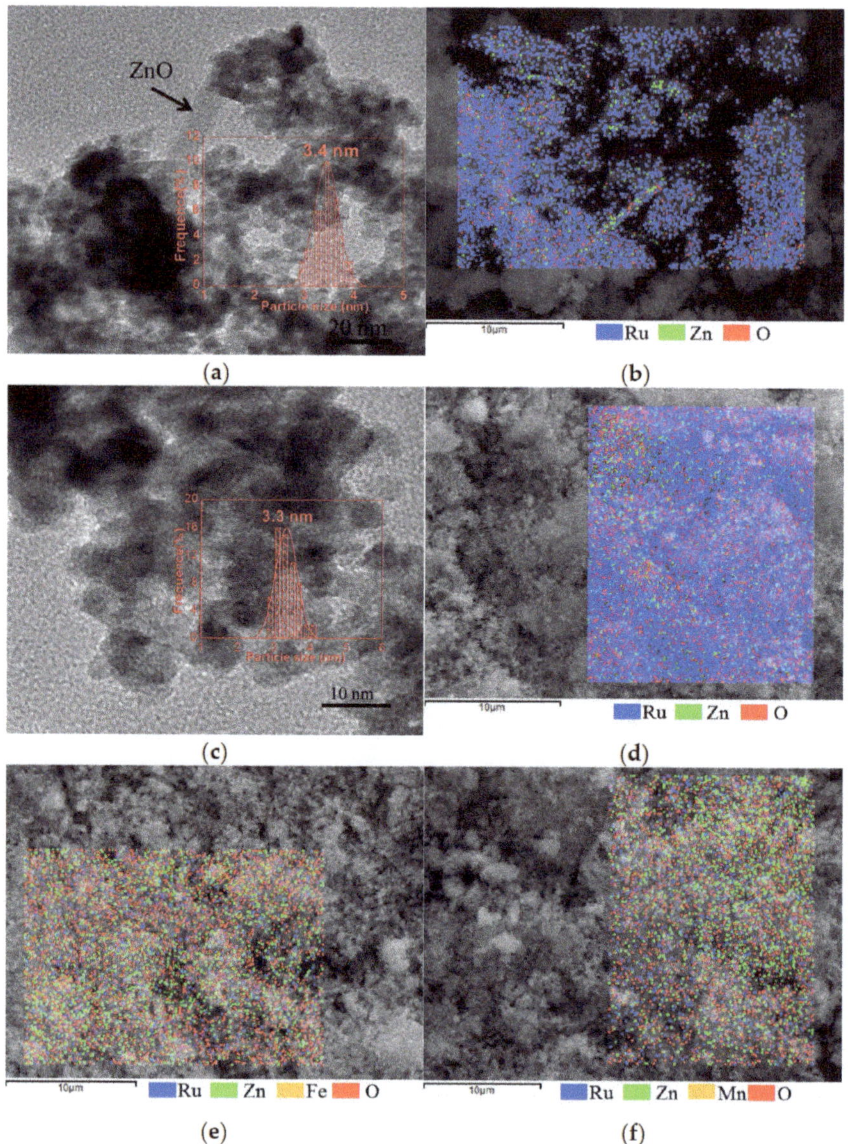

Figure 7. TEM images, SEM images, particle size distribution, and element mapping image of the spent Ru-Zn catalyst with different reaction modifiers added. (**a**) TEM image and particle size distribution (insert image) of the spent Ru-Zn catalyst without using any reaction modifier; (**b**) SEM image as well as element mapping image of the spent Ru-Zn catalyst without using any reaction modifier; (**c**) TEM image of the spent Ru-Zn catalyst applying ZnSO$_4$ (0.57 mol L^{-1}) as the reaction modifier; (**d**) SEM image as well as the element mapping image of the spent Ru-Zn catalyst using ZnSO$_4$ as a reaction modifier; (**e**) SEM image as well as the element mapping image of the spent Ru-Zn catalyst using FeSO$_4$ as a reaction modifier; (**f**) SEM image as well as the element mapping image of the spent Ru-Zn catalyst using MnSO$_4$ as a reaction modifier.

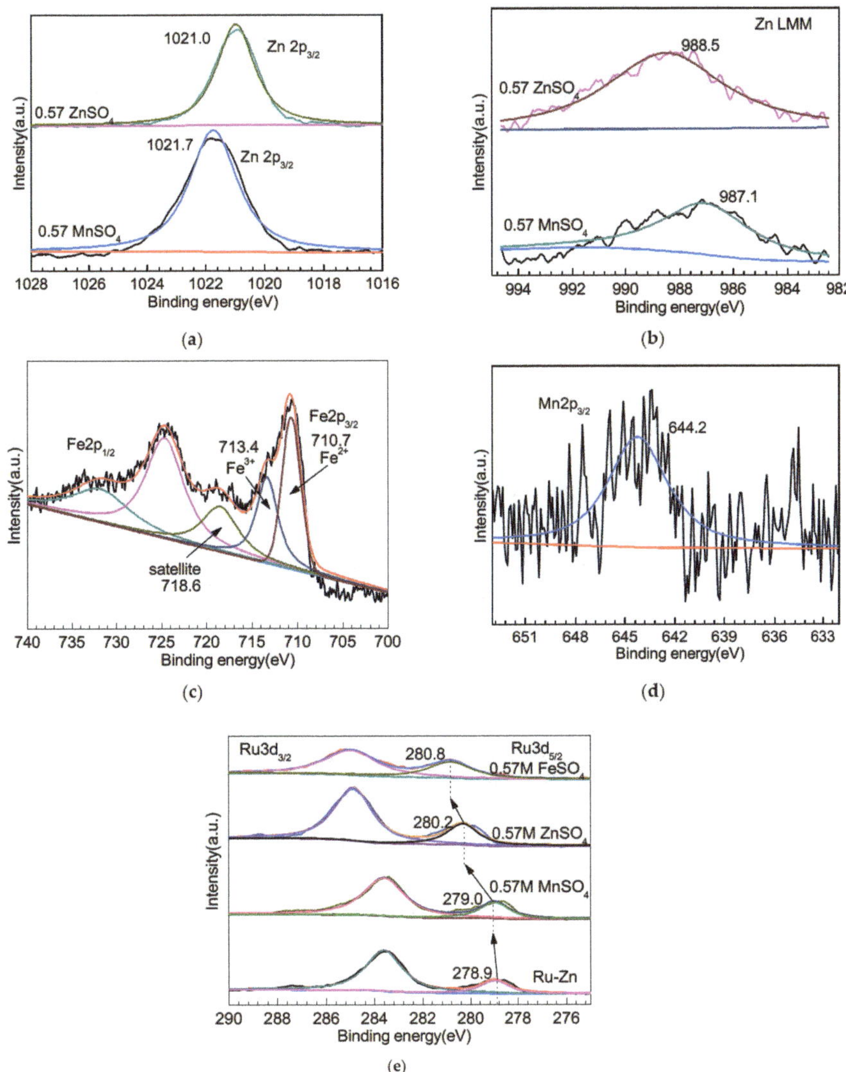

Figure 8. XPS profiles of the Ru-Zn catalyst after hydrogenation reaction applying 0.57 mol L^{-1} of ZnSO$_4$, FeSO$_4$, and MnSO$_4$ as the reaction modifiers. (**a**) Zn 2p$_{3/2}$ over the Ru-Zn catalyst applying 0.57 mol L^{-1} of ZnSO$_4$ and MnSO$_4$, respectively; (**b**) Zn LMM over the Ru-Zn catalyst applying 0.57 mol L^{-1} ZnSO$_4$ and MnSO$_4$, respectively; (**c**) Fe 2p over the Ru-Zn catalyst applying 0.57 mol L^{-1} FeSO$_4$; (**d**) Mn 2p$_{3/2}$ over the Ru-Zn catalyst applying 0.57 mol L^{-1} MnSO$_4$; and (**e**) Ru 3d$_{5/2}$ with and without reaction modifiers.

Catalytic activity towards benzene conversion and cyclohexene selectivity over the Ru-Zn catalyst with different reaction modifiers are illustrated in Figure 9. As can be observed from Figure 9a,b, in comparison to that achieved with no reaction modifier, a significant decrease of catalytic activity towards benzene conversion as well as an increase towards cyclohexene formation were obtained by using reaction modifiers. Among the used reaction modifiers, the lowest benzene conversion and the highest cyclohexene selectivity were achieved over the Ru-Zn catalyst using 0.57 mol L^{-1} of FeSO$_4$, and 0.28 mol L^{-1} of MnSO$_4$ gave the highest benzene conversion and the lowest cyclohexene selectivity within

25 min of reaction time. Notably, a lower concentration of $ZnSO_4/MnSO_4$ and a higher concentration of $FeSO_4$ were beneficial for hindering the catalytic activity towards benzene conversion and improving the selectivity towards cyclohexene generation over the Ru-Zn catalyst. Based on the characterization results and discussions, it was deemed that the adsorbed Zn^{2+}/Fe^{2+} played the essential role in the synthesis of cyclohexene from partial hydrogenation of benzene over the Ru-Zn catalyst. This can be rationalized considering that some electrons could be transferred from Ru to Zn^{2+}/Fe^{2+} to generate $Ru^{\delta+}$, which could only activate two of the π orbital of benzene, resulting in benefits for synthesis of cyclohexene via a partial hydrogenation procedure. Furthermore, Zn^{2+}/Fe^{2+} was of great capability to stabilize cyclohexene, for which the further hydrogenation of cyclohexene to cyclohexane was retarded. Additionally, the adsorbed Zn^{2+} and Fe^{2+} could cover parts of the most active sites of Ru for direct hydrogenation of benzene to cyclohexane, thus benefiting the formation of cyclohexene. Considering both catalytic activity towards benzene conversion and selectivity to cyclohexene, $ZnSO_4$ with a concentration of 0.57 mol L^{-1}, among all investigated reaction modifiers, was the most suitable for partial hydrogenation of benzene for cyclohexene production over Ru-Zn catalysts. As observed from Figure 9c,d, 48.1% of cyclohexene yield was achieved over the Ru-Zn catalyst applying 0.57 mol L^{-1} of $ZnSO_4$ as a reaction modifier within 25 min of the catalytic experiment, and the highest cyclohexene yield of 62.6% was obtained when the reaction time reached 40 min, which was one of the highest yields of cyclohexene to ever be reported [4,28].

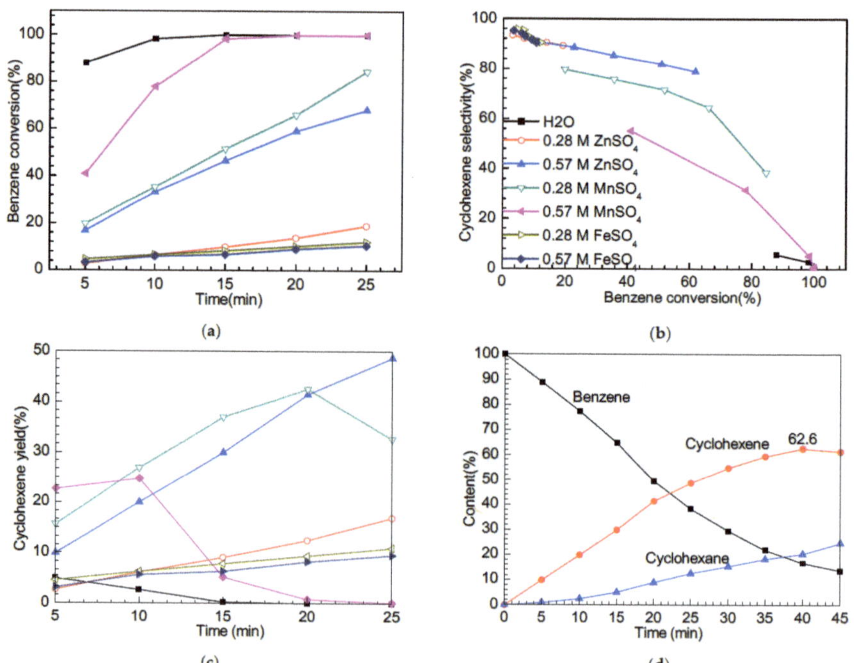

Figure 9. Catalytic activity towards benzene conversion and cyclohexene selectivity over the Ru-Zn catalyst with different reaction modifiers (m_{cat} = 1.8 g, V_{H2O} = 280 mL, $V_{benzene}$ = 140 mL, T = 423 K, p H_2 = 5.0 MPa) (**a**): benzene conversion; (**b**): cyclohexene selectivity; (**c**) cyclohexene yield; and (**d**) kinetic curve over Ru-Zn with addition of 0.57 mol L^{-1} of $ZnSO_4$.

3. Conclusions

Nano Ru-based catalysts, including monometallic Ru and Ru-Zn nanoparticles, were synthesized and evaluated on partial hydrogenation of benzene towards cyclohexene generation, during which the effect of the reaction modifiers, i.e., $ZnSO_4$, $MnSO_4$, and

$FeSO_4$, was investigated. It was found that Zn^{2+} or Fe^{2+} could be adsorbed on the surface of the monometallic Ru catalyst, which could stabilize the formed cyclohexene. This led to an enhancement of catalytic selectivity towards cyclohexene. Furthermore, electron transfer was observed from Zn^{2+} or Fe^{2+} to Ru, hindering the catalytic activity towards benzene conversion. In comparison, very few Mn^{2+} cations were adsorbed on the Ru surface, for which no cyclohexene could be detected. On the other hand, for the Ru-Zn catalyst, Zn existed as rodlike ZnO. The added $ZnSO_4$ and $FeSO_4$ could react with ZnO to generate $(Zn(OH)_2)_5(ZnSO_4)(H_2O)$ and $Fe(OH)SO_4$, respectively. This further benefited the adsorption of Zn^{2+} or Fe^{2+}, leading to the decrease of catalytic activity towards benzene conversion and the increase of selectivity towards cyclohexene synthesis. When 0.57 mol L^{-1} of $ZnSO_4$ was applied, the highest cyclohexene yield of 62.6% was achieved. When $MnSO_4$ was used as a reaction modifier, H_2SO_4 could be generated in the slurry via its hydrolysis, which reacted with ZnO to form $ZnSO_4$. The selectivity towards cyclohexene formation was then improved by the adsorbed Zn^{2+}.

4. Materials and Methods

4.1. Chemicals

All regents were analytical grade. $RuCl_3 \cdot 3H_2O$ was received from Sino-Platinum Co. Ltd. (Kunming, China). NaOH, $FeSO_4 \cdot 7H_2O$, $ZnSO_4 \cdot 7H_2O$, $MnSO_4 \cdot H_2O$, and benzene were purchased from Kemiou Chemical Reagent Co. Ltd. (Tianjin, China). All chemicals were directly used without any further purification, and deionized water was utilized in all experiments.

4.2. Preparation of Catalysts

Ru-Zn catalysts were synthesized via the following procedure: 100 mL of NaOH (0.75 mol$\cdot L^{-1}$) aqueous solution was added into 100 mL aqueous solution of $RuCl_3$ (0.18 mol$\cdot L^{-1}$) and $ZnSO_4$ (0.06 mol L^{-1}) with constant stirring at 353 K for 30 min. Then, the solid part was washed until reaching pH = 7. After that, the solid was dispersed in 200 mL of deionized water, followed by a reducing procedure under 5.0 MPa of H_2 in a 1000 mL Hastelloy autoclave with a stirring speed of 800 rpm at 423 K. After 3 h of reduction and cooling to 298 K, the fresh catalysts were gained by washing to neutral and were vacuum-dried. The obtained Ru-Zn catalysts were denoted as Ru-Zn. Similarly, monometallic Ru catalysts were synthesized from the same procedure without introducing $ZnSO_4 \cdot 7H_2O$.

4.3. Catalytic Experimental Procedure

All catalytic experiments took place in a 1000 mL GS-1 type Hastelloy autoclave. In a typical reaction, a Ru-Zn catalyst with 0.57 mol$\cdot L^{-1}$ $ZnSO_4$ as an instance as well as a 1.8 g Ru-Zn catalyst and a 280 mL aqueous solution of 0.57 mol$\cdot L^{-1}$ $ZnSO_4$ were poured into the autoclave. Then, the reactor was purified with N_2 by increasing the pressure to 5 MPa and depressurizing to 0.2 MPa 4 times. This was followed by purification using H_2 another 4 times with the same procedure. Then, the autoclave was heated under 5.0 MPa of H_2 with a stirring speed of 800 rpm. When the temperature reached 423 K, 140 mL of benzene was added into the reactor, and the stirring speed was adjusted to be 1400 rpm to eradicate the effect of mass transfer limitation. Subsequently, the liquid samples were taken from the autoclave every 5 min. All withdrawn samples were analyzed by GC-FID. The benzene conversion and the selectivity towards cyclohexene were calculated with the calibration area normalization method. After each reaction, the catalyst sample was washed until neutral and dried in an Ar atmosphere at 373 K, then stored in ethanol for further characterization. Monometallic Ru and Ru-Zn catalysts with different reaction modifiers were evaluated with the same procedure. In addition, it is important to mention that the mass balance of the reaction was 100% closed.

4.4. Catalysts Characterization

X-ray diffraction (XRD) patterns were obtained at the range of 2θ from $5°$ to $90°$, with a step size of $0.03°$ using diffracted intensity of Cu-Kα radiation ($\lambda = 0.15418$ nm) via an X'Pert Pro instrument (PAN Nallytical, Almelo, The Netherlands). Scherrer's equation was applied for the calculation of the particle size of tested samples. In addition, elemental analysis was conducted via X-ray fluorescence (XRF) (S4 Pioneer instrument from Bruker AXS, Karlsruhe, Germany). Furthermore, the morphology of the catalyst surface as well as the element scanning was tested by transmission electron microscope (TEM), a JEOL JEM 2100 from Akishima, Japan. Surface composition of the selected samples was identified by energy dispersive spectrometer (EDS). Moreover, the valence states of Ru, Zn, Fe, and Mn as well as the kinetic energy of Zn LMM electrons on the catalyst surface were analyzed by X-ray photoelectron spectroscopy (XPS) using a PHI Quantera SXM instrument from ULVAC-PHI, Japan. Al Kα (Eb = 1486.6 eV) was chosen as the radiation source, and the vacuum degree was adjusted to be 6.7×10^{-8} Pa. The C1s (Eb = 284.8 eV) line as the binding energy reference was used for calibrating and correcting the energy scale.

4.5. Theoretical Calculation

The B3LYP density functional method in conjunction with the 6-31G(d,p) basis set for cyclohexene and Lanl2DZ for the metal ions was used to optimize the complexes formed between cyclohexene and the metal ions of Fe^{2+}, Mn^{2+}, and Zn^{2+}. The calculations were all performed with the Gaussian 09 software package. Vibrational frequency analyses were performed for the optimized geometries.

Author Contributions: H.S. and Z.C. were responsible for designing the experiments and the manuscript preparation. Y.F. and X.S. were responsible for conducting the experiments and the data analysis. Z.P. was responsible for making figures. H.L. was responsible for data calculation. Z.L. was responsible for all the chemical purchases and the characterization of the samples. All authors have read and agreed to the published version of the manuscript.

Funding: This research was funded by Key Science and Technology Program of Henan Province, grant number 192102210139; The Training Plan of Young Backbone Teachers in Colleges and Universities of Henan Province, grant number 2019GGJS252; Innovation and Entrepreneurship Training Program for college students in Henan Province, grant number 202112949001; Key Scientific Research Projects of Colleges and Universities in Henan Province, grant number 18A150018; National Natural Science Foundation of China, grant number 21908203; China Postdoctoral Science Foundation, grant number 2019T120637.

Informed Consent Statement: Not applicable.

Data Availability Statement: The data presented in this study are available in this work.

Conflicts of Interest: The authors declare no conflict of interest.

References

1. Pu, J.C.; Dari, M.D.; Tang, X.Q.; Yuan, P.Q. Diffusion of benzene through water film confined in silica mesopores: Effect of competitive adsorption of solvent. *Chem. Eng. Sci.* **2020**, *224*, 115793. [CrossRef]
2. Zhou, G.B.; Jiang, L.; He, D.P. Nanoparticulate Ru on TiO_2 exposed the {100} facets: Support facet effect on selective hydrogenation of benzene to cyclohexene. *J. Catal.* **2019**, *369*, 352–362. [CrossRef]
3. Rochin, L.; Toniolo, L. Selective hydrogenation of benzene to cyclohexene using a suspended Ru catalyst in a mechanically agitated tetraphase reactor. *Catal. Today* **1999**, *48*, 255–264. [CrossRef]
4. Gonçalves, A.H.A.; Soares, J.C.S.; Araújo, L.R.R.; Zotin, F.M.Z.; Mendes, F.M.T.; Gaspar, A.B. Surface investigation by X-ray photoelectron spectroscopy of Ru-Zn catalysts for the partial hydrogenation of benzene. *Mol. Catal.* **2020**, *483*, 110710. [CrossRef]
5. Sun, H.J.; Chen, Z.H.; Li, C.G.; Chen, L.X.; Peng, Z.K.; Liu, Z.Y.; Liu, S.C. Selective hydrogenation of benzene to cyclohexene over Ru-Zn catalysts: Mechanism investigation on NaOH as a reaction additive. *Catalysts* **2018**, *8*, 104. [CrossRef]
6. Sun, H.J.; Chen, Z.H.; Li, C.G.; Chen, L.X.; Li, Y.; Peng, Z.K.; Liu, Z.Y.; Liu, S.C. Selective hydrogenation of benzene to cyclohexene over monometallic Ru catalysts: Investigation of ZnO and $ZnSO_4$ as reaction additives as well as particle size effect. *Catalysts* **2018**, *8*, 172. [CrossRef]

7. Sun, H.J.; Chen, Z.H.; Chen, L.X.; Li, H.J.; Peng, Z.K.; Liu, Z.Y.; Liu, S.C. Selective hydrogenation of benzene to cyclohexene over Ru-Zn catalysts: Investigations on the Effect of Zn content and ZrO_2 as the support and dispersant. *Catalysts* **2018**, *8*, 513. [CrossRef]

8. Liu, X.; Chen, Z.; Sun, H.; Chen, L.; Peng, Z.; Liu, Z. Investigation on Mn_3O_4 coated Ru nanoparticles for partial hydrogenation of benzene towards cyclohexene production using $ZnSO_4$, $MnSO_4$ and $FeSO_4$ as reaction additives. *Nanomaterials* **2020**, *10*, 809. [CrossRef] [PubMed]

9. Struijk, J.; Moene, R.; Kamp, T.V.D.; Scholten, J.J.F. Partial liquid phase hydrogenation of benzene to cyclohexene over ruthenium catalysts in the presence of an aqueous salt solution II. Influence of various salts on the performance of the catalyst. *Appl. Catal. A Gen.* **1992**, *89*, 77–102. [CrossRef]

10. Costa, G.P.; Gonçalves, A.H.A.; Viana, L.A.V.; Soares, J.C.S.; Passos, F.B.; Mends, F.M.T.; Gaspar, A.B. Role of $ZnNb_2O_6$ in ZnO-promoted amorphous-Nb_2O_5 supported Ru catalyst for the partial hydrogenation of benzene. *Mater. Today Chem.* **2021**, *19*, 100397. [CrossRef]

11. Spod, H.; Lucas, M.; Claus, P. Selective hydrogenation of benzene to cyclohexene over $2Ru/La_2O_3$-ZnO catalyst without additional modifiers. *ChemCatChem* **2016**, *8*, 1–9. [CrossRef]

12. Wang, J.Q.; Wang, Y.Z.; Xie, S.H.; Qiao, M.H.; Li, H.X.; Fan, K.N. Partial hydrogenation of benzene to cyclohexene on a Ru-Zn/m-ZrO_2 nanocomposite catalyst. *Appl. Catal. A Gen.* **2004**, *272*, 29–36. [CrossRef]

13. Zhang, P.; Wu, T.B.; Jiang, T.; Wang, W.T.; Liu, H.Z.; Fan, H.L.; Zhang, Z.F.; Han, B.X. Ru–Zn supported on hydroxyapatite as an effective catalyst for partial hydrogenation of benzene. *Green Chem.* **2013**, *15*, 152–159. [CrossRef]

14. Hu, S.C.; Chen, Y.W. Partial hydrogenation of benzene on Ru-Zn/SiO_2 catalysts. *Ind. Eng. Chem. Res.* **2001**, *40*, 6099–6104. [CrossRef]

15. Yuan, P.Q.; Wang, B.Q.; Ma, Y.M.; He, H.M.; Cheng, Z.M.; Yuan, W.K. Hydrogenation of cyclohexene over Ru-Zn/Ru(0001) surface alloy: A first principles density functional study. *J. Mol. Catal. A* **2009**, *301*, 140–145. [CrossRef]

16. Reddy, A.S.; Kim, J. An efficient g-C_3N_4-decorated CdS-nanoparticle-doped Fe_3O_4 hybrid catalyst for an enhanced H_2 evolution through photoelectrochemical water splitting. *Appl. Surf. Sci.* **2020**, *513*, 145836. [CrossRef]

17. Sun, H.J.; Wang, H.X.; Jiang, H.B.; Li, S.H.; Liu, S.C.; Liu, Z.Y.; Yuan, X.M.; Yang, K.J. Eeffect of $(Zn(OH)_2)_3(ZnSO_4)(H_2O)_5$ on the performance of Ru-Zn catalyst for benzene selective hydrogenation to cyclohexene. *Appl. Catal. A* **2013**, *450*, 160–168. [CrossRef]

18. Sun, H.J.; Jiang, H.B.; Dong, Y.Y.; Wang, H.X.; Pan, Y.J.; Liu, S.C.; Tang, M.S.; Liu, Z.Y. Effect of alcohols as additives on the performance of a nano-sized Ru-Zn(2.8%) catalyst for selective hydrogenation of benzene to cyclohexene. *Chem. Eng. J.* **2013**, *218*, 415–424. [CrossRef]

19. Sun, H.J.; Chen, L.X.; Huang, Z.X.; Liu, S.C.; Liu, Z.Y. Particle size effect of Ru-Zn catalysts on selective hydrogenation of benzene to cyclohexene. *Chem. J. Chin. Univ.* **2015**, *36*, 1969–1976.

20. Sun, H.J.; Li, Y.Y.; Li, S.H.; Zhang, Y.X.; Liu, S.C.; Liu, Z.Y.; Ren, B.Z. $ZnSO_4$ and La_2O_3 as Co-modifier of the monoclinic ru catalyst for selective hydrogenation of benzene to cyclohexene. *Acta Phys. Chim. Sin.* **2014**, *30*, 1332–1340.

21. Liu, J.L.; Zhu, L.J.; Pei, Y.; Zhuang, J.H.; Li, H.; Li, H.X.; Qiao, M.H.; Fan, K.N. Ce-promoted Ru/SBA-15 catalysts prepared by a "two solvent" impregnation method for selective hydrogenation of benzene to cyclohexene. *Appl. Catal. A Gen.* **2009**, *353*, 282–287. [CrossRef]

22. Sun, H.J.; Qin, H.A.; Huang, Z.X.; Su, M.F.; Li, Y.Y.; Liu, S.C.; Liu, Z.Y. Effect of reaction modifier $ZnSO_4$ and pretreatment on performance of Ru-Zn catalyst for selective hydrogenation of benzene to cyclohexene. *Chin. J. Inorg. Chem.* **2017**, *33*, 73–80.

23. Sun, H.J.; Pan, Y.J.; Jiang, H.B.; Li, S.H.; Zhang, Y.X.; Liu, S.C.; Liu, Z.Y. Effect of transition metals (Cr, Mn, Fe, Co, Ni, Cu and Zn) on the hydrogenation properties of benzene over Ru-based catalyst. *Appl. Catal. A Gen.* **2013**, *464–465*, 1–9. [CrossRef]

24. Yan, X.H.; Zhang, Q.; Zhu, M.Q.; Wang, Z.B. Selective hydrogenation of benzene to cyclohexene over Ru-Zn/ZrO_2 catalysts prepared by a two step impreganation method. *J. Mol. Catal. A Chem.* **2016**, *413*, 85–93. [CrossRef]

25. Ramos-Fernández, E.V.; Ferreira, A.F.P.; Sepúlveda-Escribano, A.; Kapteijn, F.; Rodríguez-Reinoso, F. Enhancing the catalytic performance of Pt/ZnO in the selective hydrogenation of cinnamaldehyde by Cr addition to the support. *J. Catal.* **2008**, *258*, 52–60. [CrossRef]

26. Ma, S.B.; Zhao, X.Y.; Li, Y.S.; Zhang, T.R.; Yuan, F.L.; Niu, X.Y.; Zhu, Y.J. Effect of W on the acidity and redox performance of the $Cu_{0.02}Fe_{0.2}W_aTiO_x$ (a = 0.01, 0.02, 0.03) catalysts for NH_3-SCR of NO. *Appl. Catal. B Environ.* **2019**, *248*, 226–238. [CrossRef]

27. Liu, L.J.; Xu, K.; Su, S.; He, L.M.; Qing, M.X.; Chi, H.Y.; Liu, T.; Hu, S.; Wang, Y.; Xiang, J. Efficient Sm modified Mn/TiO_2 catalysts for selective catalytic reduction of NO with NH_3 at low temperature. *Appl. Catal. A Gen.* **2020**, *592*, 117413. [CrossRef]

28. Azevedo, P.V.C.; Dias, M.V.; Gonçalves, A.H.A.; Borges, L.E.P.; Gaspar, A.B. Influence of cadmium on Ru/xCd/Al_2O_3 catalyst for benzene partial hydrogenation. *Mol. Catal.* **2021**, *499*, 111288. [CrossRef]

International Journal of
Molecular Sciences

MDPI

Article

Enhanced Sensing Ability of Brush-Like Fe$_2$O$_3$-ZnO Nanostructures towards NO$_2$ Gas via Manipulating Material Synergistic Effect

Yuan-Chang Liang * and Yu-Wei Hsu

Department of Optoelectronics and Materials Technology, National Taiwan Ocean University, Keelung 20224, Taiwan; david4250557@gmail.com
* Correspondence: yuanvictory@gmail.com; Tel.: +886-224622192

Abstract: Brush-like α-Fe$_2$O$_3$–ZnO heterostructures were synthesized through a sputtering ZnO seed-assisted hydrothermal growth method. The resulting heterostructures consisted of α-Fe$_2$O$_3$ rod templates and ZnO branched crystals with an average diameter of approximately 12 nm and length of 25 nm. The gas-sensing results demonstrated that the α-Fe$_2$O$_3$–ZnO heterostructure-based sensor exhibited excellent sensitivity, selectivity, and stability toward low-concentration NO$_2$ gas at an optimal temperature of 300 °C. The α-Fe$_2$O$_3$–ZnO sensor, in particular, demonstrated substantially higher sensitivity compared with pristine α-Fe$_2$O$_3$, along with faster response and recovery speeds under similar test conditions. An appropriate material synergic effect accounts for the considerable enhancement in the NO$_2$ gas-sensing performance of the α-Fe$_2$O$_3$–ZnO heterostructures.

Keywords: synthesis; microstructure; composite; sensing ability; enhanced mechanism

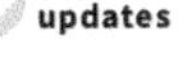
check for
updates

Citation: Liang, Y.-C.; Hsu, Y.-W. Enhanced Sensing Ability of Brush-Like Fe$_2$O$_3$-ZnO Nanostructures towards NO$_2$ Gas via Manipulating Material Synergistic Effect. *Int. J. Mol. Sci.* **2021**, *22*, 6884. https://doi.org/10.3390/ijms22136884

Academic Editor: Raghvendra Singh Yadav

Received: 24 May 2021
Accepted: 24 June 2021
Published: 26 June 2021

Publisher's Note: MDPI stays neutral with regard to jurisdictional claims in published maps and institutional affiliations.

1. Introduction

Increasing awareness of harmful gases as an environmental problem has resulted in the vital improvement of gas sensor technology. Especially, NO$_2$ gas from fuel and vehicle exhausts, even if the concentration is low, is hazardous to the human body and the ecosystem. The design and development of various semiconductors with diverse architectures to detect NO$_2$ gas molecules with high efficiency are in high demand in material technology. Hematite (α-Fe$_2$O$_3$) and zinc oxide (ZnO) are n-type semiconductors widely researched in gas sensor applications owing to their superior stability, low cost, and preparation simplicity [1,2]. Approaches to synthesizing low-dimensional α-Fe$_2$O$_3$ and ZnO structures are diverse, and they both exhibit potent sensitivity toward various volatile and toxic gases [1–5]. Notably, a large-scale growth and low-cost chemical hydrothermal method has been proposed to prepare low-degree α-Fe$_2$O$_3$ and ZnO materials for the high reproducibility and uniformity of sample preparation [2,3]. Despite the encouraging development of low-degree α-Fe$_2$O$_3$ and ZnO nanostructures, the improvement of their gas sensitivity and selectivity to target gases remains a challenge. Considerable attention has been paid to the synthesis of oxide heterostructures. Promising research into oxide heterostructures consisting of Fe$_2$O$_3$ and ZnO has indicated that a Fe$_2$O$_3$–ZnO composite system can improve the detection ability toward various harmful gases in comparison with that of single-constituent counterparts. However, most work on Fe$_2$O$_3$–ZnO heterostructures has been conducted on independent variables to improve their gas sensitivity; for example, controlling the thickness of the decorated ZnO shell for comparison with its Debye length [6], and controlling the geometrical shape of the composite structure [7]. How to integrate multiple factors for the synergetic optimal gas sensitivity of Fe$_2$O$_3$–ZnO heterostructures is still a scientific concern in this research field. Notably, low-concentration NO$_2$ gas emitted from fuel and vehicles and chemical plants threatens the environment and human health and safety. Studies on the NO$_2$ gas detection capabilities and effective sensing mechanism of Fe_2O_3–ZnO composite-based sensors are limited. In this study, the low-concentration NO_2 gas-sensing behavior and mechanism of a brush-like α-Fe_2O_3–ZnO heterostructure synthesized through an integrated sputtering ZnO seed-assisted hydrothermal growth method were investigated. The substantially improved low-concentration NO_2 gas-sensing performance of the Fe_2O_3 nanorod template through branched ZnO decoration was achieved based on the control of proper material synergistic effects, including the surface morphology, surface area, heterogeneous barrier, and the Debye length of branched crystals. The material design and fundamental synergetic effects on the NO_2 gas-sensing characteristic of brush-like Fe_2O_3–ZnO heterostructures are discussed.

2. Results

2.1. Microstructural Analysis

Figure 1a presents a scanning electron microscopic image of pristine α-Fe_2O_3 nanorods grown vertically aligned on the substrate. The average diameter of the nanorods was approximately 50 nm, with lengths of several micrometers. Figure 1b,c illustrate the morphology of as-synthesized Fe_2O_3–ZnO nanostructures. The surface of the Fe_2O_3–ZnO heterostructure was built from compactly aggregated tiny rods in the branch structures of the brush-like Fe_2O_3–ZnO heterostructures. Comparatively, the surface of the pristine Fe_2O_3 nanorods was smooth, and the as-synthesized brush heterostructures had more applicable surface area. The brush Fe_2O_3–ZnO heterostructures are more suitable for gas sensor application than individual component nanostructures are because of the rough, abundant pipes between branches and well-aligned surfaces, including the stem and branch parts of the heterostructures. Figure 1d presents the XRD patterns of the as-prepared Fe_2O_3–ZnO heterostructures. Diffraction peaks of α-Fe_2O_3 and well-defined diffraction peaks associated with hexagonal ZnO appeared in the XRD pattern according to the reference JCPDS card data in rhombohedral α-Fe_2O_3 JCPDS card (No. 33-0664) and hexagonal ZnO (No. 36-1451), respectively. Notably, the ZnO (002) is the most intensive ZnO crystallographic plane, which is in agreement with the crystallographic feature of the ZnO nanorods grown along its c-axis direction, synthesized via several approaches [7,8]. The XRD result proves the good crystalline quality of the Fe_2O_3–ZnO heterostructures herein.

2.2. UV–vis Optical Characterization

The optical properties of the as-synthesized samples were investigated by recording UV–vis absorbance spectra. Figure 2a shows the UV–Vis absorbance spectra of pristine Fe_2O_3 and a Fe_2O_3–ZnO composite. It is evident that a steep absorption edge appears at the visible light region of approximately 500–600 nm for the Fe_2O_3. This is attributed to the band-to-band transition in the α-Fe_2O_3 phase [9] Notably, the optical absorbance spectrum of the Fe_2O_3–ZnO has two obvious step absorption edges located at UV and visible light regions. The appearance of the absorption edge at the UV region is associated with the decorated ZnO, which has an UV region band energy feature [10]. The analysis of the optical absorbance spectra herein demonstrates the Fe_2O_3–ZnO heterostructure has successfully constructed through multiple hydrothermal routes with the assistance of sputtering ZnO seed layer engineering. The bandgap energies of the Fe_2O_3 and ZnO are also evaluated by converting the Kubelka–Munk equation, as displayed in Figure 2b [11]. From Figure 2b, the Fe_2O_3 and ZnO have band gap energies of approximately 2.2 eV and 3.37 eV, respectively; this bandgap energy information was further used to construct the band diagram for discussing the gas-sensing mechanism of the as-synthesized composite.

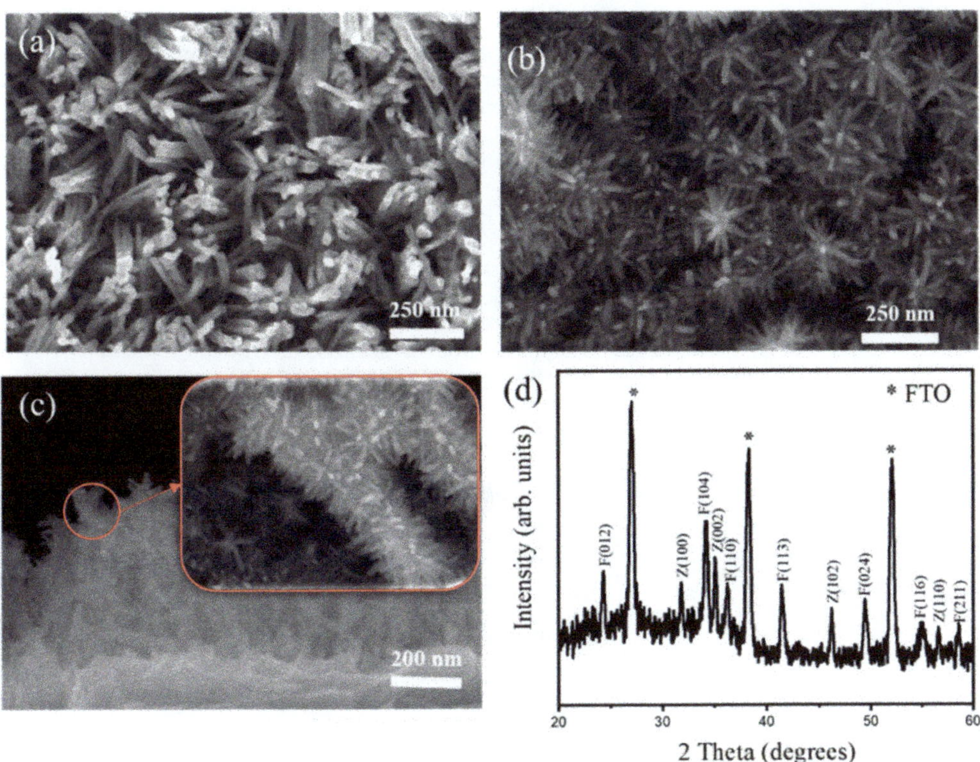

Figure 1. SEM top view image: (**a**) pristine Fe$_2$O$_3$ nanorod template and (**b**) Fe$_2$O$_3$–ZnO. (**c**) Cross-section SEM image of Figure (**b**) and the subfigure is an enlarged image. (**d**) XRD pattern of Fe$_2$O$_3$–ZnO heterostructures. The asterisk, Z, and F denote FTO, ZnO, and Fe$_2$O$_3$, respectively.

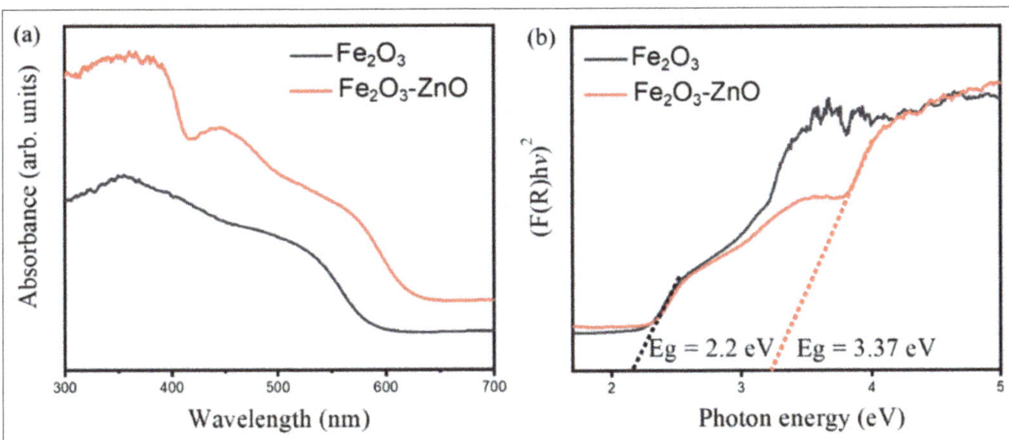

Figure 2. UV–vis absorption spectra of (**a**) pristine Fe$_2$O$_3$ nanorod template and Fe$_2$O$_3$–ZnO. (**b**) Evaluation of bandgap energies of Fe$_2$O$_3$ and Fe$_2$O$_3$–ZnO.

2.3. Transmission Electron Microscopy Analysis

Figure 3a presents the TEM image of a Fe_2O_3–ZnO heterostructure. A good brush structure was visibly displayed, and the branched rods had an average diameter and length of 12 nm and 25 nm, respectively. Figure 3b,c display the HRTEM images of the heterostructure taken from various local regions. The ordered lattice fringes with interval distances of 0.28 nm and 0.26 nm were corresponded to the interplanar spacings of α-Fe_2O_3 (110) and ZnO (002), respectively. Figure 3d demonstrates the selected area electron diffraction (SAED) pattern taken from several Fe_2O_3–ZnO composites, revealing that the concentric rings could be attributed to α-Fe_2O_3 (104), (110), and (113) and the ZnO (100), (002). The SAED analysis responds to the XRD result, confirming that crystalline Fe_2O_3–ZnO heterostructures were successfully formed. Figure 3e shows the energy dispersive spectroscopy (EDS) elemental mapping analysis of an individual Fe_2O_3–ZnO heterostructure. The Fe signals could be only detected in the core region, while Zn signals were predominant in the outer region, and the distribution of Zn element further confirmed that the surface of the Fe_2O_3 was decorated by ZnO nanorods. The O signals can be recognized from the whole heterostructure. The EDS mapping analyses confirmed that the as-synthesized product formed a compositionally homogeneous composite.

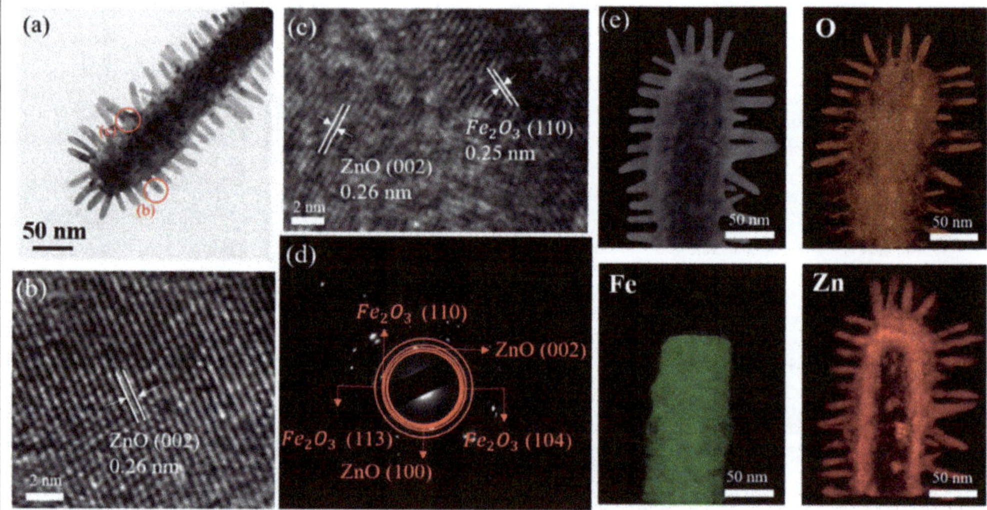

Figure 3. TEM image of Fe_2O_3–ZnO heterostructure: (**a**) low-magnification image. (**b,c**) Local HRTEM images of the heterostructure. (**d**) SEAD pattern taken from several composites. (**e**) EDS elemental mapping analysis of an individual Fe_2O_3–ZnO heterostructure.

2.4. Surface Active Site Analysis

Surface area is an important factor that affects the gas sensing characteristic, which should be investigated for the heterostructure sample. The electrochemical catalyst surface activity (ECSA) of the as-fabricated α-Fe_2O_3 and Fe_2O_3–ZnO heterostructure sample are estimated with the cyclic voltammetry (CV) from the double-layer capacitance (C_{dl}). As shown in Figure 4a,b, the CV curves of α-Fe_2O_3 and Fe_2O_3–ZnO, respectively, at non-Faradaic potential regions (-0.3–0.1 V vs. NHE) separated with different scan rates, from 0.1 mV/s to 0.5 mV/s. The ECSA can be calculated from the C_{dl} divided by the specific capacitance (C_s), namely ECSA = C_{dl}/C_s, and is proportional to the C_{dl} value. The correlation between the double-layer capacitance (C_{dl}) and the double-layer charging current (i_c) follows the equation: $C_{dl} = i_c /v$. The $i_c = (j_a - j_c)$, in which j_a is anodic current and j_c is cathodic current at the middle potential (-0.2 V) against the CV scan rate. Figure 4c shows that the plot of i_c vs. ν has a linear characteristic and the slope corresponds to the C_{dl}. Notably, for a comparison, the i_c vs. ν plot of the pristine Fe_2O_3 was also included in Figure 4c. Obviously, the Fe_2O_3–ZnO has a larger ECSA size, which is approximately 3.8 times higher than that of the pristine Fe_2O_3. The larger ESCA provided more active sites for the interface reaction between photoelectrode and electrolyte, and this is attributable to higher diffusion space between the numerous tiny branched ZnO nanorods of the heterostructure, as confirmed by the aforementioned structural analysis [12,13]. The ECSA analysis reveals that the brush heterostructure produced more surface area for reaction species adsorb or desorb on it, which might further benefit its gas-sensing ability.

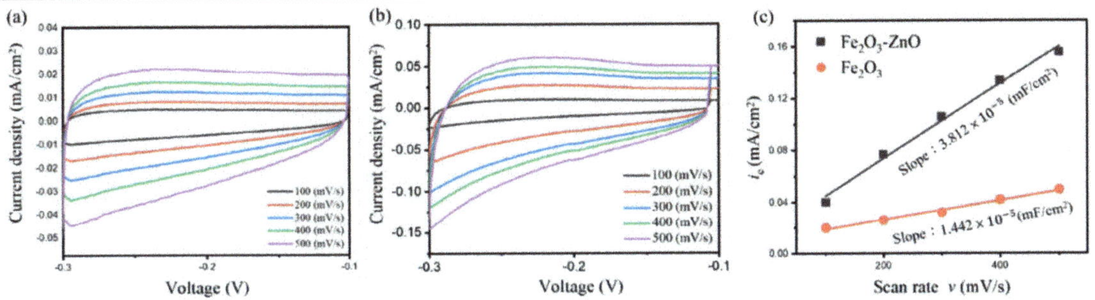

Figure 4. (**a**) Cyclic voltammetry plot of pristine Fe_2O_3 and (**b**) Fe_2O_3–ZnO at different scan rates. (**c**) ECSA analysis of Fe_2O_3 and Fe_2O_3–ZnO.

2.5. Gas Sensing Performance

As presented in Figure 5a, the pristine Fe_2O_3 and Fe_2O_3–ZnO heterostructure-based gas-sensing responses toward 10 ppm NO_2, as a function of the operating temperature, were studied. The dependence of gas-sensing responses versus operating temperature was the same for both gas sensors. A resultant equilibrium between the sensor material surface reaction with NO_2 gas molecules and the diffusion of these molecules to the material surface occurred at 300 °C [14]. Figure 5b,d present the dynamic response and recovery curves of the pristine Fe_2O_3, ZnO, and Fe_2O_3–ZnO sensors toward different NO_2 gas concentrations at 300 °C. Concentration-dependent cycling test curves for both sensors demonstrated that the Fe_2O_3 and Fe_2O_3–ZnO sensors were effective and suitably sensitive to NO_2 at various concentrations. The gas-sensing responses of the Fe_2O_3 sensor to 1, 2.5, 5, 7.5, and 10 ppm NO_2 were 1.26, 1.36, 1.54, 1.67, and 1.71, respectively. The responses of ZnO to1, 2.5, 5, 7.5, and 10 ppm NO_2 were 1.19, 1.42, 1.59, 1.75, and 1.83. Moreover, the sensing responses of the Fe_2O_3–ZnO sensor to 1, 2.5, 5, 7.5, and 10 ppm NO_2 were 1.59, 2.38, 3.17, 3.97, and 6.34, respectively. The summarized responses versus NO_2 concentration results in Figure 5e indicate that the Fe_2O_3–ZnO sensor exhibits a higher gas-sensing response than that of the Fe_2O_3 sensor under the same test conditions, demonstrating the benefit of ZnO decoration on Fe_2O_3. Furthermore, these two sensors exhibited different response and recovery speeds to NO_2 gas. The response speeds of the Fe_2O_3 sensor ranged 23 to 62 s for 1 to 10 ppm NO_2; those of the Fe_2O_3–ZnO ranged 15 to 26 s for 1 to 10 ppm NO_2. The Fe_2O_3 recovery speeds were approximately 180 to 600 s for 1 to 10 ppm NO_2, and those of the Fe_2O_3–ZnO were 85 to 185 s for 1 to 10 ppm NO_2. Substantially increased NO_2 gas-sensing responses and recovery speeds were observed for the Fe_2O_3 template decorated with abundant ZnO nanorods. The formation of the unique brush-like morphology of the Fe_2O_3–ZnO composite substantially improved the response and recovery speeds in comparison to the Fe_2O_3 sensor. The improved specific surface area and numerous heterojunctions of the Fe_2O_3–ZnO composite accounted for the observed results. A similar combined effect of hierarchical morphology and heterojunction that enhances gas-sensing performance has been proposed in an Fe_2O_3–TiO_2 composite system towards trimethylamine gas [15]. For practical usage, a gas sensor with selectivity must detect a target gas when exposed to a multicomponent gas environment; high sensor reliability within the sensing environment is a key concern. Figure 5f presents the cross-sensitivity of the Fe_2O_3–ZnO observed upon exposure to 10 ppm NO_2, 100 ppm ethanol vapor, 100 ppm ammonia, and 100 ppm hydrogen gas at 300 °C. When the comparative target gases of ethanol vapor, ammonia gas, and hydrogen gas had a higher concentration of 100 ppm, the sensing responses of the Fe_2O_3–ZnO sensor to these gases were still substantially lower than the response to 10 ppm of NO_2, revealing the superior gas-sensing selectivity of Fe_2O_3–ZnO toward NO_2 gas. The reasons for the selectivity of oxide sensing materials toward a specific target gas are complicated and no consistent statements are proposed in the literature. It is proposed that that the chemical properties of various oxidizing or reducing gases could cause a difference in the adsorption ability and reaction strength with adsorbed O species at the given sensor's operating temperature [16]. Therefore, higher selectivity toward NO_2 gas for the Fe_2O_3–ZnO herein might be associated with the higher electron affinity of NO_2 gas (2.28 eV) as compared with pre-adsorbed oxygen (0.43 eV) and other test gases [17]. The unpaired electron in the N atom of the NO_2 gas molecule easily forms bond with the oxygen species presented on the Fe_2O_3–ZnO surface, which increases the chemisorption size of NO_2 as compared to other target gases. Figure 5g presents the cycling gas-sensing curves for the Fe_2O_3–ZnO sensor to 10 ppm NO_2, with no clear sensing response deterioration after five cycling tests, demonstrating the stability and reproducibility of the Fe_2O_3–ZnO sensor. Furthermore, as exhibited in Table 1, most NO_2 gas sensors have an operating temperature above 300 °C to obtain detectable responses. Based on the NO_2 concentration and gas-sensing responses of various sensors in Table 1, the as-synthesized α-Fe_2O_3–ZnO composite sensor exhibits decent gas-sensing performance among various reported Fe_2O_3-based or ZnO-based composite sensors.

Table 1. NO_2 gas-sensing performance of various Fe_2O_3- and ZnO-based composites prepared using various methods [18–21].

Material	Synthesis Method	Temperature/ NO_2 Concentration	Response	Response Time/ Recover Time (s)	Ref.
α-Fe_2O_3-TiO_2	Solvothermal method	300 °C/5 ppm	2 (R_g/R_a)	N/A	[18]
α-Fe_2O_3-SnO_2	Hydrolysis method	300 °C/1 ppm	<0.5 (($R_g - R_a$)/R_a)	N/A	[19]
ZnO-Fe_2O_3	Co-precipitation method	400 °C/250 ppm	10.53 (R_g/R_a)	1000/4000	[20]
ZnO-CuO	Screen printing method	300 °C/29 ppm	5.57 (R_g/R_a)	N/A	[21]
α-Fe_2O_3-ZnO	Hydrothermal method	300 °C/10 ppm	6.34 (R_g/R_a)	26/185	this work

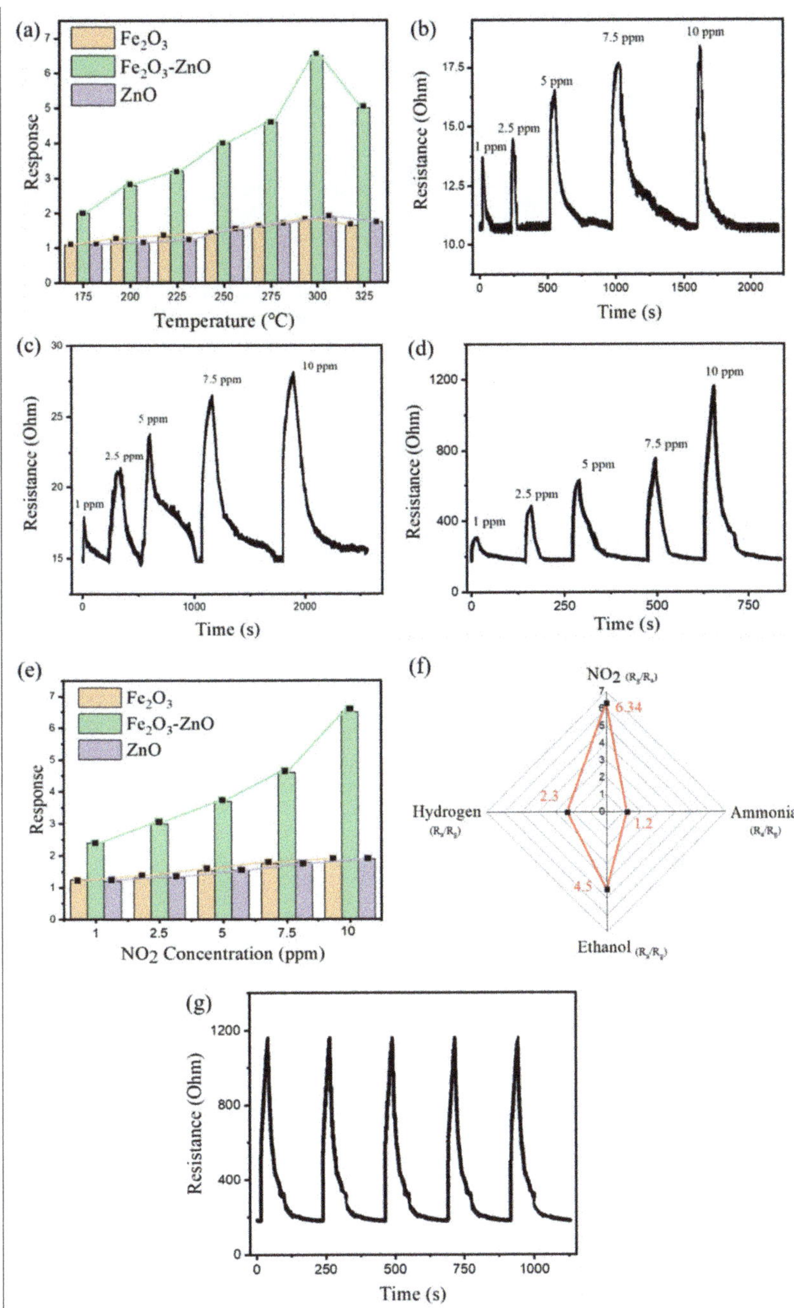

Figure 5. (**a**) Operating temperature-dependent 10 ppm NO$_2$ gas-sensing response for Fe$_2$O$_3$, ZnO, and Fe$_2$O$_3$-ZnO sensors. (**b–d**) Dynamic gas-sensing curves of Fe$_2$O$_3$, ZnO, and Fe$_2$O$_3$-ZnO sensors towards different NO$_2$ gas concentrations at 300 °C, respectively. (**e**) Gas-sensing response vs. NO$_2$ gas concentration for various sensors at 300 °C. (**f**) Cross-selectivity of Fe$_2$O$_3$-ZnO sensor. The responses are marked with red. (**g**) Cycling gas-sensing curves for the Fe$_2$O$_3$-ZnO sensor towards 10 ppm NO$_2$ gas.

3. Discussion

Figure 6a indicates that the electron transport channel size inside the pristine Fe_2O_3 nanorod can be modulated by the surface electron depletion layer, which surrounds the nanorod's surface. Initially formed through the adsorption of oxygen molecules on the surface, the electrons are extracted from the conduction band of the nanorods to form adsorbed oxygen ions (O_{ads}^{2-}) [22,23]. As presented in Figure 6b, further exposure of the Fe_2O_3 nanorod template to NO_2 gas resulted in the marked extraction of its surface electrons through the surface reaction with the NO_2 molecules, forming surface-adsorbed NO_2^- ions. The conducting channel of the Fe_2O_3 nanorod was further narrowed, thereby increasing the bulk resistance of the Fe_2O_3 sensor. For the NO_2 gas-sensing mechanism of the brush-like Fe_2O_3–ZnO heterostructure in Figure 6c,d, the influence of the sample morphology for highly accessible NO_2 gas molecule adsorption should be considered. The appreciable morphological difference between the composite and pristine Fe_2O_3 may result in several concomitant effects in the composite, such as high reactive surface area size, high airflow channel effect, and the high surface trapping and low reaction time with the target gas molecules. The aforementioned ECSA result demonstrated the substantially enhanced surface reaction of the Fe_2O_3–ZnO compared with that of the pristine Fe_2O_3. The brush-like configuration is also beneficial in the formation of pathways between the branches. The target gas molecules can effectively react with the sample surface, increasing the number of captured electrons from the Fe_2O_3 and ZnO conduction bands and effectively reducing the response and recovery time during gas-sensing tests [24,25]. The diameter of decorated ZnO branches plays a vital role in the gas-sensing performance of the heterostructure. The average diameter of the ZnO branches was approximately 12 nm, and the Debye length of the ZnO at increased temperatures was approximately 20 nm [26]. The decorated ZnO nanorods should be fully depleted at the operating temperature herein. The gas-sensing response abruptly increases when the oxide semiconductor particle size becomes smaller than the Debye length [27]. Moreover, the Fe_2O_3–ZnO sensor's gas-sensing response to NO_2 gas was not only influenced by the depletion layer from the surfaces of each ZnO branch and Fe_2O_3 nanorod template, but also by the formation of abundant Fe_2O_3/ZnO heterogeneous junctions. These form because the different work functions of Fe_2O_3 and ZnO in the composite contribute varying degrees of bulk resistance upon exposure to NO_2 gas [7]. The percolation network of electrons pass through potential barriers at heterogeneous junctions between the ZnO branches and Fe_2O_3 template. This results in high initial bulk resistance (189 Ohm) in the Fe_2O_3–ZnO sensor, almost 18 times that of the pristine Fe_2O_3 (10.5 Ohm). The initially formed Fe_2O_3/ZnO interfacial potential barriers effectively modulate the electron transport between the two constituents by adsorbing or desorbing target gas molecules [28]. These junctions can be considered additional active sites, resulting in the enhancement of the gas-sensing response when the sensor material is exposed to NO_2 gas. When the Fe_2O_3–ZnO was further exposed to NO_2 gas, as presented in Figure 6d, the Fe_2O_3/ZnO interfacial depletion layer size was appreciably widened, resulting in the rugged, narrow conduction channel of the Fe_2O_3 template. The band emery variations of the Fe_2O_3/ZnO at the surface and interface regions also account for the improved NO_2 gas-sensing performance of the Fe_2O_3-ZnO composite. Figure 6e shows the band diagram of the Fe_2O_3/ZnO system at the equilibrium state which is constructed with the band energy parameters from previous bandgap value evaluation and literature [7]. Figure 6f displays the band structure variation of the Fe_2O_3-ZnO exposed to NO_2 gas. The Fe_2O_3/ZnO heterogeneous interface due to the electron trapping by interface states might dominate the gas-sensing behavior of the composite toward NO_2 gas. The potential barrier-controlled carrier transport mechanism might be predominant over the surface depletion mechanism for Fe_2O_3-ZnO; this is due to the fact that no electrons are available in the depleted decorated ZnO (size smaller than Debye length) to react with the NO_2 molecules at the given test condition. Comparing the schematic configurations of NO_2 gas-sensing mechanisms in Fe_2O_3 and Fe_2O_3-ZnO, the degree of bulk resistance variation associated with ZnO and Fe_2O_3 surface depletion and the Fe_2O_3/ZnO interfacial junctions in the heterostructural system were expected to be substantially higher than that of the pristine Fe_2O_3 template, which was merely modulated by the surface depletion layer size variation before and after exposure to NO_2 gas. Based on the proposed NO_2 gas-sensing model of the Fe_2O_3–ZnO heterostructure, the low-concentration NO_2 gas-sensing performance of the pristine Fe_2O_3 nanorods was considerably improved through the suitable control of material synergistic effects through the formation of brush-like Fe_2O_3–ZnO heterostructures.

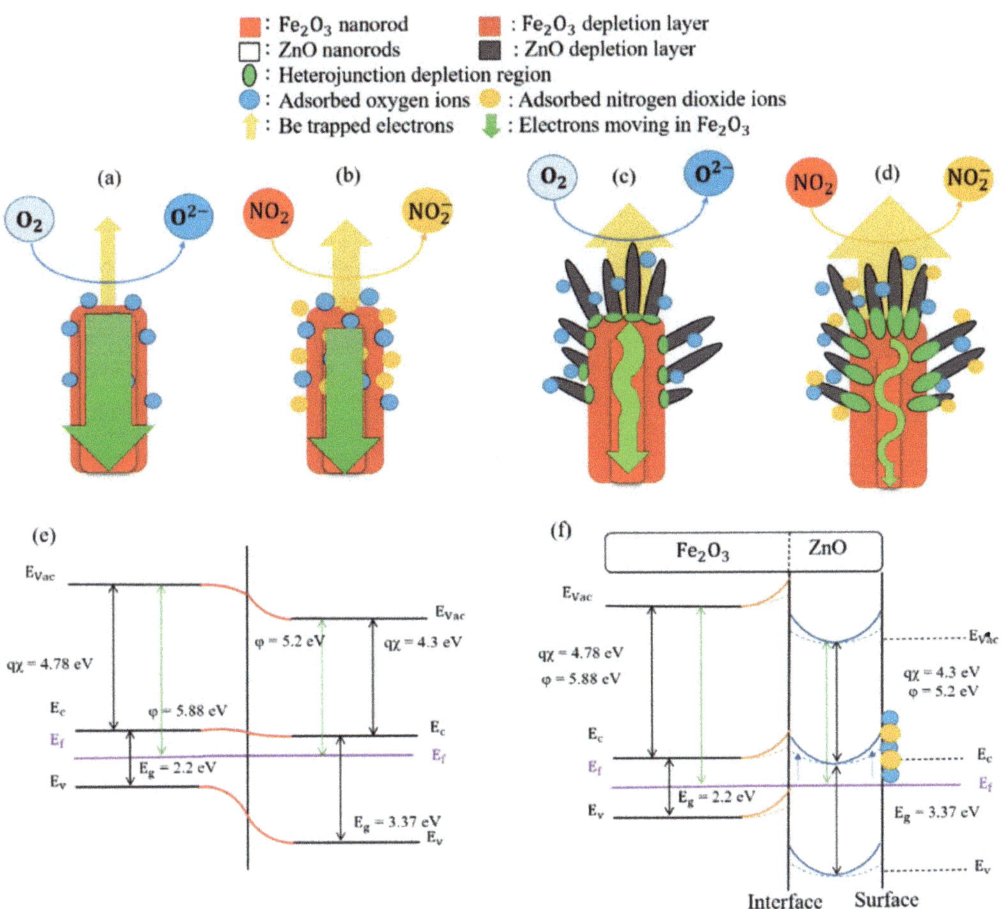

Figure 6. Schematic diagrams illustrating the gas sensing mechanisms of Fe_2O_3 template: (**a**) in air, (**b**) in NO_2 gas, and those of Fe_2O_3-ZnO: (**c**) in air, (**d**) in NO_2 gas. (**e**) Energy band diagram of Fe_2O_3/ZnO after equilibrium. (**f**) Band bending variation of Fe_2O_3/ZnO toward NO_2 gas (from dashed line to solid line).

4. Materials and Methods

4.1. Materials Synthesis

The α-Fe_2O_3 nanorod template was synthesized through a hydrothermal method. The 3.028 g of iron chloride hexahydrate ($FeCl_3 \cdot 6H_2O$) and 0.72 g of urea ($CO(NH_2)_2$) were dissolved into deionized water to form a 60 mL mixture solution as a reaction solution. The F-doped tin oxide (FTO) substrates were perpendicularly leant in the aforementioned solution, and then tightly sealed and maintained at 100 °C for 24 h in an electric oven. Finally, the products were annealed at 500 °C for 30 min in the muffle furnace. The detailed hydrothermal reaction parameters have also been described elsewhere [3]. The Fe_2O_3 nanorod template was coated with a ZnO seed layer via radio-frequency sputtering at room temperature with a ZnO ceramic target. The sputtering process was conducted in an Ar/O_2 mixed atmosphere with an Ar/O_2 ratio of 4/1. The sputtering working pressure was 20 mTorr and the sputtering power was ZnO 90 W in this study [10]. The sputtering duration of the ZnO seed layer is 60 min. After that, the ZnO seed layer-coated Fe_2O_3 nanorod template on FTO substrates was perpendicularly suspended in a homogeneous mixture solution containing 100 mL of deionized water, 0.7 g of Hexamethylenetetramine $(C_6H_{12}N_4)$, and 1.4869 g of Zinc Nitrate Hexahydrate $(Zn(NO_3)_2)$. The hydrothermal reaction was conducted at 95 °C for 1.5 h to synthesize brush-like Fe_2O_3-ZnO samples.

4.2. Characterization Analysis

The surface morphology of the as-synthesized samples was investigated by scanning electron microscopy (SEM). The analysis of samples' crystal structures was conducted by X-ray diffractometer (XRD) with Cu Kα radiation in the two theta ranges of 20–60°. Moreover, the optical absorption spectra of the samples were investigated via a diffuse-reflectance mode with an ultraviolet–visible spectrophotometer (UV–vis). Furthermore, the detailed microstructures of the composite were characterized by high-resolution transmission electron microscopy (HRTEM) equipped energy-dispersive X-ray spectroscopy (EDS). The electrochemical catalyst surface activity (ECSA) was performed in a three-electrode electrochemical system, where the as-synthesized sample was used as the working electrode, a Pt wire was used as the counter electrode, and an Ag/AgCl electrode was used as the reference electrode in an aqueous solution containing 0.5 M Na_2SO_4.

4.3. Gas Sensing Experiments

The surface of gas sensors made from the pure Fe_2O_3 and the Fe_2O_3-ZnO composite were coated with patterned platinum film for electric contacts with probes during gas-sensing measurements. Various concentrations of NO_2 gas (1~10 ppm) were introduced into the test chamber, using dry air as the carrier gas. A resistance meter was used to observe the resistance changes of the sensor devices before and after the supply of the target gas. The responses of sensors toward NO_2 herein are defined as Rg/Ra, Ra and Rg are the sensor resistances in air and target gas, respectively. By contrast, the responses of sensors toward other reducing gases are defined as Ra/Rg. The response and recovery speeds of sensors are defined as the duration required to reach the 90% maximum resistance toward NO_2 gas and the duration required to decrease 90% resistance with the removal of NO_2 gas, respectively.

5. Conclusions

In summary, the integrated controllable preparation of Fe_2O_3–ZnO brush-like hierarchical nanostructures through a combination two-step hydrothermal method with sputtering ZnO seed layer support was proposed. The NO_2 gas-sensing performance of the Fe_2O_3–ZnO heterostructures was substantially improved compared with that of the pristine Fe_2O_3. The improved performance of the Fe_2O_3–ZnO sensor was attributed to synergistic effects, including a ZnO branch size smaller than the Debye length, the branched morphology of the heterostructure inducing abundant air flow channels, and the formation of abundant heterojunction barriers. The experimental results serve as a sound reference for the design of a Fe_2O_3–ZnO heterogeneous sensor based on the control of appropriate material synergistic effects, with the effective detection of low-concentration NO_2 gas.

Author Contributions: Conceptualization: Y.-C.L. and Y.-W.H.; formal analysis and investigation: Y.-W.H.; writing—original draft preparation: Y.-C.L. and Y.-W.H.; writing—review and editing: Y.-C.L.; supervision: Y.-C.L. All authors have read and agreed to the published version of the manuscript.

Funding: This research was funded by Ministry of Science and Technology of Taiwan. (Grant No. MOST 108-2221-E-019-034-MY3).

Data Availability Statement: The data presented in this study are available in this article.

Conflicts of Interest: The authors declare no conflict of interest.

References

1. Sun, P.; He, X.; Wang, W.; Ma, J.; Sun, Y.; Lu, G. Template-free synthesis of monodisperse α-Fe$_2$O$_3$ porous ellipsoids and their application to gas sensors. *CrystEngComm* **2012**, *14*, 2229–2234. [CrossRef]
2. Liang, Y.-C.; Lin, T.-Y.; Lee, C.-M. Crystal growth and shell layer crystal feature-dependent sensing and photoactivity performance of zinc oxide–indium oxide core–shell nanorod heterostructures. *CrystEngComm* **2015**, *17*, 7948–7955. [CrossRef]
3. Liang, Y.-C.; Hung, C.-S. Design of Hydrothermally Derived Fe$_2$O$_3$ Rods with Enhanced Dual Functionality via Sputtering Decoration of a Thin ZnO Coverage Layer. *ACS Omega* **2020**, *5*, 16272–16283. [CrossRef] [PubMed]
4. Liang, Y.C.; Chang, C.W. Improvement of Ethanol Gas-Sensing Responses of ZnO–WO$_3$ Composite Nanorods through Annealing Induced Local Phase Transformation. *Nanomaterials* **2019**, *9*, 669–680. [CrossRef]
5. Rackauskas, S.; Barbero, N.; Barolo, C.; Viscardi, G. ZnO Nanowire Application in Chemoresistive Sensing: A Review. *Nanomaterials* **2017**, *7*, 381. [CrossRef]
6. Zhu, C.; Chen, Y.; Wang, R.; Wang, L.; Cao, M.-S.; Shi, X. Synthesis and enhanced ethanol sensing properties of α-Fe$_2$O$_3$/ZnO heteronanostructures. *Sens. Actuators B Chem.* **2009**, *140*, 185–189. [CrossRef]
7. Zhang, J.; Liu, X.; Wang, L.; Yang, T.; Guo, X.; Wu, S.; Wang, S.; Zhang, S. Synthesis and gas sensing properties of α-Fe$_2$O$_=$@ ZnO core–shell nanospindles. *Nanotechnology* **2011**, *22*, 185501–185507. [CrossRef]
8. Liang, Y.-C.; Hu, C.-Y.; Zhong, H.; Wang, J.-L. Crystal synthesis and effects of epitaxial perovskite manganite underlayer conditions on characteristics of ZnO nanostructured heterostructures. *Nanoscale* **2013**, *5*, 2346–2351. [CrossRef]
9. Maji, S.K.; Mukherjee, N.; Mondal, A.; Adhikary, B. Synthesis, characterization and photocatalytic activity of α-Fe$_2$O$_3$ nanoparticles. *Polyhedron* **2012**, *33*, 145–149. [CrossRef]
10. Liang, Y.-C.; Zhao, W.-C. Morphology-dependent photocatalytic and gas-sensing functions of three-dimensional TiO$_2$–ZnO nanoarchitectures. *CrystEngComm* **2020**, *22*, 7575–7589. [CrossRef]
11. Liang, Y.C.; Lung, T.W. Growth of Hydrothermally Derived CdS-Based Nanostructures with Various Crystal Features and Photoactivated Properties. *Nanoscale Res. Lett.* **2016**, *11*, 264–274. [CrossRef]
12. Zhong, Y.; Yang, S.; Zhang, S.; Cai, X.; Gao, Q.; Yu, X.; Xu, Y.; Zhou, X.; Peng, F.; Fang, Y. CdS branched TiO$_2$: Rods-on-rods nanoarrays for efficient photoelectrochemical (PEC) and self-bias photocatalytic (PC) hydrogen production. *J. Power Sources* **2019**, *430*, 32–42. [CrossRef]
13. Long, X.; Gao, L.; Li, F.; Hu, Y.; Wei, S.; Wang, C.; Wang, T.; Jin, J.; Ma, J. Bamboo shoots shaped FeVO$_4$ passivated ZnO nanorods photoanode for improved charge separation/transfer process towards efficient solar water splitting. *Appl. Catal. B Environ.* **2019**, *257*, 117813. [CrossRef]
14. Liang, Y.C.; Chao, Y. Crystal phase content-dependent functionality of dual phase SnO$_2$–WO$_3$ nanocomposite films via cosputtering crystal growth. *RSC Adv.* **2019**, *9*, 6482–6493. [CrossRef]
15. Lou, Z.; Li, F.; Deng, J.; Wang, L.; Zhang, T. Branch-like Hierarchical Heterostructure (α-Fe$_2$O$_3$/TiO$_2$): A Novel Sensing Material for Trimethylamine Gas Sensor. *ACS Appl. Mater. Interfaces* **2013**, *5*, 12310–12316. [CrossRef]
16. Xu, K.; Zhao, W.; Yu, X.; Duan, S.; Zeng, W. Enhanced ethanol sensing performance using Co$_3$O$_4$–ZnSnO$_3$ arrays prepared on alumina substrates. *Phys. E Low-Dimens. Syst. Nanostruct.* **2020**, *117*, 113825. [CrossRef]
17. Shaikh, S.; Ganbavale, V.; Mohite, S.; Patil, U.; Rajpure, K. ZnO nanorod based highly selective visible blind ultra-violet photodetector and highly sensitive NO$_2$ gas sensor. *Superlattices Microstruct.* **2018**, *120*, 170–186. [CrossRef]
18. Kheel, H.; Sun, G.J.; Lee, J.K.; Lee, S.; Dwivedi, R.P.; Lee, C. Enhanced H2S sensing performance of TiO$_2$-decorated α-Fe$_2$O$_3$ nanorod sensors. *Ceram. Int.* **2016**, *42*, 18597–18604. [CrossRef]
19. Kotsikau, D.; Ivanovskaya, M.; Orlik, D.; Falasconi, M. Gas-sensitive properties of thin and thick film sensors based on Fe$_2$O$_3$–SnO$_2$ nanocomposites. *Sens. Actuators B Chem.* **2004**, *101*, 199–206. [CrossRef]
20. Zhang, B.; Huang, Y.; Vinluan, R.; Wang, S.; Cui, C.; Lu, X.; Peng, C.; Zhang, M.; Zheng, J.; Gao, P.X. Enhancing ZnO nanowire gas sensors using Au/Fe$_2$O$_3$ hybrid nanoparticle decoration. *Nanotechnology* **2020**, *31*, 325505. [CrossRef]
21. Yang, L.; Xie, C.; Zhang, G.; Zhao, J.; Yu, X.; Zeng, D.; Zhang, S. Enhanced response to NO$_2$ with CuO/ZnO laminated heterostructured configuration. *Sens. Actuators B Chem.* **2014**, *195*, 500–508. [CrossRef]
22. Yan, S.; Ma, S.; Li, W.; Xu, X.; Cheng, L.; Song, H.; Liang, X. Synthesis of SnO$_2$–ZnO heterostructured nanofibers for enhanced ethanol gas-sensing performance. *Sens. Actuators B Chem.* **2015**, *221*, 88–95. [CrossRef]
23. Wang, X.; Ren, P.; Tian, H.; Fan, H.; Cai, C.; Liu, W. Enhanced gas sensing properties of SnO$_2$: The role of the oxygen defects induced by quenching. *J. Alloys Compd.* **2016**, *669*, 29–37. [CrossRef]
24. Mohammad, S.M.; Hassan, Z.; Talib, R.A.; Ahmed, N.M.; Al-Azawi, M.A.; Abd-Alghafour, N.M.; Chin, C.W.; Al-Hardan, N.H. Fabrication of a highly flexible low-cost H2 gas sensor using ZnO nanorods grown on an ultra-thin nylon substrate. *J. Mater. Sci. Mater. Electron.* **2016**, *27*, 9461–9469. [CrossRef]

25. Ge, Y.; Kan, K.; Yang, Y.; Zhou, L.; Jing, L.; Shen, P.; Li, L.; Shi, K. Highly mesoporous hierarchical nickel and cobalt double hydroxide composite: Fabrication, characterization and ultrafast NOx gas sensors at room temperature. *J. Mater. Chem. A* **2014**, *2*, 4961–4969. [CrossRef]
26. Chen, Y.; Zhu, C.L.; Xiao, G. Reduced-temperature ethanol sensing characteristics of flower-like ZnO nanorods synthesized by a sonochemical method. *Nanotechnology* **2006**, *17*, 4537–4541. [CrossRef]
27. Lee, J.-H. Gas sensors using hierarchical and hollow oxide nanostructures: Overview. *Sens. Actuators B Chem.* **2009**, *140*, 319–336. [CrossRef]
28. Sysoev, V.V.; Goschnick, J.; Schneider, T.; Strelcov, E.; Kolmakov, A. A Gradient Microarray Electronic Nose Based on Percolating SnO2Nanowire Sensing Elements. *Nano Lett.* **2007**, *7*, 3182–3188. [CrossRef]

International Journal of
Molecular Sciences

MDPI

Article

Cu and Cu-SWCNT Nanoparticles' Suspension in Pulsatile Casson Fluid Flow via Darcy–Forchheimer Porous Channel with Compliant Walls: A Prospective Model for Blood Flow in Stenosed Arteries

Amjad Ali [1,*], Zainab Bukhari [1], Muhammad Umar [1], Muhammad Ali Ismail [2] and Zaheer Abbas [3]

[1] Centre for Advanced Studies in Pure and Applied Mathematics, Bahauddin Zakariya University, Multan 60800, Pakistan; zainabbukhari@student.bzu.edu.pk (Z.B.); muhammadumar@bzu.edu.pk (M.U.)

[2] Department of Computer and Information System Engineering, NED University of Engineering and Technology, Karachi 75270, Pakistan; maismail@neduet.edu.pk

[3] Department of Mathematics, The Islamia University of Bahawalpur, Bahawalpur 63100, Pakistan; zaheer.abbas@iub.edu.pk

* Correspondence: amjadali@bzu.edu.pk

Citation: Ali, A.; Bukhari, Z.; Umar, M.; Ismail, M.A.; Abbas, Z. Cu and Cu-SWCNT Nanoparticles' Suspension in Pulsatile Casson Fluid Flow via Darcy–Forchheimer Porous Channel with Compliant Walls: A Prospective Model for Blood Flow in Stenosed Arteries. *Int. J. Mol. Sci.* **2021**, *22*, 6494. https://doi.org/10.3390/ijms22126494

Academic Editor: Raghvendra Singh Yadav

Received: 24 May 2021
Accepted: 5 June 2021
Published: 17 June 2021

Publisher's Note: MDPI stays neutral with regard to jurisdictional claims in published maps and institutional affiliations.

Abstract: The use of experimental relations to approximate the efficient thermophysical properties of a nanofluid (NF) with Cu nanoparticles (NPs) and hybrid nanofluid (HNF) with Cu-SWCNT NPs and subsequently model the two-dimensional pulsatile Casson fluid flow under the impact of the magnetic field and thermal radiation is a novelty of the current study. Heat and mass transfer analysis of the pulsatile flow of non-Newtonian Casson HNF via a Darcy–Forchheimer porous channel with compliant walls is presented. Such a problem offers a prospective model to study the blood flow via stenosed arteries. A finite-difference flow solver is used to numerically solve the system obtained using the vorticity stream function formulation on the time-dependent governing equations. The behavior of Cu-based NF and Cu-SWCNT-based HNF on the wall shear stress (WSS), velocity, temperature, and concentration profiles are analyzed graphically. The influence of the Casson parameter, radiation parameter, Hartmann number, Darcy number, Soret number, Reynolds number, Strouhal number, and Peclet number on the flow profiles are analyzed. Furthermore, the influence of the flow parameters on the non-dimensional numbers such as the skin friction coefficient, Nusselt number, and Sherwood number is also discussed. These quantities escalate as the Reynolds number is enhanced and reduce by escalating the porosity parameter. The Peclet number shows a high impact on the microorganism's density in a blood NF. The HNF has been shown to have superior thermal properties to the traditional one. These results could help in devising hydraulic treatments for blood flow in highly stenosed arteries, biomechanical system design, and industrial plants in which flow pulsation is essential.

Keywords: compliant walls; constricted channel; pulsatile flow; blood flow; nanoparticles; hybrid nanofluid

1. Introduction

The study of blood flow through an artery that has stenosis plays an important role in understanding cardiovascular diseases. The stenosis created is due to the undesired growth of lumen inside the blood vessels, which results in a reduction of normal blood flow. In severe cases, the blockage of arteries may lead to stroke, heart attack, or other cardiovascular diseases. The human blood flow is frequently modeled as a non-Newtonian fluid. Moreover, the geometry of channels with compliant walls, i.e., having constrictions on the walls models, has been used to model arteries that have stenosis. Young [1] studied in detail the behavior of blood flow in arteries that have stenosis. He explained the effect of stenosis on blood flow, pressure distribution, and wall shearing stress distribution along with the stenosis, velocity, and flow separation region near the stenosis. The Casson model is a non-Newtonian liquid model that is used to simulate shearing thinning and stress. This model will be shown in terms of mathematical expression in Section 2. As a result of the unique characteristics, the Casson fluid is considered as a promising rheological fluid model for human blood. The rheological behavior of blood in microcirculation is mostly dictated by red blood cells. As a result of their peculiar features, they are responsible for blood's non-Newtonian behavior. Casson fluid flow in a channel using the Brinkman model was investigated by Chaturani and Upadhya [2]. They concluded that increasing yield stress and decreasing the value of permeability reduces the flow rate. These results can be used to treat the blood clots in a coronary artery. The pulsating motion study has significant importance while analyzing the blood flow. Due to the cyclic nature of the pulse, the blood supply in arteries is intrinsically erratic and pulsating. Its analysis is challenging in both clinical and analytical contexts. Shit and Roy [3] analyzed the effects of magnetic field and heat on the pulsatile flow of blood in the arteries with stenosis. They concluded that the heat transfer rate has a direct relation toward the magnetic parameter. Elahi et al. [4] investigated the heat and mass transfer of non-Newtonian fluid flow in arteries with mild stenosis having permeable walls. Haghighi and Asl [5] presented the mathematical modeling of the pulsatile blood flow via overlapping constricted tapered vessels. Amjad et al. [6] investigated the influence of the magnetic field on the flow behavior of Casson fluid moving through a constricted channel using Darcy's law for the steady and pulsatile flows. They deduced that the magnetic parameter could be used to control the flow separation region. Zainab et al. [7] expanded the study to analyze the heat transfer of Casson fluid flow in a constricted channel. They concluded that temperature profile has a direct relation toward the thermal radiation parameter and inverse relation toward the Prandtl number. Amjad et al. [8] studied the pulsatile flow of non-Newtonian micropolar fluid flow in a constricted channel, and Umar et al. [9] expanded the study to analyzed the heat transfer effects. They concluded that the Nusselt number is an increasing function of the Reynolds number and Prandtl number. Amjad et al. [10] examined the pulsatile flow behavior of micropolar-Casson fluid in a constricted channel as an exemplar problem of blood flow. They reported the direct relation of the Casson fluid parameter toward the wall shear stress. Some other relevant studies can also be seen in [11–13].

Heat transfer analysis has gained importance among researchers due to its vast applications in industries and biomedical engineering. Different kinds of fluids are used as heat carriers. The common fluids usually used as a heat/energy carrier to improve the product quality in industries are water, ethylene glycol, vegetable oil, paraffin oil, etc. [14]. The thermophysical properties of such base fluids are enhanced by the suspension of nanoparticles (NPs) of highly conductive solids. Common examples of nanoparticles (NPs) include those of copper (Cu), silver (Ag), copper oxide (CuO), aluminum oxide (Al_2O_3), titanium oxide (TiO_2), single-wall carbon nanotubes (SWCNTs), and multi-wall carbon nanotubes (MWCNTs). The literature reveals that remarkable efforts have been made by scientists to enhance heat transfer using different NFs. Recently, nanofluids (NFs) have been used in cancer therapeutics, hyperthermia, and drug carriers in biomedical engineering. The science of cardiovascular mechanics, in which an irregular plaque forms in the arterial lumen, as well as other fields such as biomedical engineering, chemical engineering, environmental degradation, and so on, benefit greatly from the dispersion of solute into vessels. The NFs formed using more than one type of NP are so-called hybrid nanofluids (HNFs). HNF might serve the desired purpose by combining the properties of its constituent materials. As a result of the synergistic impact, HNFs are found to exhibit better thermophysical characteristics compared with individual NFs [14]. Waini et al. [15] discussed the influence of transpiration on HNF flow and heat transfer for uniform shear flow over a stretching/shrinking surface. Lund et al. [16] analyzed the effects of different parameters on the flow profiles of an unsteady magnetohydrodynamic (MHD) flow of HNF mixture of copper–aluminum oxide/water over the stretching shrinking sheet in the existence of thermal radiation. Khan et al. [17] numerically studied impact variable viscosity in inclined MHD Williamson NF flow over a nonlinearly stretching sheet. They reported that the velocity profile declined with the rise of inclination angle, Hartmann number, and variable viscosity. The 3D flow of Cu-Al$_2$O$_3$/water HNF on an expanding surface was studied by Devi and Anjali [18] using the RK–Fehlberg integration technique. The numerical results demonstrated a higher heat transfer rate for the HNF than the copper-based NF. Vasua et al. [19] simulated the 2D rheological laminar hemodynamics via diseased tapered artery with mild stenosis theoretically and computationally. They considered the effect of different metallic NPs homogeneously suspended in the blood motivated by pharmacology applications. Priyadharshini and Ponalagusamy [20] investigated the blood flow having magnetic NPs in a stenosed artery. Some examples of relevant studies can also be found in [21–24].

Aman et al. [25,26] considered a novel HNF model with advanced thermophysical properties for an MHD Casson fluid in a porous vertical medium. They used alumina and copper NPs and analyzed the heat and mass transfer effects. Kasim et al. [27] considered the heat transfer effects over an unsteady stretching sheet. They used blood fluid-based copper and alumina NPs. Nadeem et al. [28] analyzed the impact of chemical reaction in a Casson NF flow. Jamshed and Aziz [29] carried out a study to analyze the entropy generation and heat transfer analysis of a Casson HNF under the impact of transverse magnetic field thermal radiation and Cattaneo-Christov heat flux model. Aside from that reported, the following recent studies, such as those by Souayeh et al. [30], Ullah et al. [31], and Aziz and Afify [32], can be referred to as additional knowledge correlated with the Casson NF flows. Dinarvand et al. [33] demonstrated steady laminar mixed convection incompressible viscous and electrically conducting HNF (CuO- Cu/blood) flow over the plane stagnation point over a horizontal porous stretching layer with an external magnetic field, taking into account the induced magnetic field effect, which can be used in biomedical fields, especially in drug delivery. Reddy [34] investigated the model for electro-MHD flow over a stagnation point flow of HNF with non-linear thermal radiation and non-uniform heat source/sink that is implemented to blood-based NFs for two different NPs. For bacterial development in the heart valve, Elelamy et al. [35] examined a mathematical model using numerical simulation. NPS was employed for antibacterial and antibody characteristics. They employed non-Newtonian fluid of Casson micropolar blood flow in the heart valve for 2D, since antibiotics are widely believed to be homogeneously dispersed via the blood. El Kot et al. [36] investigated the behavior of a gold–titanium oxide NPs combination suspended in blood as a base fluid in a damaged coronary artery. Their major purpose is to examine the shed light on the HNF flows via a vertical diseased artery in the presence of the catheter tube with heat transfer.

The current research adds into the literature the heat as well as mass transfer analysis of non-Newtonian Casson HNF flow with Cu and SWCNT as the two types of constituting nanoparticles influenced by the Lorentz force in channel with compliant walls and Darcian porous medium. The unsteady governing equations are transformed using the vorticity-stream function technique, and the resulting model is solved using a finite difference method (FDM). The objective is to examine the cumulative impact of the applied magnetic field and thermal radiation on the wall shear stress (WSS), velocity, temperature, and concentration profiles. The flow regulating parameters for the present study include the Hartmann number (the magnetic parameter), Strouhal number (the pulsation parameter), radiation parameter, Darcy's number (the porosity parameter), Casson fluid parameter, Reynolds number, Soret number, and Peclet number (the heat or mass diffusion parameter). By definition, the Peclet number encompasses the effects of the Prandtl number for heat diffusion. Furthermore, the influence of the flow parameters on the non-dimensional numbers such as the skin friction coefficient, Nusselt number, and Sherwood number is also discussed. The results of the current study would provide insight to the relevant researchers and engineers to design concerning products. The non-Newtonian pulsatile nanofluid flows help in understanding the influence of various metallic NPs homogeneously suspended in the blood, which is driven by drug trafficking (pharmacology) applications. A thorough interpretation with the computations of particular significance to pharmacological NP-mediated rheological blood flow in stenosed arteries may be considered in future research.

The remaining parts of the article are organized in the following manner. The mathematical modeling and its transformation into a solvable form, specifically using the stream and vorticity functions, is presented in Section 2. Section 3 discusses the numerical scheme and validation. Section 4 discusses the findings and related discussions. Section 5 contains the concluding remarks.

2. Mathematical Model and Formulation

2.1. Problem Statement

A two-dimensional pulsatile flow of a non-Newtonian Casson HNF in a porous channel with symmetrical constrictions under the impact of a uniform magnetic field perpendicular to the fluid flow and thermal radiation is considered. Along with it, the chemical reaction is also considered. NF based on Cu NPs and HNF based on Cu-SWCNT NPs are two different kinds of NFs considered in this study to enhance the heat and mass transfer. Furthermore, thermal equilibrium among the base fluid and NPs and no-slip conditions are considered for the current study. The constriction on the channel walls is formed using the following expression

$$y_1(x) = \begin{cases} \frac{h_1}{2}\left[1 + \cos\left(\frac{\pi x}{x_0}\right)\right], & |x| \le x_0 \\ 0, & |x| > x_0 \end{cases}$$

$$y_2(x) = \begin{cases} 1 - \frac{h_2}{2}\left[1 + \cos\left(\frac{\pi x}{x_0}\right)\right], & |x| \le x_0 \\ 1, & |x| > x_0 \end{cases} \tag{1}$$

The mathematical model for the Casson fluid is given as [10]

$$\tau_{ij} = \begin{cases} 2\left(\mu_\beta + p_y/\sqrt{2\pi}\right)e_{ij} & \pi > \pi_c \\ 2\left(\mu_\beta + p_y/\sqrt{2\pi_c}\right)e_{ij} & \pi < \pi_c \end{cases} \tag{2}$$

2.2. Governing Equations

The walls of the channel have a pair of symmetric constriction bumps (see Figure 1). We suppose a Cartesian coordinate system $(\tilde{x}, \tilde{y})$ such that the flow direction is along the $\tilde{x}$-axis and the direction of **B** is along the $\tilde{y}$-axis. The resulting electric field **J** is normal to the plane of flow. In the transformed coordinate system (to be discussed later on), the constrictions are spanned from $x = -x_0$ to $x = x_0$ with its center at $x = 0$, as shown in Figure 1. Thus, the length of the constriction is $2x_0$.

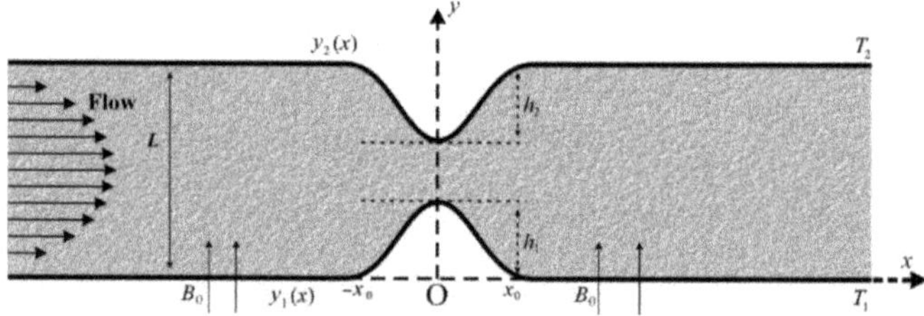

Figure 1. A schematic diagram of the flow channel with symmetrical constriction on both walls.

The flow phenomenon is represented by the unsteady incompressible viscous flow equations as follows.

The continuity equation:

$$\frac{\partial \tilde{u}}{\partial \tilde{x}} + \frac{\partial \tilde{v}}{\partial \tilde{y}} = 0 \tag{3}$$

The momentum equation:

$$\frac{\partial \tilde{u}}{\partial \tilde{t}} + \tilde{u}\frac{\partial \tilde{u}}{\partial \tilde{x}} + \tilde{v}\frac{\partial \tilde{u}}{\partial \tilde{y}} = -\frac{1}{\rho_{hnf}}\frac{\partial \tilde{p}}{\partial \tilde{x}} + \frac{\mu_{hnf}}{\rho_{hnf}}\left(1+\frac{1}{\beta}\right)\nabla^2\tilde{u} + \frac{1}{\rho_{hnf}}(\mathbf{J}\times\mathbf{B})_x - \frac{v_f}{k}\tilde{u} \tag{4}$$

$$\frac{\partial \tilde{v}}{\partial \tilde{t}} + \tilde{u}\frac{\partial \tilde{v}}{\partial \tilde{x}} + \tilde{v}\frac{\partial \tilde{v}}{\partial \tilde{y}} = -\frac{1}{\rho_{hnf}}\frac{\partial \tilde{p}}{\partial \tilde{y}} + \frac{\mu_{hnf}}{\rho_{hnf}}\left(1+\frac{1}{\beta}\right)\nabla^2\tilde{v} - \frac{v_f}{k}\tilde{v} \tag{5}$$

where $\mathbf{J} \equiv (J_x, J_y, J_z)$, $\mathbf{B} \equiv (0, B_0, 0)$.

The energy equation:

$$\frac{\partial \tilde{T}}{\partial \tilde{t}} + \tilde{u}\frac{\partial \tilde{T}}{\partial \tilde{x}} + \tilde{v}\frac{\partial \tilde{T}}{\partial \tilde{y}} = \frac{k_{hnf}}{(\rho C_p)_{hnf}}\nabla^2\tilde{T} - \frac{1}{(\rho C_p)_{hnf}}\frac{\partial q}{\partial \tilde{y}} \tag{6}$$

where the radiative heat flux q defined as $q = -\left(\frac{4\sigma}{3k^*}4T_\infty^3\frac{\partial \tilde{T}}{\partial \tilde{y}}\right)$ and $\tilde{T}^4 \cong 4T_\infty^3\tilde{T} - 3T_\infty^4$ allows the expansion of T^4 using Taylor series about T_∞ (free stream temperature); then, $q = -\left(\frac{4\sigma}{3k^*}4T_\infty^3\frac{\partial \tilde{T}}{\partial \tilde{y}}\right)$ so $\frac{\partial q}{\partial \tilde{y}} = -\left(\frac{16\sigma}{3k^*}T_\infty^3\frac{\partial^2\tilde{T}}{\partial \tilde{y}^2}\right)$. Equation (6) becomes

$$\frac{\partial \tilde{T}}{\partial \tilde{t}} + \tilde{u}\frac{\partial \tilde{T}}{\partial \tilde{x}} + \tilde{v}\frac{\partial \tilde{T}}{\partial \tilde{y}} = \frac{k_{hnf}}{(\rho C_p)_{hnf}}\nabla^2\tilde{T} + \frac{16\sigma T_\infty^3}{3k^*(\rho C_p)_{hnf}}\frac{\partial^2\tilde{T}}{\partial \tilde{y}^2} \tag{7}$$

The concentration equation:

$$\frac{\partial \tilde{C}}{\partial \tilde{t}} + \tilde{u}\frac{\partial \tilde{C}}{\partial \tilde{x}} + \tilde{v}\frac{\partial \tilde{C}}{\partial \tilde{y}} = D\nabla^2\tilde{C} + \frac{DK_T}{T_m}\nabla^2\tilde{T} \tag{8}$$

The expression $\mathbf{J} \times \mathbf{B}$ in Equation (4) is simplified using Ohm's law and Maxwell's equation. By Ohm's law

$$J_x = 0, \quad J_y = 0, \quad J_z = \sigma(E_z + \tilde{u}B_0) \tag{9}$$

Maxwell's equation $\nabla \times \mathbf{E} = 0$ for steady flow implies that $E_z = a$, where a is a constant number. We can assume a to be zero. Then, Equation (9) gives, $J_z = \sigma\tilde{u}B_0$. Therefore, $\mathbf{J} \times \mathbf{B} = -\sigma\tilde{u}B_0^2$. So, Equation (3) becomes

$$\frac{\partial \tilde{u}}{\partial \tilde{t}} + \tilde{u}\frac{\partial \tilde{u}}{\partial \tilde{x}} + \tilde{v}\frac{\partial \tilde{u}}{\partial \tilde{y}} = -\frac{1}{\rho_{hnf}}\frac{\partial \tilde{p}}{\partial \tilde{x}} + \frac{\mu_{hnf}}{\rho_{hnf}}\left(1+\frac{1}{\beta}\right)\nabla^2\tilde{u} - \frac{1}{\rho_{hnf}}\sigma_f\tilde{u}B_0^2 - \frac{v_f}{k}\tilde{u} \tag{10}$$

The following dimensionless variables are considered to convert the governing Equations (3), (5), (7), (8) and (10) into the dimensionless form:

$$x = \frac{\tilde{x}}{L}, y = \frac{\tilde{y}}{L}, u = \frac{\tilde{u}}{U}, v = \frac{\tilde{v}}{U}, t = \frac{\tilde{t}}{T}, St = \frac{L}{UT}, p = \frac{\tilde{p}}{\rho_f U^2}, Re = \frac{\rho_f UL}{\mu_f},$$

$$M = B_0 L\sqrt{\frac{\sigma_f}{\rho_f v_f}}, Da = \frac{v_f}{U\sqrt{k}}, Pr = \frac{\mu_f C_{p,f}}{k_f}, Rd = \frac{16\sigma T_\infty^3}{3k^* k_f}, \varphi = \frac{\tilde{C}-C_2}{C_1-C_2}, \theta = \frac{\tilde{T}-T_2}{T_1-T_2}, \tag{11}$$

$$Sc = \frac{\mu_f}{\rho_f D}, Sr = \frac{\rho_f DK_T(T_1-T_2)}{\mu_f T_m(C_1-C_2)}, Peh = Pr\cdot Re, Pem = Sc\cdot Re$$

Here, D is the coefficient of mass diffusivity, K_T is the thermal diffusion ratio, and T_m is the mean temperature. The effective thermophysical properties of HNF [16,37] are presented as follows.

The HNF's effective dynamic viscosity μ_{hnf} is given as

$$\mu_{hnf} = \frac{\mu_f}{(1-\phi_1)^{2.5}(1-\phi_2)^{2.5}} \tag{12}$$

Here, ϕ_1 and ϕ_2 show the volume fractions of NPs Cu and SWCNT, respectively. The NF's effective density ρ_{hnf} is given as

$$\rho_{hnf} = (1 - \phi_2)\left[(1 - \phi_1)\rho_f + \phi_1\rho_{s_1}\right] + \phi_2\rho_{s_2} \tag{13}$$

Here, s_1 and s_2 in subscripts correspond to the property of Cu and SWCNT, respectively. Moreover, hnf, nf, and f in the subscripts correspond to the property of the HNF, NF, and base fluid water, respectively.

The NF's heat capacitance $\left(\rho C_p\right)_{hnf}$ is given as

$$\left(\rho C_p\right)_{hnf} = (1 - \phi_2)\left[(1 - \phi_1)\left(\rho C_p\right)_f + \phi_1\left(\rho C_p\right)_{s_1}\right] + \phi_2\left(\rho C_p\right)_{s_2}. \tag{14}$$

The NF's effective thermal conductivity k_{hnf} is given as

$$\frac{k_{hnf}}{k_{nf}} = \frac{k_{s_2} + 2k_{nf} - 2\phi_2\left(k_{nf} - k_{s_2}\right)}{k_{s_2} + 2k_{nf} + 2\phi_2\left(k_{nf} - k_{s_2}\right)} \tag{15}$$

where

$$\frac{k_{nf}}{k_f} = \frac{k_{s_1} + 2k_f - 2\phi_1\left(k_f - k_{s_1}\right)}{k_{s_1} + 2k_f + 2\phi_1\left(k_f - k_{s_1}\right)} \tag{16}$$

where ϕ shows the volume fraction of NF, and the subscripts f and s correspond to the fluid and solid-state properties, respectively. For the present work, the nanoparticles of the following materials are considered: Cu and SWCNT.

The thermophysical properties of the base fluid water and the NPs under consideration are shown in Table 1. The properties in Table 1 are given at 24.6 °C. The viscosity of water at 24.6 °C is approximately 0.0000009009 m^2/s. Thus, according to Equation (12), $\mu_{hnf} = 7.940348428 \times 10^{-7}$ m^2/s with $\phi_1 = \phi_2 = 0.03$.

Table 1. Thermophysical properties of the base fluid water and the two kinds of NPs [38,39].

Physical Property	Base Fluid	Nanoparticles	
	H$_2$O	Cu	SWCNT
C_p (J/kgK)	4179	385	235
ρ (kg/m^3)	997.1	8933	10500
k (W/mK)	0.613	401	429

Equations (3), (5), (7), (8) and (10) after transformation become

$$\frac{\partial u}{\partial x} + \frac{\partial u}{\partial y} = 0 \tag{17}$$

$$St\frac{\partial u}{\partial t} + u\frac{\partial u}{\partial x} + v\frac{\partial u}{\partial y} = -\frac{1}{\varnothing_1}\frac{\partial p}{\partial x} + \frac{1}{Re\varnothing_3}\left(1 + \frac{1}{\beta}\right)\nabla^2 u - \frac{1}{\varnothing_1}\frac{M^2}{Re}u - ReD_a^2 u \tag{18}$$

$$St\frac{\partial v}{\partial t} + u\frac{\partial v}{\partial x} + v\frac{\partial v}{\partial y} = -\frac{1}{\varnothing_1}\frac{\partial p}{\partial y} + \frac{1}{Re\varnothing_3}\left(1 + \frac{1}{\beta}\right)\nabla^2 v - ReD_a^2 v \tag{19}$$

$$St\frac{\partial \theta}{\partial t} + u\frac{\partial \theta}{\partial x} + v\frac{\partial \theta}{\partial y} = \frac{1}{Peh}\frac{\varnothing_5}{\varnothing_4}\left(\frac{\partial^2 \theta}{\partial x^2} + \left(1 + \frac{Rd}{\varnothing_5}\right)\frac{\partial^2 \theta}{\partial y^2}\right) \tag{20}$$

$$St\frac{\partial \varphi}{\partial t} + u\frac{\partial \varphi}{\partial x} + v\frac{\partial \varphi}{\partial y} = \frac{1}{Pem}\left(\frac{\partial^2 \varphi}{\partial x^2} + \frac{\partial^2 \varphi}{\partial y^2}\right) + \frac{Sr}{Re}\left(\frac{\partial^2 \theta}{\partial x^2} + \frac{\partial^2 \theta}{\partial y^2}\right) \tag{21}$$

where

$$\varnothing_1 = (1 - \phi_2)\left[(1 - \phi_1) + \phi_1\frac{\rho_{s_1}}{\rho_f}\right] + \phi_2\frac{\rho_{s_2}}{\rho_f}$$

$$\varnothing_2 = \frac{1}{(1 - \phi_1)^{2.5}(1 - \phi_2)^{2.5}}$$

$$\varnothing_3 = (1 - \phi_1)^{2.5}(1 - \phi_2)^{2.5}\left((1 - \phi_2)\left[(1 - \phi_1) + \phi_1\frac{\rho_{s_1}}{\rho_f}\right] + \phi_2\frac{\rho_{s_2}}{\rho_f}\right)$$

$$\varnothing_4 = (1 - \phi_2)\left[(1 - \phi_1) + \phi_1\frac{(\rho C_p)_{s_1}}{(\rho C_p)_f}\right] + \phi_2\frac{(\rho C_p)_{s_2}}{(\rho C_p)_f}$$

$$\varnothing_5 = \frac{k_{hnf}}{k_f}$$

2.3. Stream and Vorticity Functions

The following transformation is used to transform the governing equations into the stream (ψ) and vorticity (ω) functions:

$$u = \frac{\partial\psi}{\partial y}, \quad v = -\frac{\partial\psi}{\partial x}, \quad \omega = \frac{\partial v}{\partial x} - \frac{\partial u}{\partial y} \tag{22}$$

The WSS (τ_w) is calculated using the vorticity at the wall, as both are orthogonal to each other [40]. Applying Equation (22) to Equations (18) and (19), and rearranging result:

$$St\frac{\partial}{\partial t}\left(\frac{\partial v}{\partial x} - \frac{\partial u}{\partial y}\right) + u\frac{\partial}{\partial x}\left(\frac{\partial v}{\partial x} - \frac{\partial u}{\partial y}\right) + v\frac{\partial}{\partial y}\left(\frac{\partial v}{\partial x} - \frac{\partial u}{\partial y}\right)$$
$$= \frac{1}{Re\varnothing_3}\left(1 + \frac{1}{\beta}\right)\left[\frac{\partial^2}{\partial x^2}\left(\frac{\partial v}{\partial x} - \frac{\partial u}{\partial y}\right) + \frac{\partial^2}{\partial y^2}\left(\frac{\partial v}{\partial x} - \frac{\partial u}{\partial y}\right)\right] + \frac{1}{\varnothing_1}\frac{M^2}{Re}\frac{\partial u}{\partial y} + ReD_a^2\omega \tag{23}$$

$$St\frac{\partial\omega}{\partial t} + \frac{\partial\psi}{\partial y}\frac{\partial\omega}{\partial x} - \frac{\partial\psi}{\partial x}\frac{\partial\omega}{\partial y} = \frac{1}{\varnothing_3}\frac{1}{Re}\left(1 + \frac{1}{\beta}\right)\left[\frac{\partial^2\omega}{\partial x^2} + \frac{\partial^2\omega}{\partial y^2}\right] + \frac{1}{\varnothing_1}\frac{M^2}{Re}\frac{\partial^2\psi}{\partial y^2} + ReD_a^2\omega \tag{24}$$

The Equation for the stream function ψ (Poisson equation) is given by

$$\frac{\partial^2\psi}{\partial x^2} + \frac{\partial^2\psi}{\partial y^2} = -\omega \tag{25}$$

Here u, v, θ, and φ are the primitive variables, whereas ω and ψ are the non-primitive variables.

2.4. Boundary Conditions

The Equation of motion (4) for the steady case, in the presence of electric and magnetic fields, becomes

$$\mu_{hnf}\left(1 + \frac{1}{\beta}\right)\frac{\partial^2\tilde{u}}{\partial\tilde{y}^2} - \sigma_f\tilde{u}B_0^2 - \rho_{hnf}\frac{v_f}{k}\tilde{u} = \frac{\partial\tilde{p}}{\partial\tilde{x}} + \sigma_f B_0 E_z \tag{26}$$

Again, for the steady flow, Maxwell's equation, $\nabla \times \mathbf{E} = 0$ gives $E_z = a$, where a is a constant number, and all the other variables depend only on $\tilde{y}$, except for the pressure gradient, i.e., $\frac{\partial\tilde{p}}{\partial\tilde{x}} = $ constant.

Using the thermophysical properties and Equation (11), Equation (26) becomes

$$\varnothing_2\left(1 + \frac{1}{\beta}\right)\frac{d^2u}{dy^2} - \left(M^2 + \varnothing_1 Re^2 D_a^2\right)u = \frac{L^2}{\rho_f v_f U}\left(\frac{\partial\tilde{p}}{\partial\tilde{x}} + \sigma_f B_0 E_z\right) \tag{27}$$

For $M \neq 0$,

$$\psi = \left[\frac{M^2 \sqrt{g} \cosh\left(\frac{M}{2}\right)\left(\frac{M_1}{\sqrt{g}}\cosh\left(\frac{M_1}{2\sqrt{g}}\right)\eta - \sinh\left(\frac{M_1}{\sqrt{g}}\left(\eta - \frac{1}{2}\right)\right)\right)}{8M_1^3 \sinh^2\left(\frac{M}{4}\right)\cosh\left(\frac{M_1}{2\sqrt{g}}\right)} \right] \times [1 + \epsilon \sin(2\pi t)] \tag{28}$$

For $M = 0$ and $D_a = 0$,

$$\psi = \frac{1}{6g}\left(3\eta^2 - 2\eta^3\right)[1 + \epsilon \sin(2\pi t)] \tag{29}$$

For $M = 0$ and $D_a \neq 0$,

$$\psi = \frac{2}{Re^2 D_a^2}\left[\eta - \frac{\sqrt{g}}{ReD_a}\left(\sinh(g_1) - \cosh(g_1)\tanh\left(\frac{g_1}{2}\right)\right)\right] \times [1 + \epsilon \sin(2\pi t)] \tag{30}$$

Here,

$$g = \varnothing_2\left(1 + \frac{1}{\beta}\right), \quad M_1^2 = (M^2 + \varnothing_1 Re^2 D_a^2), \quad g_1 = \frac{\sqrt{\varnothing_1}ReD_a\eta}{\sqrt{g}}$$

where ϵ denotes the pulsating amplitude, $\epsilon = 0$ indicates the steady flow conditions, and $\epsilon = 1$ indicates the pulsatile flow conditions. For the temperature, the boundary conditions after transformation are given by, $\theta = 1$, at $\eta = 0$; $\theta = 0$, at $\eta = 1$. By substituting $\eta = 0$ and $\eta = 1$ in (28)–(30), the wall boundary conditions for ψ at the upper and lower walls are obtained.

The initial and boundary conditions for the variables (stream function ψ, the vorticity ω, temperature θ, and concentration φ) are given in the new coordinate system:

$$\psi = \omega = \theta = \varphi = 0, \quad \text{for all } t \leq 0$$

The wall boundary conditions for ω in the (ξ, η) system are given by:

$$\omega = -\left[\left(Q^2 + D^2\right)\frac{\partial^2 \psi}{\partial \eta^2}\right]_{\eta=0,1} \tag{31}$$

The wall boundary conditions for θ and φ in the (ξ, η) system are given by:

$$\theta = \varphi = 1|_{\eta=0}, \quad \theta = \varphi = 0|_{\eta=1} \tag{32}$$

At the outlet, the boundary conditions for all the variables are used as the fully developed flow. The flow is defined as sinusoidal for the pulsating flow:

$$u(y, t) = u(y)[1 + \sin(2\pi t)], \quad v = 0 \tag{33}$$

Furthermore, no-slip conditions are considered on the walls.

2.5. Transformation of Coordinates

The constricted part of the channel wall is mapped to a straight line. Consider the following coordinate transformation

$$\xi = x, \quad \eta = \frac{y - y_1(x)}{y_2(x) - y_1(x)} \tag{34}$$

After this transformation, the lower and the upper walls of the channel are represented by $\eta = 0$ and $\eta = 1$, respectively, and Equations (20) and (21), (24) and (25) in the new coordinate system (ξ, η) are given as

$$St\frac{\partial \omega}{\partial t} + u\left(\frac{\partial \omega}{\partial \xi} - Q\frac{\partial \omega}{\partial \eta}\right) + vD\frac{\partial \omega}{\partial \eta}$$
$$= \frac{1}{\varnothing_3}\frac{1}{Re}\left(1 + \frac{1}{\beta}\right)\left[\frac{\partial^2 \omega}{\partial \xi^2} - (P - 2QR)\frac{\partial \omega}{\partial \eta} - 2Q\frac{\partial^2 \omega}{\partial \xi \partial \eta} + (Q^2 + D^2)\frac{\partial^2 \omega}{\partial \eta^2}\right] + \frac{1}{\varnothing_1}\frac{M^2}{Re}D^2\frac{\partial^2 \psi}{\partial \eta^2} + ReD_a^2\omega \tag{35}$$

$$\frac{\partial^2 \psi}{\partial \xi^2} - (P - 2QR)\frac{\partial \psi}{\partial \eta} - 2Q\frac{\partial^2 \psi}{\partial \xi \partial \eta} + \left(Q^2 + D^2\right)\frac{\partial^2 \psi}{\partial \eta^2} = -\omega \tag{36}$$

$$St\frac{\partial \theta}{\partial t} + u\left(\frac{\partial \theta}{\partial \xi} - Q\frac{\partial \theta}{\partial \eta}\right) + vD\frac{\partial \theta}{\partial \eta} = \frac{1}{Peh}\frac{\varnothing_5}{\varnothing_4}\left[\frac{\partial^2 \theta}{\partial \xi^2} - (P - 2QR)\frac{\partial \theta}{\partial \eta} - 2Q\frac{\partial^2 \theta}{\partial \xi \partial \eta} + \left(Q^2 + \left(1 + \frac{Rd}{\varnothing_5}\right)D^2\right)\frac{\partial^2 \theta}{\partial \eta^2}\right] \tag{37}$$

$$St\frac{\partial \varphi}{\partial t} + u\left(\frac{\partial \varphi}{\partial \xi} - Q\frac{\partial \varphi}{\partial \eta}\right) + vD\frac{\partial \varphi}{\partial \eta} = \frac{1}{Pem}\left[\frac{\partial^2 \varphi}{\partial \xi^2} - (P - 2QR)\frac{\partial \varphi}{\partial \eta} - 2Q\frac{\partial^2 \varphi}{\partial \xi \partial \eta} + \left(Q^2 + D^2\right)\frac{\partial^2 \varphi}{\partial \eta^2}\right]$$
$$+ \frac{Sr}{Re}\left[\frac{\partial^2 \theta}{\partial \xi^2} - (P - 2QR)\frac{\partial \theta}{\partial \eta} - 2Q\frac{\partial^2 \theta}{\partial \xi \partial \eta} + \left(Q^2 + D^2\right)\frac{\partial^2 \theta}{\partial \eta^2}\right] \tag{38}$$

where

$$P = P(\xi, \eta) = \frac{\eta y_2''(\xi) + (1-\eta)y_1''(\xi)}{y_2(\xi) - y_1(\xi)}, \quad Q = Q(\xi, \eta) = \frac{\eta y_2'(\xi) + (1-\eta)y_1'(\xi)}{y_2(\xi) - y_1(\xi)},$$
$$R = R(\xi) = \frac{y_2'(\xi) - y_1'(\xi)}{y_2(\xi) - y_1(\xi)}, \quad D = D(\xi) = \frac{1}{y_2(\xi) - y_1(\xi)}$$

where the velocity components u and v in terms of (ξ, η) take the forms

$$u = D(\xi)\frac{\partial \psi}{\partial \eta}, \quad v = Q(\xi, \eta)\frac{\partial \psi}{\partial \eta} - \frac{\partial \psi}{\partial \xi}$$

The non-dimensional physical quantities, Nusselt number (Nu), Sherwood (Sh), and skin friction coefficient $\left(C_f\right)$, are defined as

$$Nu = -\left(\frac{L}{k_f(T_1 - T_2)}\right)\left(k_{hnf} + \frac{16\sigma T_\infty^3}{3k^*}\right)\frac{\partial \widetilde{T}}{\partial \widetilde{y}}\Big|_{\widetilde{y}=0}, \quad C_f = \frac{\tau_w}{\rho_f u_w^2}, \quad Sh = \frac{-hc_w}{(C_1 - C_2)}$$

where τ_w, and c_w are defined as

$$\tau_w = \left[\left(\mu_{hnf}\left(1 + \frac{1}{\beta}\right)\right)\frac{\partial \widetilde{u}}{\partial \widetilde{y}}\right]_{\widetilde{y}=0}, \quad c_w = \left[\frac{\partial \widetilde{C}}{\partial \widetilde{y}}\right]_{\widetilde{y}=0}$$

After using dimensionless variables from Equation (10) and the coordinate transformation from Equation (34), we get

$$C_f = \frac{1}{u_w^2}\left[-\frac{1}{Re\varnothing_3}\left(1 + \frac{1}{\beta}\right)D\frac{\partial u}{\partial \eta}\right]_{\eta=0}$$

$$Nu = -\left[\frac{k_{hnf}}{k_f}(1 + Rd)\theta'(0)\right] = -\left[\varnothing_5(1 + Rd)\theta'(0)\right]$$

$$Sh = \left[-D\frac{\partial \varphi}{\partial \eta}\right]_{\eta=0}$$

where $Rd = \frac{16\sigma T_\infty^3}{3k^* k_{hnf}}$ and $\varnothing_5 = \frac{k_{hnf}}{k_f}$.

3. Numerical Scheme and Validation

The system (35)–(38) subject to the relevant boundary conditions is solved numerically as in [6–10,41,42]. We consider the computation domain as $-10 \leq \xi \leq 10$ and $0 \leq \eta \leq 1$. For a stable solution, the domain is discretized using the step size of $\Delta\xi = 0.05$ and $\Delta\eta = 0.02$ in ξ and η directions, thus forming a Cartesian grid of 400×50 elements. The height of constriction at both walls is considered as 0.35, and the Reynolds number is taken as 800 unless specified otherwise. Moreover, time step size $\Delta t = 0.00005$ is considered for the flow with pulsation. The solution at time level l is specified, but the solution at each time level $l + 1$, for $l = 0, 1, 2, \cdots$, is computed. Using the well-known TDMA (Tri-Diagonal Matrix Algorithm) and central differences for the discretization of the space derivatives,

Equation (36) is solved for $\psi = \psi(\xi, \eta)$. Whereas, using the ADI (Alternating Direction Implicit) scheme, Equations (35) and (37), (38) are solved for $\omega = \omega(\xi, \eta)$, $\theta = \theta(\xi, \eta)$, and $\varphi = \varphi(\xi, \eta)$, respectively. The forward/backward difference is used to discretize the time derivative, and the central differences are used to discretize the space derivatives. The computations for the present study are performed in a sequential fashion. The results can be found by parallel computing for time-efficient solutions; see Ali and Syed [43].

The pulsatile flows were simulated over a long enough period of time to avoid transition effects on the solution; i.e., a steady periodicity in the flow is established. For this purpose, we choose the sixth time interval that has been used to compute these data. The results are displayed at the four time instants of the pulse cycle; i.e., $t = 0$, 0.25, 0.5, 0.75. At $t = 0$, the pulse cycle is started. For $0 < t < 0.25$, the flow accelerates, and its peak is at $t = 0.25$. For $0.25 < t < 0.5$, the flow rate decreases and reaches its minimum at $t = 0.5$. At $t = 0.75$, the instantaneous flow rate becomes zero [6,10,42]. Moreover, to analyze the profiles at the prominent x locations, the results are displayed at selected time instants and axial locations. We perform simulations for a long enough time; however, in most cases, the results are displayed at selected time instants and axial locations, especially $x = 0$ (throat of the constriction) and $x = 2$ (in the lee of the constriction) where the fluid enters the low-pressure zone from the high-pressure zone.

For validation, the present results for the pulsatile flow of the base fluid (i.e., with $\phi_1 = 0$, $\phi_2 = 0$) are compared with those obtained by Amjad et al. [6] in the case of Casson fluid flow. Figure 2 shows a good agreement of the present results, specifically the WSS, with [6] for $M = 0$, 5, 10, 15 at $t = 0.25$.

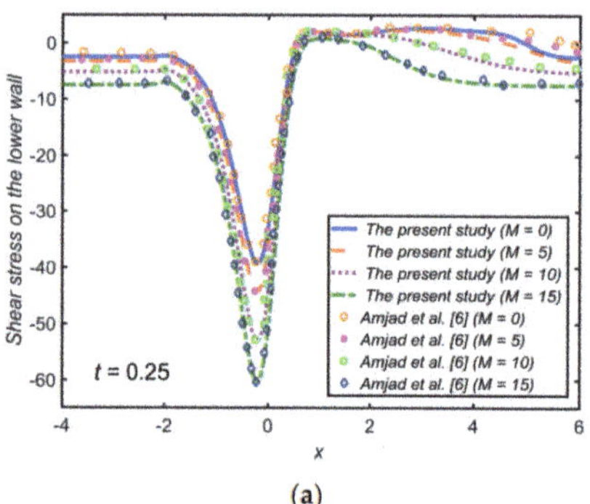

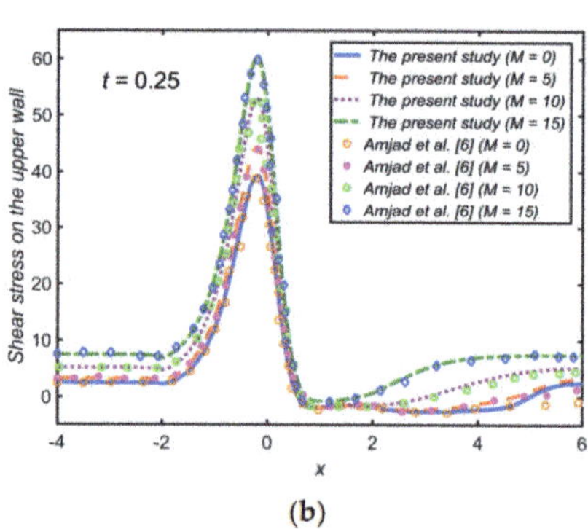

Figure 2. The WSS distribution for distinct values of the Hartmann number M (**a**) on the lower wall; (**b**) on the upper wall.

4. Results and Discussion

This section deals with the physical influence of constraints on the NF with Cu NPs and HNF with Cu-SWCNT NPs when $M = 5$, $St = 0.02$, $\beta = 0.5$, $D_a = 0.002$, $Re = 800$, $Rd = 0.2$, $Peh = 490$, $Pem = 420$, $Sr = 0.8$, and $\phi_1 = \phi_2 = 0.03$. The foregoing values are kept constant in the entire article except when we explicitly mention otherwise. The initial guesses for dependent variables are selected using the hit and trial method until the correct asymptotic behavior of the solution for Cu-based NF is achieved.

4.1. The Magnetic Parameter Effect

Figure 3 depicts the impact of Cu-based NF and Cu-SWCNT-based HNF for Hartman number $M = 0$, 5, 10, 15 on (a) the WSS distribution on the upper wall, (b) u profile at $x = 0$, (c) u profile at $x = 2$, (d) the temperature (θ) profile at $x = 0$, (e) the temperature (θ) profile at $x = 2$, and (f) the concentration (φ) profile at $x = 0$. Figure 3a shows that as M estimations intensify, the WSS on the upper wall escalates, and it is maximum at $t = 0.25$. The WSS for HNF is higher as compared to NF. Figure 3b,c show that the velocity field's boundary layer thickness escalates as M grows. This is due to the fact that the Lorentz force measures the presence of M, so a delaying force is conceived in the velocity field. As Hartmann number estimations grow, the restricting force increases, and the velocity field reduces. The HNF velocity profile is slightly lower than that of the NF at $x = 0$. The profiles are not parabolic, as some backflow in the vicinity of the walls is observed at $x = 2$, so in this case, the velocity profile for HNF is higher as compared to NF. The backflow reduces with an increase in M. Figure 3d–f show the opposite effects of M on the temperature and concentration fields. The Hartmann number has an inverse relation toward the density of HNF. So, it is found that as the estimations of M grow, the fluid density shrinks and the temperature increases. Hence, escalation in M shrinks the fluid density, which causes temperature rises. The HNF temperature profile is slightly higher than that of the NF.

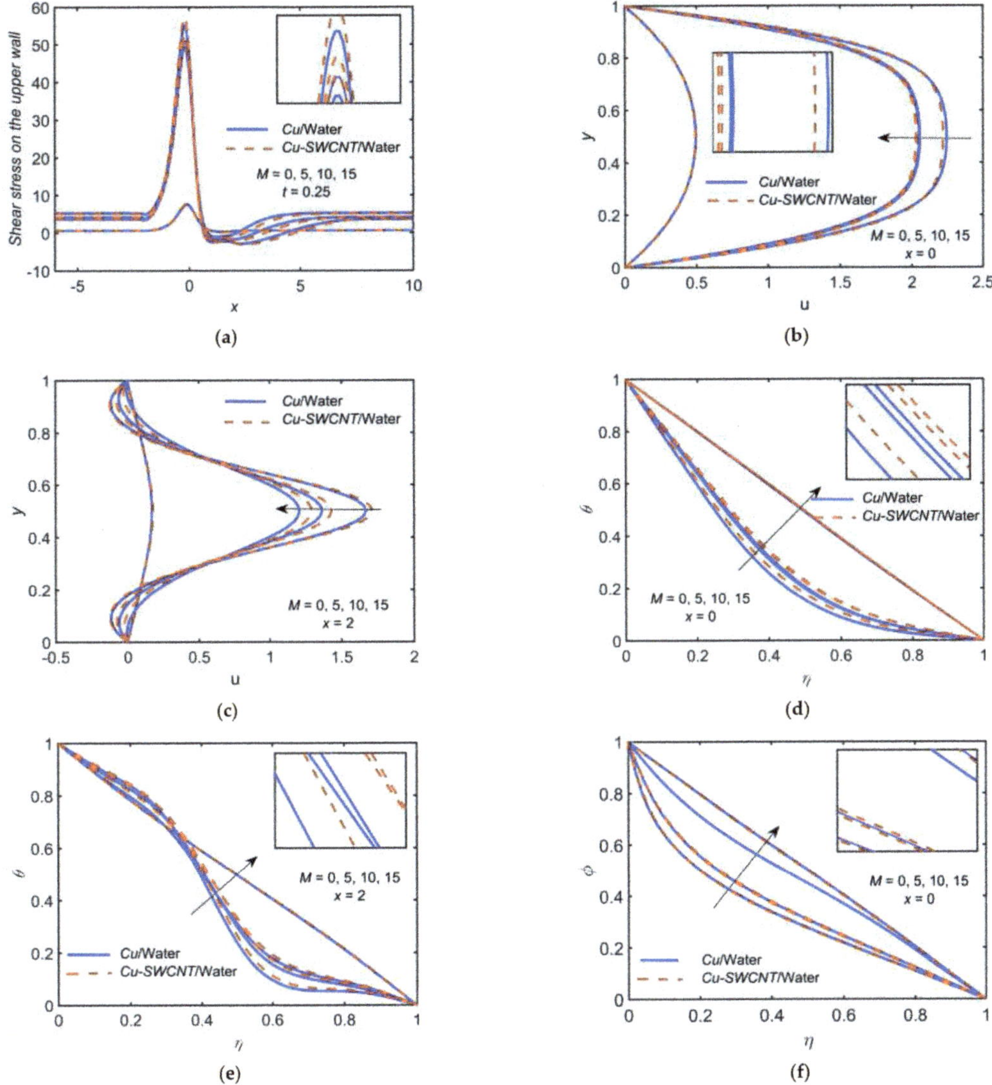

Figure 3. (a) The WSS distribution, (b) u profile at $x = 0$, (c) u profile at $x = 2$, (d) θ profile at $x = 0$, (e) θ profile at $x = 2$, and (f) φ profile at $x = 0$ for distinct values of M.

4.2. The Pulsation Parameter Effect

Figure 4 depicts the impact of Cu-based NF and Cu-SWCNT-based HNF for Strouhal number (pulsation parameter) $St = 0.02, 0.04, 0.06, 0.08$ on (a) the WSS distribution on the upper wall, (b) u profile at $x = 0$, (c) u profile at $x = 2$, (d) the temperature (θ) profile at $x = 0$, (e) the temperature (θ) profile at $x = 2$, and (f) the concentration (φ) profile at $x = 0$. Figure 4a shows that as the St estimations intensify, the WSS on the upper wall escalates, and it is maximum at $t = 0.25$. The WSS for HNF is higher as compared to NF. The velocity field's boundary layer thickness escalates as St grows, and the velocity field reduces, as shown in Figure 4b,c. The u profile coincides for all the values of St, and the HNF velocity profile is slightly lower than that of the NF when $x = 0$. The profiles are not parabolic, as some backflow in the vicinity of the walls is observed at $x = 2$, so in this case, the velocity profile for HNF is higher as compared to NF. Figure 4d–f show the opposite effects of St on the temperature and concentration fields. It is found that as the estimations of St grow, the temperature at the surface and the thermal boundary layer thickness reduces. The HNF θ profile is slightly higher than that of the NF for temperature profiles, but it is opposite in the case of the φ profile.

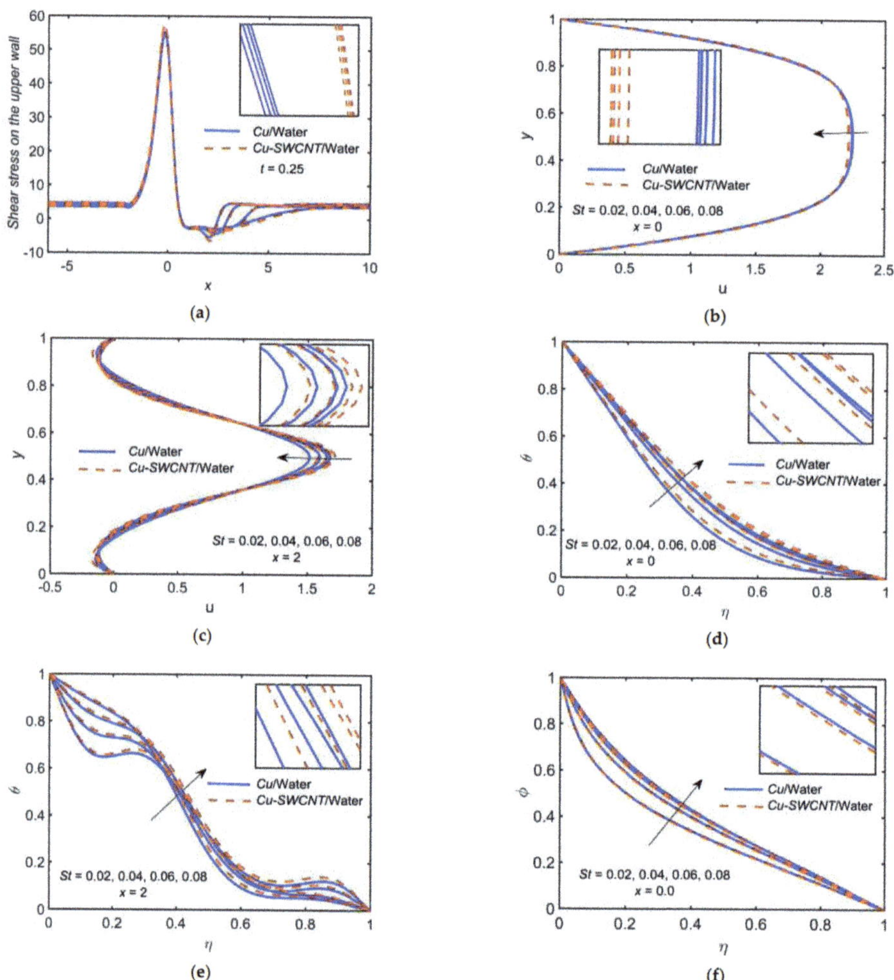

Figure 4. (a) The WSS distribution, (b) u profile at $x = 0$, (c) u profile at $x = 2$, (d) θ profile at $x = 0$, (e) θ profile at $x = 2$, and (f) φ profile at $x = 0$ for distinct values of St.

4.3. The Casson Fluid Parameter Effect

Figure 5 depicts the impact of Cu-based NF and Cu-SWCNT-based HNF for Casson fluid parameter $\beta = 0.5$, 1, 1.5, 2 on (a) the WSS distribution on the upper wall, (b) the u profile at $x = 0$, (c) the u profile at $x = 2$, (d) the temperature (θ) profile at $x = 0$, (e) the temperature (θ) profile at $x = 2$, and (f) the concentration (φ) profile at $x = 0$. Figure 5a shows that as β estimations intensify, the WSS on the upper wall escalates, and it is maximum at $t = 0.25$. The WSS for HNF is higher as compared to NF. The velocity field's boundary layer thickness reduces as β grows, and the velocity field escalates, as shown in Figure 5b,c. The reason for this behavior is that as the value of β grows, the elasticity of HNF rises, causing the HNF to become more viscous. Physically, the boundary layer thickness in such a case reduces as β grows. The HNF velocity profile is slightly lower than that of the NF when $x = 0$, but an opposite behavior was noticed when $x = 2$. Figure 5d–f show that the θ and φ profiles have a declining behavior toward β. The HNF temperature profile is slightly higher than that of the NF for temperature profiles but opposite in the case of concentration profile.

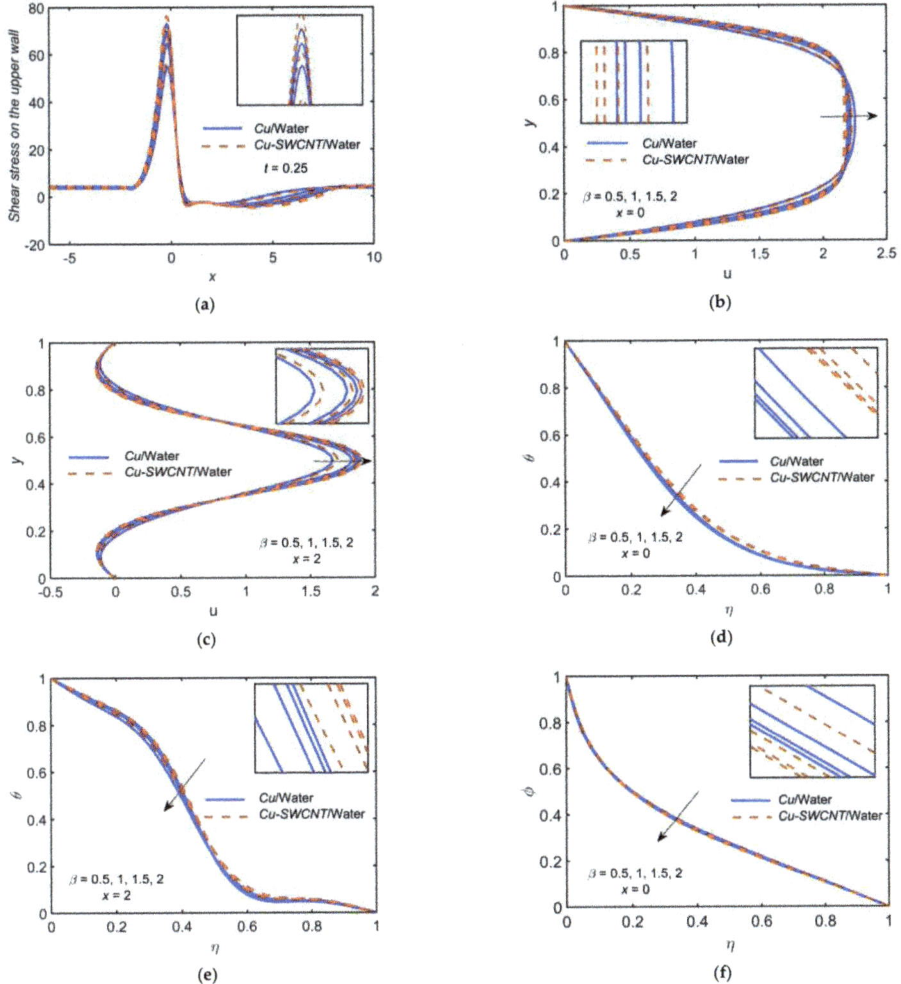

Figure 5. (a) The WSS distribution, (b) u profile at $x = 0$, (c) u profile at $x = 2$, (d) θ profile at $x = 0$, (e) θ profile at $x = 2$, and (f) φ profile at $x = 0$ for distinct values of β.

4.4. The Porosity Parameter Effect

Figure 6 depicts the impact of Cu-based NF and Cu-SWCNT-based HNF for porosity parameter $D_a = 0.002, 0.004, 0.006, 0.008$ on (a) the WSS distribution on the upper wall, (b) u profile at $x = 0$, (c) u profile at $x = 2$, (d) the temperature (θ) profile at $x = 0$, (e) the temperature (θ) profile at $x = 2$, and (f) the concentration (φ) profile at $x = 0$. Figure 6a shows that as D_a estimations intensify, the WSS on the upper wall reduces. The WSS for HNF is higher as compared to NF. Figure 6b,c show that the velocity field's boundary layer thickness escalates as D_a grows and the velocity field reduces. The HNF velocity profile is slightly lower than that of the NF when $x = 0$, but opposite behavior is noticed when $x = 2$. Figure 6d–f show that the θ and φ profiles escalate as D_a grows. The HNF temperature profile is slightly higher than that of the NF for temperature profiles but opposite in the case of concentration profile.

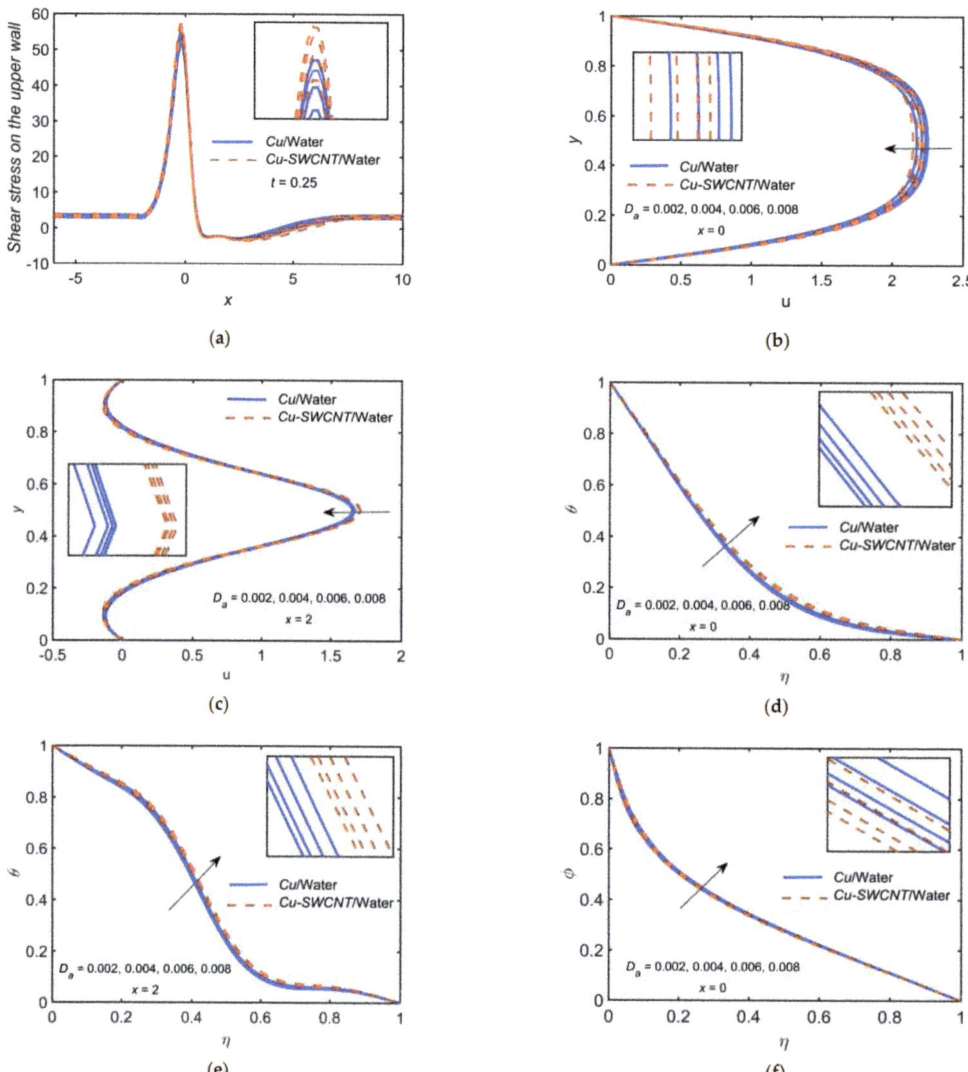

Figure 6. (**a**) The WSS distribution, (**b**) u profile at $x = 0$, (**c**) u profile at $x = 2$, (**d**) θ profile at $x = 0$, (**e**) θ profile at $x = 2$, and (**f**) φ profile at $x = 0$ for distinct values of D_a.

4.5. The Reynolds Number Effect

Figure 7 depicts the impact of Cu-based NF and Cu-SWCNT-based HNF for Reynolds number $Re = 800, 1000, 1200, 1400$ on (a) the WSS distribution on the upper wall, (b) the u profile at $x = 0$, (c) the u profile at $x = 2$, (d) the temperature (θ) profile at $x = 0$, (e) the temperature (θ) profile at $x = 2$, and (f) the concentration (φ) profile at $x = 0$. Figure 7a shows that as Re estimations intensify, the WSS on the upper wall escalates. As expected, the magnitudes of the peak values of the profiles are lower for the lower Re. The WSS for HNF is higher compared to NF. Figure 7b,c show that the velocity field's boundary layer thickness escalates as Re grows and the velocity field reduces. The reason for this condition is the formation of a thinner boundary layer with the increase of Re. The HNF velocity profile is slightly lower than that of the NF when $x = 0$, but the opposite behavior is noticed when $x = 2$. Figure 7d–f shows that the θ profile escalates but the φ profile reduces as Re grows. The HNF temperature profile is slightly higher than that of the NF for temperature profiles but opposite in the case of concentration profile. This may be due to the shear thinning nature of HNF, where effective viscosity becomes almost constant at a high shear rate. It may be noted that at higher Re, any shear thinning fluid behaves similar to Newtonian fluid. In fact, escalating values of Re reduce the internal energy of NPs, which lowers the concentration profile.

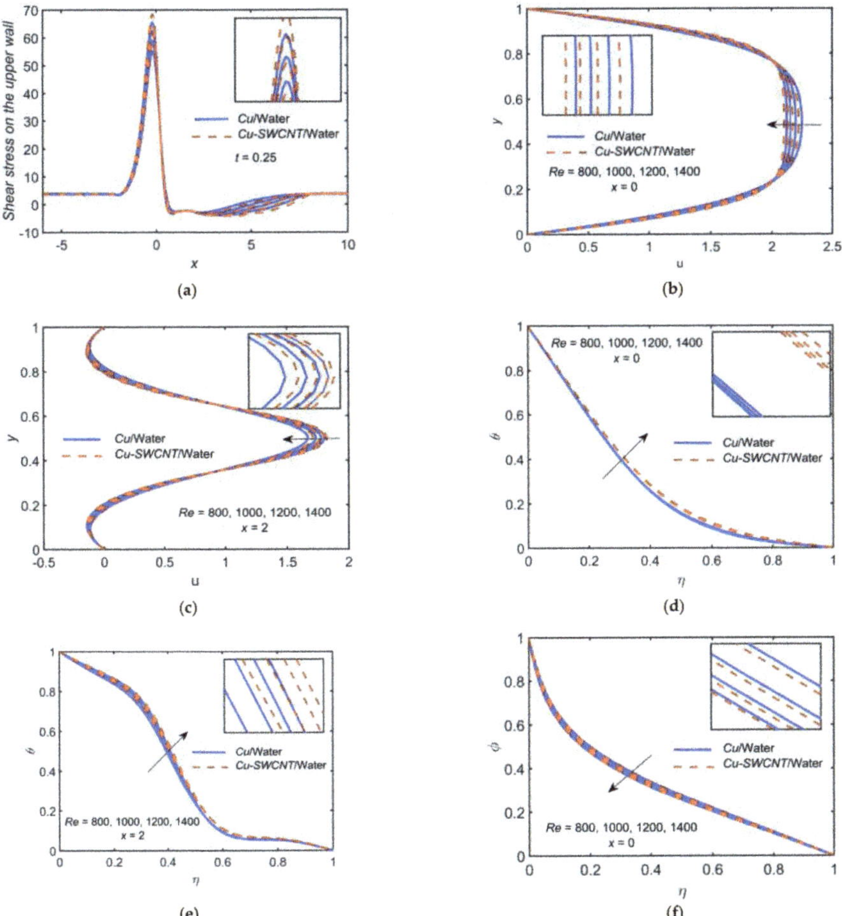

Figure 7. (a) The WSS distribution, (b) u profile at $x = 0$, (c) u profile at $x = 2$, (d) θ profile at $x = 0$, (e) θ profile at $x = 2$, and (f) φ profile at $x = 0$ for distinct values of Re.

4.6. The Varying Time Effect

Figure 8 depicts the impact of Cu-based NF and Cu-SWCNT-based HNF for time $t = 0$, 0.25, 0.50, 0.75 on (a) the u profile at $x = 0$, (b) the u profile at $x = 2$, (c) the θ profile at $x = 0$, and (d) the φ profile at $x = 0$. Figure 8a,b show that for the HNF, the particle concentration in the fluid is even higher than that of the NF, which causes an enhancement of the density and dynamic viscosity. The flow velocity experiences reductions accordingly. At $x = 0$, the maximum value of u grows, and the curves appear to be parabolic for $0 \leq t \leq 0.5$. At $t = 0.75$, u is dropped substantially. At $x = 2$, which is the vicinity of the limiting point of the constriction downstream, backflow occurs near the walls. The HNF velocity profile is slightly lower than that of the NF when $x = 0$, but opposite behavior is noticed when $x = 2$. Figure 8c shows that the temperature profile escalates for HNF compared to NF as the values of t upsurge. A rapid growth in temperature is due to the hybrid nature of NF, since it escalates the thermal conductivity. Figure 8d shows the opposite behavior in the case of concentration profile.

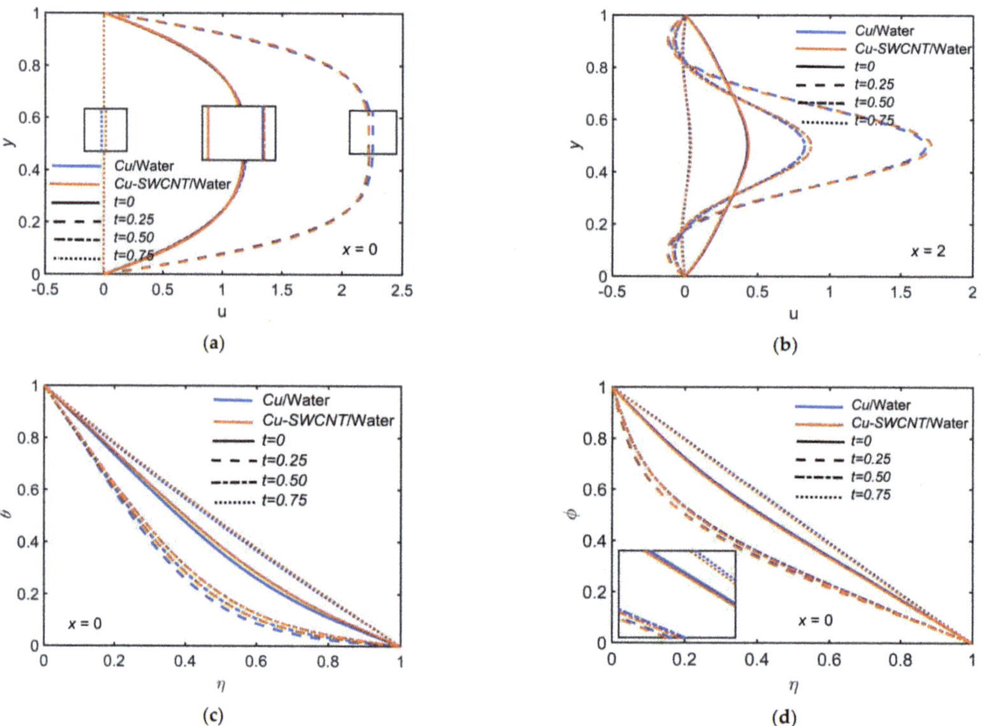

Figure 8. (a) u profile at $x = 0$, (b) u profile at $x = 2$, (c) θ profile at $x = 0$, and (d) φ profile at $x = 0$ for distinct values of t.

4.7. The Radiation Parameter Effect

Figure 9 depicts the impact of Cu-based NF and Cu-SWCNT-based HNF for Radiation parameter $Rd = 0.2$, 0.6, 1, 1.4 on (a) the temperature (θ) profile at $x = 0$, (b) the temperature (θ) profile at $x = 2$, (c) the concentration (φ) profile at $x = 0$, and (d) the concentration (φ) profile at $x = 2$. Figure 9a,b show that as Rd estimations intensify, the HNF thermal distribution is improved. Basically, the values of Rd provide extra heat to NF, which results in the rise of the temperature profile. Theoretically, thermal radiation causes an escalation in the amount of heat on the surface. This means that the fluid absorbs more heat from the radiation. Heat is dissipated away from the surface, raising the temperature profile of HNFs. In general, as Rd estimations intensify, it escalates the electromagnetic radiation and, as a result, the fluid's hotness. Figure 9c,d show that the concentration profiles reduce as Rd escalates. The HNF temperature profile is slightly higher than that of the NF for temperature profiles but opposite in the case of concentration profile.

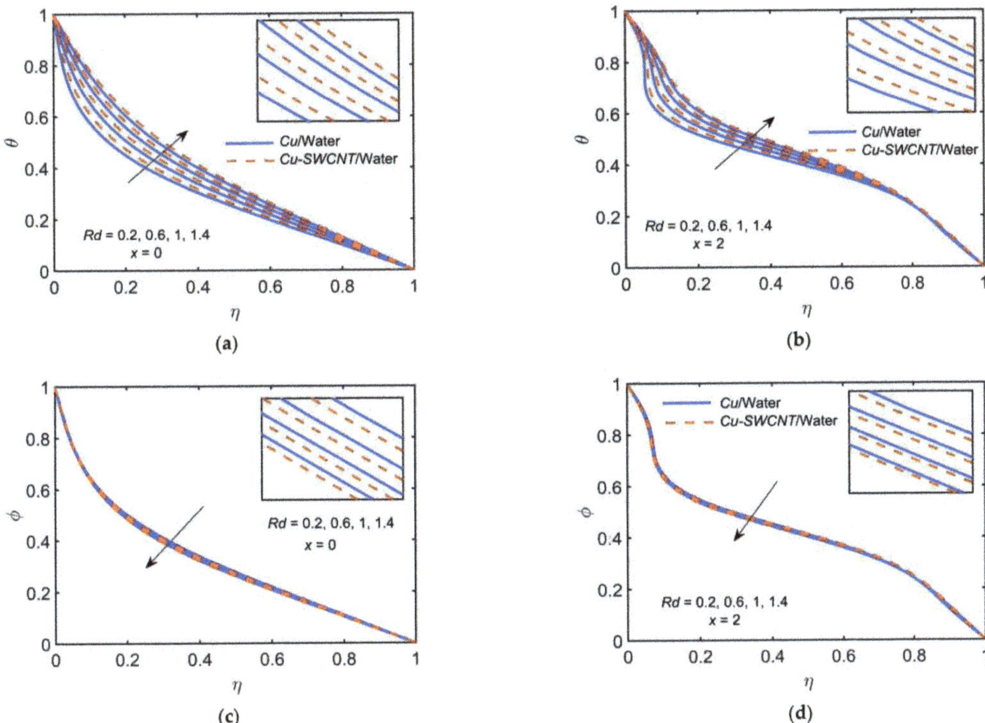

Figure 9. (**a**) θ profile at $x = 0$, (**b**) θ profile at $x = 2$, (**c**) φ profile at $x = 0$, and (**d**) φ profile at $x = 2$ for distinct values of Rd.

4.8. The Solid Volume Fraction Effect

Figure 10 depicts the impact of Cu-based NF and Cu-SWCNT-based HNF for the solid volume fraction $\phi = 0$, 0.03, 0.06, 0.09 on (a) the WSS distribution on the upper wall, (b) the u profile at $x = 0$, (c) the u profile at $x = 2$, (d) the temperature (θ) profile at $x = 0$, and (e) the concentration (φ) profile at $x = 0$. Figure 7a shows that as ϕ estimations intensify, the WSS on the upper wall reduces. There is no significant impact observed of varying ϕ by comparing HNF and NF. Figure 7b,c show that the velocity field reduces. Physically, the momentum boundary layer thickness tends to decrease with upsurging ϕ. Thus, the velocity field's boundary layer thickness escalates as ϕ grows. The HNF velocity profile is slightly lower than that of the NF when $x = 0$, but the opposite behavior is noticed when $x = 2$ where backflow is observed. Figure 7d–f show that the temperature and concentration profiles escalate as ϕ grows. When looking at the graphs, it is clear that the convection heat transfer coefficient escalates as the volume fraction of NPs grows in all HNFs. This is due to the fact that the fraction of NPs affects the thermophysical properties of the base fluid and improves the convective heat transfer performance. In addition, the thermal conductivity and thermal boundary layer thickness reduce with escalating ϕ values. The HNF temperature profile is slightly higher than that of the NF for temperature profiles but opposite in the case of concentration profile.

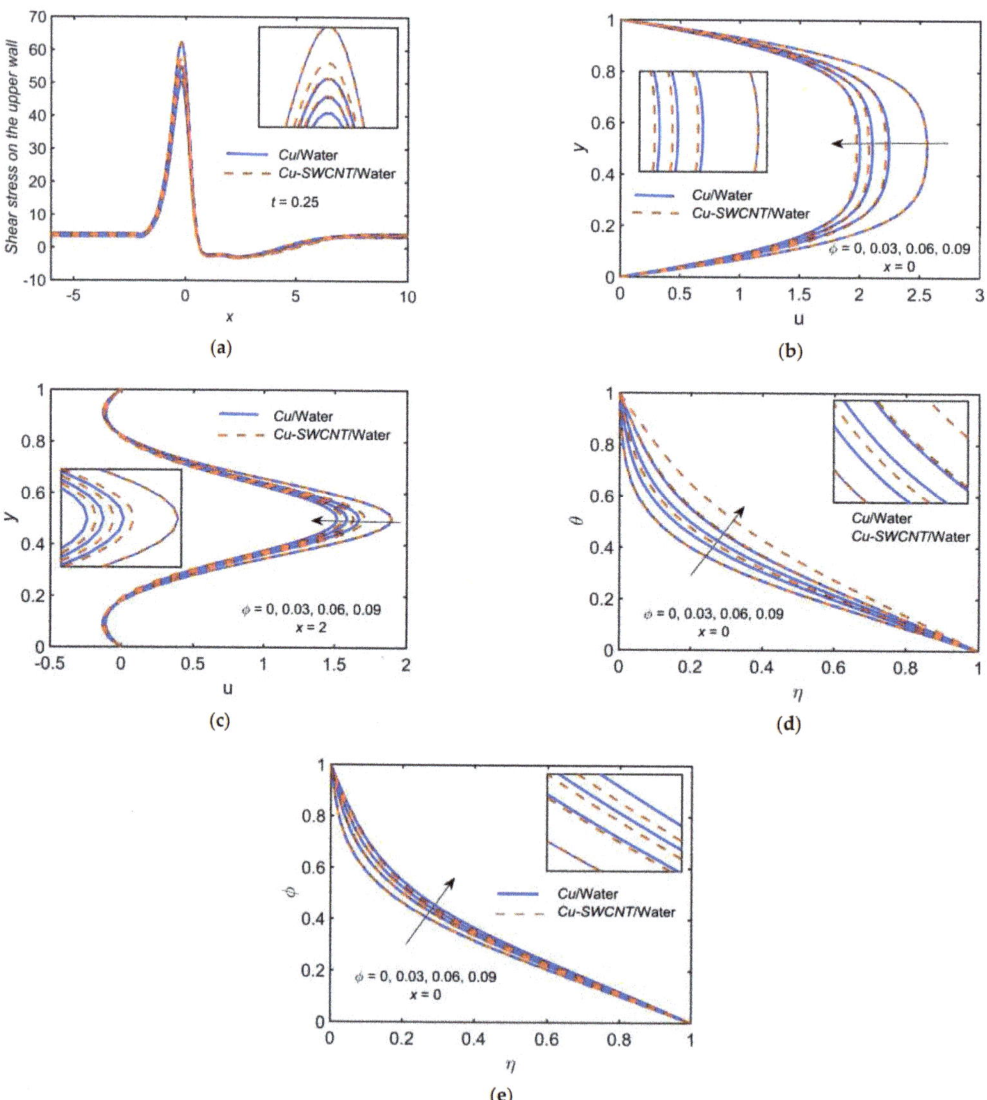

Figure 10. (**a**) The WSS distribution, (**b**) u profile at $x = 0$, (**c**) u profile at $x = 2$, (**d**) θ profile at $x = 0$, and (**e**) φ profile at $x = 0$ for distinct values of ϕ.

4.9. Some Other Physical Parameters Effect

Figure 11 depicts the impact of Cu-based NF and Cu-SWCNT-based HNF for the heat diffusion parameter $Peh = 210, 350, 490, 630$ on (a) the temperature (θ) profile at $x = 0$, and (b) the concentration (φ) profile at $x = 0$. This figure shows that as Peh estimations intensify, the temperature profile reduces, and the concentration profile escalates. The Peclet number shows a high impact on the microorganism's density in a blood NF. The HNF temperature profile is slightly higher than that of the NF for temperature profiles but opposite in the case of concentration profile.

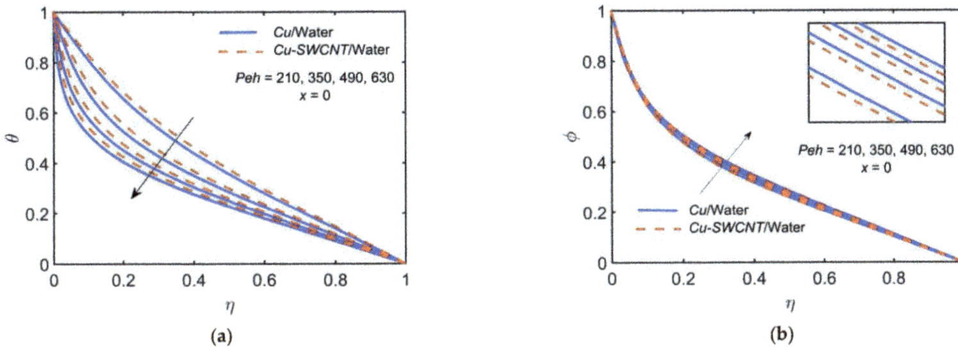

Figure 11. (**a**) The θ profile and (**b**) φ profile at $x = 0$ for distinct values of *Peh*.

Figure 12 depicts the impact of Cu-based NF and Cu-SWCNT-based HNF for the mass diffusion parameter *Pem* = 220, 420, 620 on (a) the concentration (φ) profile at $x = 0$ and (b) the concentration (φ) profile at $x = 2$. The figure shows that as *Pem* estimations intensify, the concentration profile reduces. The HNF concentration profile is slightly lower than that of the NF.

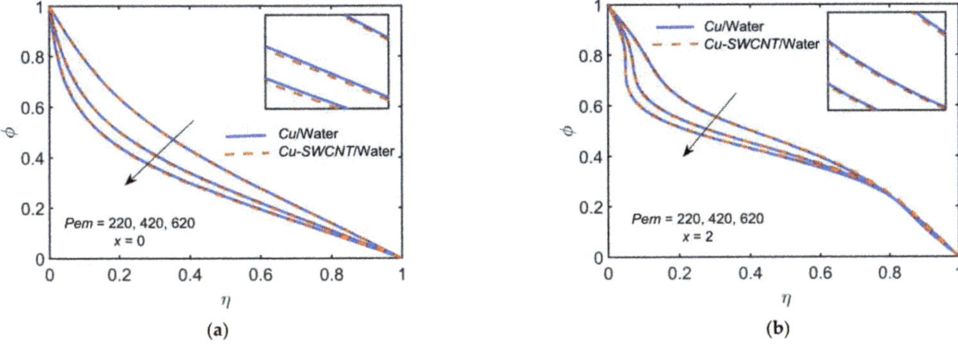

Figure 12. The φ profile at (**a**) $x = 0$ and (**b**) $x = 2$ for distinct values of *Pem*.

Figure 13 depicts the impact of Cu-based NF and Cu-SWCNT-based HNF for the Soret number *Sr* = 1, 1.5, 2, 2.5 on (a) the concentration (φ) profile at $x = 0$, and (b) the concentration (φ) profile at $x = 2$. The figure shows that as *Sr* estimations intensify, the concentration profile escalates. The HNF concentration profile is slightly higher than that of the NF.

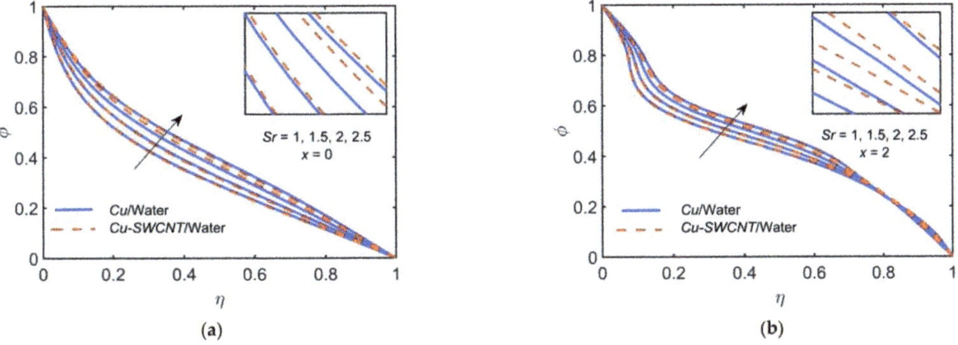

Figure 13. The φ profile at (**a**) $x = 0$ and (**b**) $x = 2$ for distinct values of *Sr*.

4.10. The Influence of Physical Parameters on the Nusselt Number, Sherwood Number, and Skin Friction Coefficient Profiles

Figure 14 depicts the effect of various flow governing parameters on Nusselt number Nu. The Nu escalates as β, Re, and Peh estimations intensify. Nu reduces upon escalating M, D_a, and Rd.

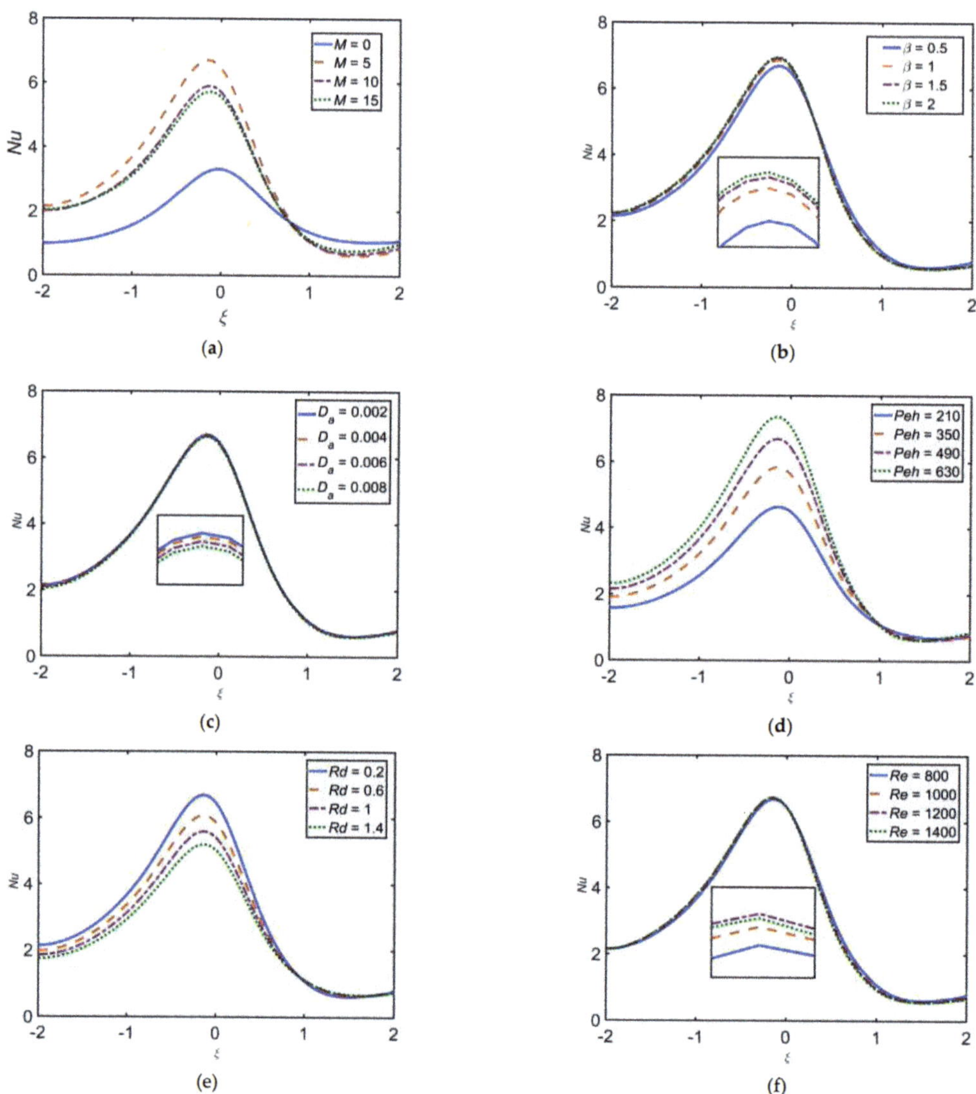

Figure 14. Impact of (**a**) M, (**b**) β, (**c**) D_a, (**d**) Peh, (**e**) Rd, and (**f**) Re on Nu.

Figure 15 depicts the influence of different flow governing parameters on the Sherwood number Sh. It is observed that Sh escalates as β, Rd, Re, and Pem estimations intensify, whereas Sh reduces upon escalating D_a, M, Peh, and Sr.

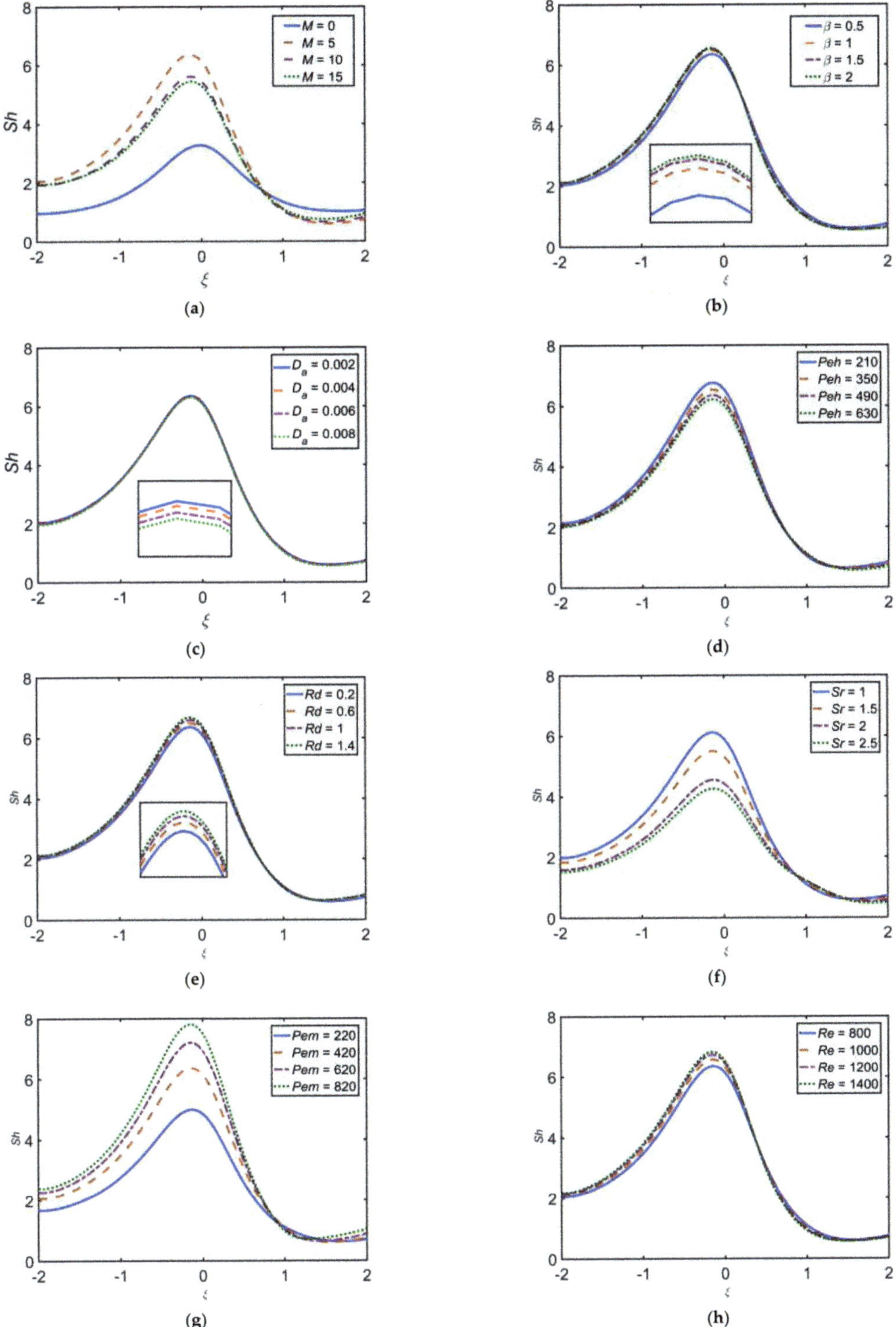

Figure 15. Impact of (**a**) M, (**b**) β, (**c**) D_a, (**d**) Peh, (**e**) Rd, (**f**) Sr, (**g**) Pem, and (**h**) Re on Sh.

Figure 16 depicts the effect of various flow governing parameters on skin friction coefficient Sk. The Sk has a direct relation with D_a and Re, whereas it has an inverse relation with M and β.

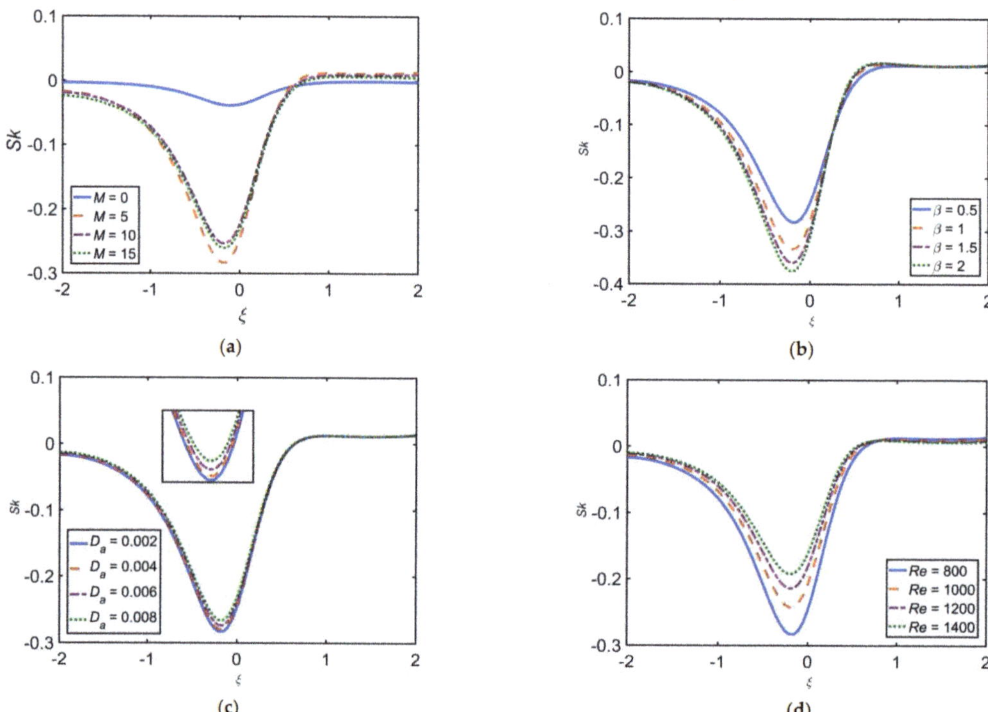

Figure 16. Impact of (a) M, (b) β, (c) D_a, and (d) Re on Sk.

The Nusselt number grows as the flow becomes more turbulent due to the rising number of collisions among the fluid particles. Since more viscous fluids have a lower Reynolds number, less heat transfer occurs, lowering the Nusselt number. The Nusselt number grows as β estimations intensify.

The streamlines, vorticity, temperature distribution, and concentration plots with $M = 5$, $St = 0.02$, $Rd = 0.2$, and $Re = 800$ are shown in Figure 17 for Cu-SWCNT-based HNF. The formation of vertical eddies in the vicinity of the walls can be observed. The eddies increase for the HNF and slowly occupy a major part of the channel downstream of the constriction. The inclusion of multiple types of NPs adds more energy. This results in higher increments in the temperature as well as the thickness of the thermal boundary layer. A rapid rise in the temperature is due to the HNF, since it increases the thermal conductivity. However, the significant impact can be seen in the case of concentration profile.

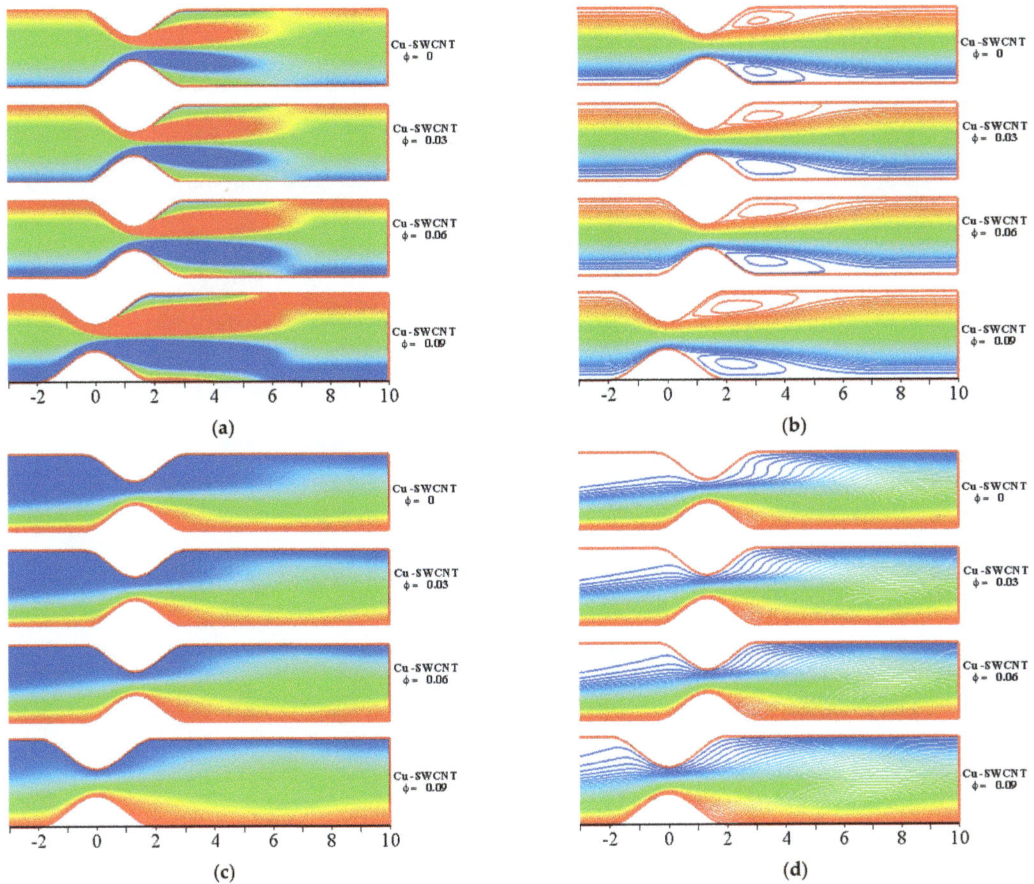

Figure 17. The impact of HNF on (**a**) vorticity, (**b**) streamlines, (**c**) temperature distributions, and (**d**) concentration profile for distinct values of ϕ with $Re = 800$, $M = 5$, $St = 0.02$, and $Rd = 0.2$, at the maximum flow rate ($t = 0.25$).

5. Concluding Remarks

In this research, heat and mass transfer analysis of the pulsatile flow of a Casson hybrid nanofluid through a constricted channel in the presence of viscous incompressible MHD pulsatile fluid flow over a rectangular channel is numerically investigated. The impact of flow governing parameters such as the magnetic parameter, Casson parameter, Reynolds number, Strouhal number, porosity parameter, radiation parameter, Peclet number for the diffusion of heat and mass, and Soret number on the WSS, u, θ, and φ profiles for the sake of comparison between the behavior of Cu-based NF and Cu-SWCNT-based HNF are analyzed in the form of graphs. The physical quantities and heat and mass transfer coefficients are also calculated for the present model, and the impact of flow controlling parameters on these quantities is graphically analyzed. The key findings can be summarized as follows.

(1) The WSS escalates with the rising value of M, St, β, and Re, whereas it reduces with rising values of Da and ϕ. The WSS for HNF is observed to be the highest, which is followed by NF.

(2) The velocity field's boundary layer thickness escalates as M, St, D_a, Re, and ϕ grow, and the velocity field reduces but shows opposite behavior for β. As Hartmann number estimations grow, the restricting force increases, and the velocity field reduces.

The HNF velocity profile is slightly lower than that of the NF when $x = 0$, but opposite behavior is noticed when $x = 2$ in the case of ϕ.

(3) The research shows that a magnetic field applied to the blood reduces the velocity of both blood and magnetic particles.

(4) Temperature profiles escalate as M, St, D_a, Re, Rd, and ϕ grow but reduce as β, and Peh estimations intensify.

(5) A rapid rise in temperature is observed due to the HNF, since it increases the thermal conductivity when comparing the NF and HNF. Hence, it is interpreted that the HNF hits higher temperatures compared to the NF.

(6) The concentration profile escalates as M, St, D_a, Peh, Sr, and ϕ grow but reduces as β, Re, Rd, Pem and Peh estimations intensify.

(7) The heat transfer rate, Sherwood number, and skin friction coefficient escalate as Re estimations intensify and reduce by escalating the porosity parameter. The Nusselt number grows as β estimations intensify.

This research might provide a good picture of the heat and mass transfer activity of blood supply in a circulatory system for treatment such as multiple hyperthermia therapies.

Author Contributions: Conceptualization, A.A. and Z.B.; methodology, A.A., Z.B. and M.U.; software, Z.B., M.U. and M.A.I.; validation, M.U. and Z.A.; formal analysis, A.A. and Z.B.; investigation, A.A., Z.B., M.A.I. and Z.A.; writing—original draft preparation, A.A. and Z.B.; writing—review and editing, A.A., M.U. and M.A.I.; visualization, Z.B. and M.U.; supervision, A.A., M.A.I. and Z.A.; project administration, A.A. and Z.A.; All authors have read and agreed to the published version of the manuscript.

Funding: This research received no external funding.

Institutional Review Board Statement: Not applicable.

Informed Consent Statement: Not applicable.

Data Availability Statement: Not applicable.

Conflicts of Interest: The authors declare no conflict of interest.

Nomenclature

y_1	constricted channel's lower wall
y_2	constricted channel's upper wall
h_1	upper wall constriction height
h_2	lower wall constriction height
π	product of deformation rate's component
π_c	critical value of π
p_y	yield stress of the fluid
μ_β	plastic dynamic viscosity
$\tilde{u}$	velocity in $\tilde{x}$-direction
$\tilde{v}$	velocity in $\tilde{y}$-direction
ρ	density
$\tilde{p}$	pressure
μ	dynamic viscosity
β	Casson fluid parameter
k	Thermal conductivity
U	characteristic flow velocity
D	coefficient of mass diffusivity
J	current density
B	magnetic field
B_0	uniform magnetic field strength

ν	kinematic viscosity
ζ	Transformed x-coordinate
η	Transformed y-coordinate
Pr	Prandtl number
c	parameter defining weak/strong concentrations/turbulence
$\mathbf{E}$	electric field
q	radiative heat flux
σ	electrical conductivity
c_p	specific heat constant
L	length among channel walls
T	flow pulsation period
St	Strouhal number
D_a	Darcy number
Re	Reynolds number
M	Hartmann number
μ	Dynamic viscosity
μ_m	medium's magnetic permeability
Rd	Radiation parameter
φ	concentration profile
θ	temperature profile
Sr	Soret number
Peh	Peclet number for heat transfer
Pem	Peclet number for mass transfer
ψ	stream function
ω	vorticity function
Sk	skin friction coefficient
Nu	Nusselt number
Sh	Sherwood number
ϵ	pulsating amplitude
Re	Reynolds number

References

1. Young, D.F. Fluid mechanics of arterial stenoses. *J. Biomech. Eng.* **1979**, *101*, 157–175. [CrossRef]
2. Chaturani, P.; Upadhya, V.S. Casson fluid model for pulsatile flow of blood under periodic body acceleration. *J. Biorheol.* **1990**, *27*, 619–630. [CrossRef] [PubMed]
3. Shit, G.C.; Roy, M. Pulsatile flow and heat transfer of a magneto-micropolar fluid through a stenosed artery under the influence of body acceleration. *J. Mech. Med. Biol.* **2011**, *11*, 643–661. [CrossRef]
4. Ellahi, R.; Rahman, S.U.; Nadeem, S.; Akbar, N.S. Influence of heat and mass transfer on micropolar fluid of blood flow through a tapered stenosed arteries with permeable walls. *J. Comput. Theor. Nanosci.* **2014**, *11*, 1156–1163. [CrossRef]
5. Haghighi, A.R.; Asl, M.S. Mathematical modelling of micropolar fluid flow through an overlapping arterial stenosis. *Int. J. Biomath.* **2015**, *8*, 1–15. [CrossRef]
6. Ali, A.; Farooq, H.; Abbas, Z.; Bukhari, Z.; Fatima, A. Impact of Lorentz force on the pulsatile flow of non-Newtonian Casson fluid in a constricted channel using Darcy's law: A numerical study. *Sci. Rep.* **2020**, *10*, 10629. [CrossRef]
7. Bukhari, Z.; Ali, A.; Abbas, Z.; Farooq, H. The pulsatile flow of thermally developed non-Newtonian Casson fluid in a channel with wall. *AIP Adv.* **2020**, *11*, 025324. [CrossRef]
8. Ali, A.; Umar, M.; Abbas, Z.; Shahzadi, G.; Bukhari, Z.; Saleem, A. Numerical investigation of MHD pulsatile flow of micropolar fluid in a channel with symmetrically constricted walls. *Mathematics* **2021**, *9*, 1000. [CrossRef]
9. Umar, M.; Ali, A.; Bukhari, Z.; Shahzadi, G.; Saleem, A. Impact of Lorentz force in thermally developed pulsatile micropolar fluid flow in a constricted channel. *Energies* **2021**, *14*, 2173. [CrossRef]
10. Ali, A.; Umar, M.; Bukhari, Z.; Abbas, Z. Pulsating flow of a micropolar-Casson fluid through a constricted channel influenced by a magnetic field and Darcian porous medium: A numerical study. *Results Phys.* **2020**, *19*, 103544. [CrossRef]
11. Chitra1, M.; Suhasini, M. Effect of the Porous Medium through a Constricted Channel under the Influence of Transverse Magnetic Field. *Int. J. Pure Appl. Math. Sci.* **2017**, *10*, 185–192.
12. Santra, A.K.; Sen, S.; Chakraborty, N. Study of heat transfer due to laminar flow of copper–water nanofluid through two isothermally heated parallel plates. *Int. J. Ther. Sci.* **2009**, *48*, 391–400. [CrossRef]
13. Shukla, J.B.; Parihar, R.S.; Rao, B.R. Effects of stenosis on non-Newtonian flow of the blood in an artery. *Bull. Math. Biol.* **1980**, *42*, 283–294. [CrossRef]
14. Huminicl, G.; Huminic, A. Hybrid nanofluids for heat transfer applications—A state of the art review. *Int. J. Heat Mass Trans.* **2018**, *125*, 82–103. [CrossRef]

15. Waini, I.; Ishak, A.; Pop, I. Transpiration effects on hybrid NF flow and heat transfer over a stretching/shrinking sheet with uniform shear flow. *Alex. Eng. J.* **2020**, *59*, 91–99. [CrossRef]
16. Lund, L.A.; Omar, Z.; Khan, I.; Sherif, E.M. Dual Solutions and Stability Analysis of a Hybrid NF over a Stretching/Shrinking Sheet Executing MHD Flow. *Symmetry* **2020**, *12*, 276. [CrossRef]
17. Khan, M.; Malik, M.; Salahuddin, T.; Hussian, A. Heat and mass transfer of Williamson nanofluid flow yield by an inclined Lorentz force over a non-linear stretching sheet. *Results Phys.* **2018**, *8*, 862–868. [CrossRef]
18. Devi, S.S.U.; Devi, S.P.A. Numerical investigation on three dimensional hybrid Cu-Al$_2$O$_3$/water nanofluid flow over a stretching sheet with effecting Lorentz force subject to Newtonian heatings. *Can. J. Phys.* **2016**, *94*, 490–496. [CrossRef]
19. Vasua, B.; Dubey, A.; Bég, O.A.; Gorla, R.S.R. Micropolar pulsatile blood flow conveying nanoparticles in a stenotic tapered artery: Non-Newtonian pharmacodynamic simulation. *Comput. Biol. Med.* **2020**, *126*, 104025. [CrossRef] [PubMed]
20. Priyadharshini, S.; Ponalagusamy, R. Mathematical modelling for pulsatile flow of Casson fluid along with magnetic nanoparticles in a stenosed artery under external magnetic field and body acceleration. *Neural Comput. Appl.* **2019**, *31*, 813–826. [CrossRef]
21. Suresh, S.; Venkitaraj, K.; Selvakumar, P.; Chandrasekar, M. Effect of Al$_2$O$_3$/water hybrid nanouid in heat transfer. *Exp. Ther. Fluid Sci.* **2012**, *8*, 54–60. [CrossRef]
22. Momin, G.G. Experimental investigation of mixed convection with Al$_2$O$_3$/water and hybrid nanofluid in inclined tube for laminar flow. *Int. J. Sci. Res. Sci.* **2013**, *2*, 195–202.
23. Prakash, M.; Devi, S.S.U. Hydromagnetic hybrid Al$_2$O$_3$–Cu/water nanofluid flow over a slendering stretching sheet with prescribed surface temperature. *Asian J. Hum. Soc. Sci.* **2016**, *6*, 1921–1936. [CrossRef]
24. Bahiraei, M.; Mazaheri, N. Application of a novel hybrid nanofluid containing grapheme platinum nanoparticles in a chaotic twisted geometry for utilization in miniature devices thermal and energy effciency considerations. *Int. J. Mech. Sci.* **2018**, *138*, 337–349. [CrossRef]
25. Aman, S.; Zokri, S.M.; Ismail, Z.; Salleh, M.Z.; Khan, I. Effect of MHD and Porosity on Exact Solutions and Flow of a Hybrid Casson-Nanofluid. *IJFMTS* **2018**, *44*, 131–139.
26. Aman, S.; Zokri, S.M.; Ismail, Z.; Salleh, M.Z.; Khan, I. Casson Model of MHD Flow of SA-Based Hybrid Nanofluid Using Caputo Time-Fractional Models. *Defect Diffus. Forum* **2019**, *390*, 83–90. [CrossRef]
27. Kamis, N.I.; Basir, M.F.M.; Shafie, S.; Khairuddin, T.K.A.; Jiann, L.Y. Suction effect on an unsteady Casson hybrid nanofluid film past a stretching sheet with heat transfer analysis. *Mater. Sci. Eng.* **2021**, *1078*, 012019.
28. Nadeem, S.; Haq, R.U.; Akbar, N.S. MHD three-dimensional boundary layer flow of Casson nanofluid past a linearly stretching sheet with convective boundary condition. *IEEE Trans. Nanotech.* **2014**, *13*, 109–115. [CrossRef]
29. Jamshed, W.; Aziz, A. Cattaneo–Christov based study of TiO$_2$TiO$_2$–CuO/EG Casson hybrid nanofluid flow over a stretching surface with entropy generation. *Appl. Nanosci.* **2018**, *8*, 685–698. [CrossRef]
30. Souayeh, B.; Reddy, M.G.; Sreenivasulu, P.; Poornima, T.; Rahimi-Gorji, M.; Alarifi, I.M. Comparative analysis on non-linear radiative heat transfer on MHD Casson nanofluid past a thin needle. *J. Mol. Liq.* **2019**, *284*, 163–174. [CrossRef]
31. Ullah, I.; Nisar, K.S.; Shafie, S.; Khan, I.; Qasim, M.; Khan, A. Unsteady free convection flow of casson nanofluid over a non-linear stretching sheet. *IEEE Access* **2019**, *7*, 93076–93087. [CrossRef]
32. Aziz, A.M.; Afify, A.A. Effect of Hall current on MHD slip flow of Casson nanofluid over a stretching sheet with zero nanoparticle mass flux. *Thermophys. Aeromech.* **2019**, *26*, 429–443. [CrossRef]
33. Dinarvand, S.; Rostami, M.N.; Dinarvand, R.; Pop, I. Improvement of drug delivery micro-circulatory system with a novel pattern of *CuO- Cu*/blood hybrid nanofluid flow towards a porous stretching sheet. *Int. J. Numer. Methods Heat Fluid Flow* **2019**, *29*, 4408–4429. [CrossRef]
34. Reddy, P.B. Biomedical aspects of entropy generation on electromagnetohydrodynamic blood flow of hybrid nanofluid with non-linear thermal radiation and non-uniform heat source/sink. *Eur. Phys. J. Plus* **2020**, *135*, 852. [CrossRef]
35. Elelamy, A.F.; Elgazery, N.; Ellahi, R. Blood flow of MHD non-Newtonian nanofluid with heat transfer and slip effects: Application of bacterial growth in heart valve. *Int. J. Numer. Methods Heat Fluid Flow* **2020**, *30*, 4883–4908. [CrossRef]
36. El Kot, M.A.; Elmaboud, Y.A. Hybrid nanofluid flows through a vertical diseased coronary artery with heat transfer. *J. Mech. Med. Biol.* **2021**, *21*, 2150012. [CrossRef]
37. Aly, E.H.; Pop, I. MHD flow and heat transfer over a permeable stretching/shrinking sheet in a hybrid nanofluid with a convective boundary condition. *Int. J. Numer. Methods Heat Fluid Flow* **2019**, *29*, 3012–3038. [CrossRef]
38. Alizadeh, M.; Dogonchi, A.S.; Ganji, D.D. Micropolar nanofluid flow and heat transfer between penetrable walls in the presence of thermal radiation and magnetic field. *Case Stud. Therm. Eng.* **2018**, *12*, 319–332. [CrossRef]
39. Sheikholeslami, M.; Ganji, D.D. Heat transfer of *Cu*-water nanofluid flow between parallel plates. *Powder Technol.* **2013**, *235*, 873–879. [CrossRef]
40. Wu, J.Z.; Ma, H.Y.; Zhou, M.D. *Vorticity and Vortex Dynamics*; Springer Science & Business Media: Cham, Switzerland, 2007.
41. Ali, A.; Fatima, A.; Bukhari, Z.; Farooq, H.; Abbas, Z. Non-Newtonian Casson pulsatile fluid flow influenced by Lorentz force in a porous channel with multiple constrictions: A numerical study. *Korea-Aust. Rheol. J.* **2020**, *33*, 79–90. [CrossRef]
42. Bandyopadhay, S.; Layek, G.C. Study of magnetohydrodynamic pulsatile flow in a constricted channel. *Commun. Nonlinear Sci. Numer. Simulat.* **2012**, *17*, 2434–2446. [CrossRef]
43. Ali, A.; Syed, K.S. An outlook of high performance computing infrastructures for scientific computing. *Adv. Comput.* **2013**, *91*, 87–118.

International Journal of
Molecular Sciences

Article

Mg,Si—Co-Substituted Hydroxyapatite/Alginate Composite Beads Loaded with Raloxifene for Potential Use in Bone Tissue Regeneration

Katarzyna Szurkowska [1], Paulina Kazimierczak [2] and Joanna Kolmas [1,*]

1   Department of Analytical Chemistry, Chair of Analytical Chemistry and Biomaterials, Faculty of Pharmacy, Medical University of Warsaw, 02-097 Warsaw, Poland; katarzyna.szurkowska@wum.edu.pl
2   Chair and Department of Biochemistry and Biotechnology, Faculty of Pharmacy, Medical University of Lublin, 20-093 Lublin, Poland; paulina.kazimierczak@umlub.pl
*   Correspondence: joanna.kolmas@wum.edu.pl; Tel.: +48-22-5720706

Abstract: Osteoporosis is a worldwide chronic disease characterized by increasing bone fragility and fracture likelihood. In the treatment of bone defects, materials based on calcium phosphates (CaPs) are used due to their high resemblance to bone mineral, their non-toxicity, and their affinity to ionic modifications and increasing osteogenic properties. Moreover, CaPs, especially hydroxyapatite (HA), can be successfully used as a vehicle for local drug delivery. Therefore, the aim of this work was to fabricate hydroxyapatite-based composite beads for potential use as local carriers for raloxifene. HA powder, modified with magnesium and silicon ions (Mg,Si-HA) (both of which play beneficial roles in bone formation), was used to prepare composite beads. As an organic matrix, sodium alginate with chondroitin sulphate and/or keratin was applied. Cross-linking of beads containing raloxifene hydrochloride (RAL) was carried out with Mg ions in order to additionally increase the concentration of this element on the material surface. The morphology and porosity of three different types of beads obtained in this work were characterized by scanning electron microscopy (SEM) and mercury intrusion porosimetry, respectively. The Mg and Si released from the Mg,Si-HA powder and from the beads were measured by inductively coupled plasma optical emission spectrometry (ICP-OES). In vitro RAL release profiles were investigated for 12 weeks and studied using UV/Vis spectroscopy. The beads were also subjected to in vitro biological tests on osteoblast and osteosarcoma cell lines. All the obtained beads revealed a spherical shape with a rough, porous surface. The beads based on chondroitin sulphate and keratin (CS/KER-RAL) with the lowest porosity resulted in the highest resistance to crushing. Results revealed that these beads possessed the most sustained drug release and no burst release effect. Based on the results, it was possible to select the optimal bead composition, consisting of a mixture of chondroitin sulphate and keratin.

Keywords: nanocrystalline hydroxyapatite; composite biomaterials; raloxifene; drug delivery system; magnesium ions; silicate ions

Citation: Szurkowska, K.; Kazimierczak, P.; Kolmas, J. Mg,Si—Co-Substituted Hydroxyapatite/Alginate Composite Beads Loaded with Raloxifene for Potential Use in Bone Tissue Regeneration. *Int. J. Mol. Sci.* **2021**, 22, 2933. https://doi.org/10.3390/ijms22062933

Academic Editor: Raghvendra Singh Yadav

Received: 17 February 2021
Accepted: 11 March 2021
Published: 13 March 2021

Publisher's Note: MDPI stays neutral with regard to jurisdictional claims in published maps and institutional affiliations.

1. Introduction

In many bone diseases involving the formation of cavities and difficulties with regeneration (such as osteoporosis or bone metastasis), bone substitutes are needed to provide structural support for cells and the newly formed osseous tissue, and to induce natural processes of tissue regeneration and development [1–3]. Simultaneously, appropriate pharmacotherapy is required to induce bone formation, inhibit osteoclast activity, relieve pain, or provide antibacterial prophylaxis [4,5].

Recently, preparation, characterization, and application of innovative multifunctional biomaterials, with the potential to deliver drugs into the bone tissue, have attracted much attention [6–8]. This is thanks to their unique properties such as the possibility for controlled release of drugs, thereby reducing the therapeutic dose and minimizing toxicity [9].

Synthetic hydroxyapatite (HA) with general formula $Ca_{10}(PO_4)_6(OH)_2$ is similar to bone apatite and has high biocompatibility [10–12]. These characteristics mean that HA has received a great deal of attention as an inorganic biomaterial in bone tissue replacement. Furthermore, thanks to a high affinity for ionic substitution, HA can be modified in order to obtain additional biological, physicochemical or mechanical properties [13,14]. Finally, due to its high loading capacity, HA could be used for local drug delivery in the treatment and prophylaxis of bone tissue disorders, primarily osteoporosis, bone tumors, and infections [15–17].

In the present work, HA modified with magnesium and orthosilicate ions (Mg,Si-HA) was used in order to prepare three-dimensional composite beads for potential use as a drug delivery system into the bone. Mg^{2+} and SiO_4^{4-} play important roles in the bone mineralization process influencing osteoblast and osteoclast activities, and in bone formation, stimulating collagen type I synthesis [18–23].

Raloxifene hydrochloride (RAL) was selected as model drug for the release studies. RAL is a second-generation selective estrogen receptor modulator (SERM) that is used to prevent and treat osteoporosis in postmenopausal women. It exhibits estrogenic effects on the skeletal system and antiestrogenic effects on the breast and endometrium [24–27]. It thereby prevents the loss of bone mass associated with osteoporosis and decreases the risk of osteoporotic fractures in elderly women. It is interesting that the drug is also used for an adjunctive treatment for schizophrenia or prevention of breast cancer, whilst the latest research reports its potential activity in reducing Covid-19 related mortality [28–30].

It is important to note that the current oral therapy with RAL is insufficient because it has poor oral bioavailability (2%) due to hepatic first-pass metabolism and poor water solubility [31]. This fact makes it attractive for controlled drug delivery, which could overcome the limitations of its application in clinical practice.

RAL is classified as a Class II drug according to the Biopharmaceutics Classification System. This means that it has low solubility and high permeability. Therefore, intensive research on new drug formulation or other administration routes to improve its pharmacokinetic properties and bioavailability is required [32–36].

So far, there have been several studies on the local delivery of RAL by calcium phosphate-based materials and there is scope for further development [37,38]. The presented research is a continuation of previous work on beads based on substituted HA, alginate, and chondroitin sulphate, and it involves the introduction of a drug substance whilst investigating the release profile of the drug and substituted ions [39].

To expand the scope of the research, three bead compositions were compared: (i) chondroitin sulphate (CS), (ii) keratin (KER), and (iii) a mixture of both of them in a 1:1 weight ratio (CS/KER). In vitro drug release profiles from beads containing raloxifene, i.e., CS-RAL, KER-RAL, and CS/KER-RAL, were compared. The resulting beads were subjected to in vitro biological tests on osteoblast hFOB and osteosarcoma Saos-2 cell lines. Biological results were compared to control beads without the addition of the drug, CS-c, KER-c, and CS/KER-c, respectively. Porosity and mechanical strength tests were also carried out.

2. Results and Discussion

2.1. Bead Morphology

Morphology of representative granules before and after drying is illustrated in Figure 1. As can be clearly seen, the beads show a very smooth surface before drying, whilst dry beads are dramatically rougher. This may be an important advantage to potentially promote the adhesion of bone cells involved in the remodeling of damaged tissue. The obtained granules are of various diameters, from 2 to 5 mm. For further studies, granules with a 3 mm diameter were selected.

Figure 1. Optical images of representative beads before drying (**A**) and after drying (**B**).

For morphological analysis, the SEM images taken from three types of beads containing RAL (CS-RAL, KER-RAL and CS/KER-RAL) are shown in Figure 2. All the beads revealed a spherical shape with a rough surface. The CS-RAL beads exhibit the most heterogeneous outer surface, whereas the cross-sectional analysis indicated their relatively dense and smooth interior with numerous fine pores near the outer surface of the spheres. The keratin-containing granules (KER-RAL and CS/KER-RAL) have small indentations on the outer surface, while their graining appears to be more homogenous.

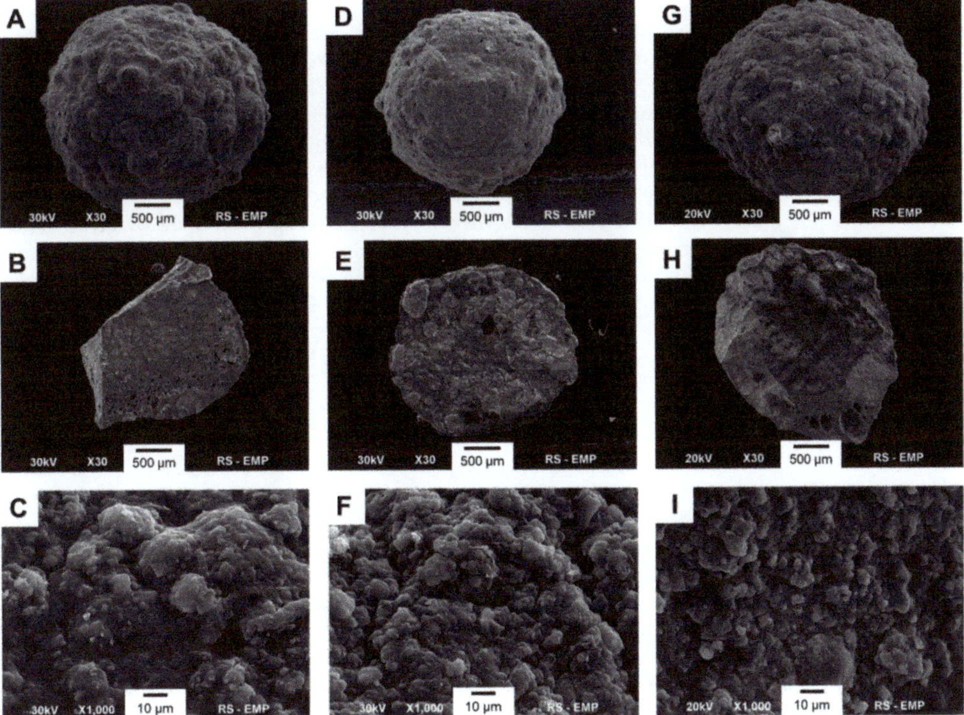

Figure 2. SEM images of CS-RAL (**A–C**), KER-RAL (**D–F**), and CS/KER-RAL (**G–I**). First line: whole bead at ×30 magnification; second line: internal cross-section at ×30 magnification; third line: outer surface at ×1000 magnification.

The same measurements were performed for control beads without the drug, but no difference in morphology was observed.

2.2. Mechanical Strength and Porosity

The results obtained by the Hg intrusion technique for the samples containing RAL are summarized in Table 1. The apparent and true density values were similar for all the samples (CS-RAL, KER-RAL and CS/KER-RAL) and were in the ranges 1.4–1.5 g/cm^3 and 1.7–1.9 g/cm^3, respectively. This indicates that all the samples achieved a high degree of densification during cross-linking.

Table 1. Porosity results of CS-RAL, KER-RAL, and CS/KER-RAL beads measured by the Hg intrusion technique.

Sample	S_{Hg} [m^2/g] ± 0.1	ρ_b [g/cm^3] ± 0.05	ρ_p [g/cm^3] ± 0.05	P [%] ± 0.1	V_t [cm^3/g] ± 0.001	V_{me} [cm^3/g] ± 0.001	%V_{me} [%] ± 0.1	CR [N/bead] ± 0.1
CS-RAL	47	1.5	1.9	19	0.12	0.08	63	139
KER-RAL	71	1.4	1.7	20	0.15	0.11	74	137
CS/KER-RAL	45	1.5	1.8	13	0.09	0.06	67	177

S_{Hg}, surface area; ρ_b, apparent density; ρ_p, true density; P, porosity; V_t, total pore volume; V_{me}, mesopores volume; %V_{me}, percentage of mesopores; CR, crush resistance.

The obtained beads have a moderate degree of pore surface development (45–71 m^2/g), while the total pore volume is in the range 0.09–0.15 cm^3/g, which confirms that relatively compact and dense materials were obtained. The percentage of mesopores between 63 and 74% allows the granules to be classified as mesoporous materials. It should be noted that the KER-RAL sample has the most developed pore surface (71 m^2/g) and at the same time the highest volume (0.09 cm^3/g).

The pore size distribution (data not presented) showed that all the obtained beads mainly contain pores up to 10 nm in size, including micropores. Pores with a larger diameter (over 0.1 mm) were also observed; however, their volume for each sample was only 0.02 cm^3/g. They may be referred to as inter-grain spaces (cavities), which are probably formed during the cross-linking of the alginate and drying process.

It should be noted that the degree of porosity and pore size distribution are crucial parameters characterizing drug delivery systems applied to bone replacement and regeneration. The osteogenic properties of the material result from the presence of pores, which facilitate migration of cells, integration with the host tissue, and ensure vascularization. The interconnected pores with a diameter of approx. 100–300 μm ensure cell adhesion and migration, while small pores are required for effective drug delivery systems [40,41].

The average values of the destructive forces for individual samples were also compared. As expected, the beads based on CS/KER-RAL with the lowest porosity (13%) resulted in the highest resistance to crushing (177 N/bead).

2.3. Ion Release

In the first step, the release of magnesium and silicon ions from the starting powder (Mg,Si-HA) for bead fabrication was carried out. Figure 3 shows the cumulative release curves for the ions over four weeks. Interestingly, there are significant differences between the release profile of magnesium and silicon. Magnesium is released gradually and in very small quantities, up to 2.4% of its total content after four weeks. Furthermore, the burst release effect can also be observed for up to 24 h, although, during this time, only slightly more than 1% of the total amount of Mg introduced into the powder is released (see Figure 3). A different release profile applies to silicon, where up to 57% of the introduced element was released within four weeks, and approximately 30% within 24 h.

In order to explain the ion release results, our earlier studies on the physicochemical properties of the Mg,Si-HA powder should be recalled [39]. The Mg,Si-HA sample was shown to be nanocrystalline with the elongated-shaped crystals with 24 ± 2 nm and 7 ± 1 nm along the *c* and *a* axes, respectively. Moreover, according to ^{31}P ssNMR studies, the Mg,Si-HA material, in addition to a well-ordered crystalline core, is characterized by the presence of a non-apatitic layer, called the hydrated surface layer, which is typical for nanocrystalline hydroxyapatite. The ^{29}Si ssNMR studies showed that a significant amount of silicon is located just in this surface layer forming "silica gel".

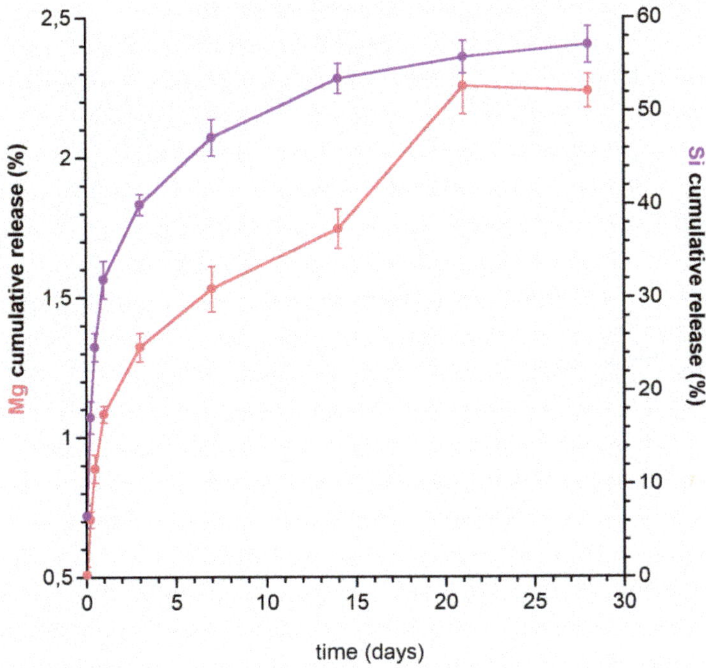

Figure 3. Magnesium and silicon release profiles from Mg,Si-HA powder.

A fairly fast release of silicon may be explained by its location on the surface. Initially, the ions, weakly bound with the crystals, pass to the medium, then the ions from the crystal interior are released.

In turn, slow release of magnesium suggests that these ions have been incorporated into the crystalline core and are released along with the slow dissolution of the substituted hydroxyapatite.

Taking into account the small amount of magnesium introduced into the Mg,Si-HA sample (0.26 wt%) (resulting from the substitution limit), we decided to increase its content in the obtained beads by using Mg^{2+} as a cross-linker for sodium alginate. It is worth adding that, conventionally, aqueous calcium solutions are used to cross-link alginates. Figure 4a shows the magnesium release from the obtained beads. As expected, the amount of Mg released from the beads increased significantly compared to that released from the powder. The CS-RAL sample exhibits the lowest amount of Mg released to the medium. It should be noted that this type of bead is also characterized by the lowest porosity. Therefore, Mg^{2+} ions from the cross-linking solution had difficulty accessing the interior of the beads and penetrated them poorly, and, in addition, were released to a lesser extent. During the formation of more porous granules (KER-RAL and CS/KER-RAL), the Mg ions from the cross-linking solution not only binded to alginate but also adsorbed on the bead porous structure. This is why the amount of release magnesium was so high and exceeded 100—100% corresponds to the average amount of cross-linked magnesium ions, however, each type of granule has a different binding capacity.

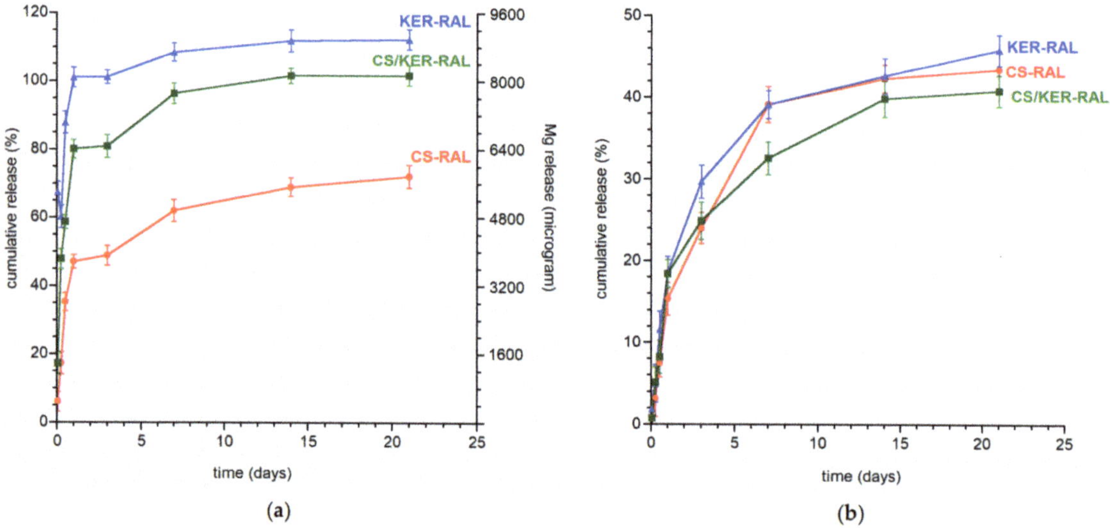

Figure 4. (**a**) The magnesium release profiles from CS-RAL, KER-RAL, and CS/KER-RAL; (**b**) the silicon release profiles from CS-RAL, KER-RAL, and CS/KER-RAL.

In conclusion, regardless of the type of beads, it was possible to obtain the initial rapid discharge of Mg^{2+} ions into the medium, followed by stabilization of the release. This is particularly beneficial since Mg ions are responsible for the biocompatibility of the implant materials and bone substitutes.

It is worth paying attention to the release of silicon from the obtained composite granules (see Figure 4b). In the first 24 h, the released silicon was only 15.3–18.6% of the total amount, unlike the release from the Mg,Si-HA powder material, where the value was 31.9%. Moreover, the further release is also slightly slower because the amount of Si released from each type of bead did not exceed 45.9% (vs. 55.9% from the Mg,Si-HA powder). Therefore, thanks to composite production, it was possible to stabilize the release of Si ions and obtain a gradual, extended-release profile. The materials may exhibit a long-lasting osteogenic effect after in vivo implantation which should be analyzed in further research.

2.4. Drug Release

After the promising preliminary results of ion release from the powders and granules, an extended three-month release study of raloxifene hydrochloride was performed. The graphs showing the drug substance release profile are presented in Figure 5a,b.

RAL is gradually released from all the obtained granules without a burst release effect, as may be evidenced by the profiles recorded during the first 24 h (see Figure 5b). The release profile of each type of granule is similar, the curves differ mainly in the amount of RAL released to the medium.

The most favorable release profile was observed with CS/KER-RAL granules containing a mixture of chondroitin sulphate and keratin. The drug was evenly distributed throughout the volume of the granules and was released gradually as the material swelled and slowly dissolved. We managed to avoid disintegration of the granules and the sudden initial burst of the drug, which often occurs in granules containing only sodium alginate [42–44]. After 12 weeks, the pellets did not disintegrate, and the amount of released RAL reached up to 60% of the total quantity and still did not plateau. Interestingly, beads containing only keratin exhibit the slowest drug release, while the addition of KER to CS accelerates the release of RAL to the medium.

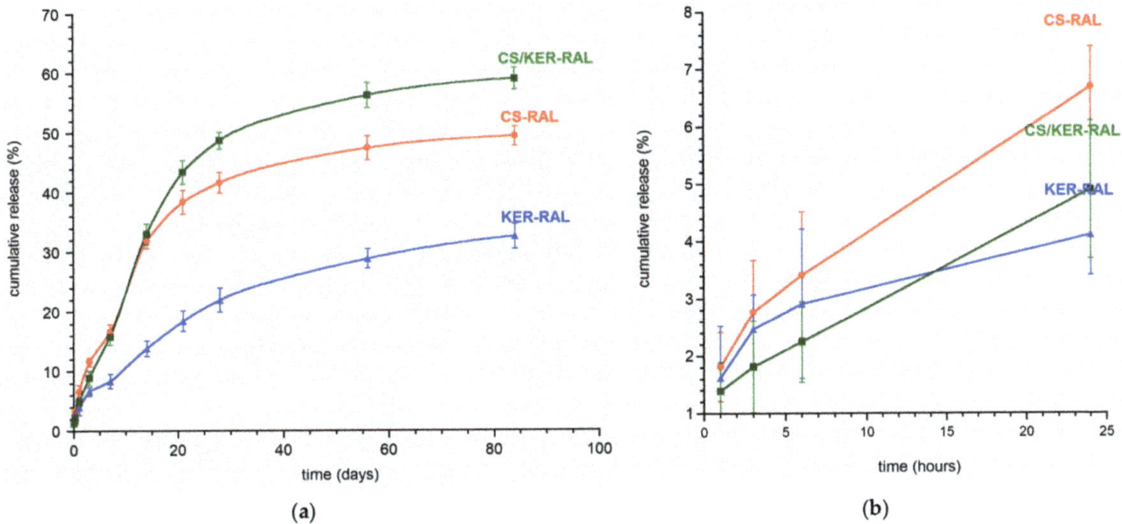

Figure 5. (**a**) Raloxifene hydrochloride release profile from CS-RAL, KER-RAL, and CS/KER-RAL over 12 weeks; (**b**) Raloxifene hydrochloride release profile from CS-RAL, KER-RAL, and CS/KER-RAL during the first 24 h.

2.5. In Vitro Cytotoxicity Assessment

In this study, an MTT assay was performed to assess the cytotoxic effect of 24-h granule extracts on the viability of hFOB 1.19 and Saos-2 cells after a 48-h exposure (see Figure 6). In vitro cytotoxicity evaluation showed that an extract of CS-c and CS/KER-c granules slightly decreased hFOB 1.19 cells viability to 81.83% and 71.27%, respectively. Nevertheless, it should be noted that, according to the ISO 10993-5 procedure, 100% extract of the biomaterial should be considered as non-toxic when the percentage of cell viability is higher than 70%. Importantly, the addition of raloxifene to CS-RAL and CS/KER-RAL granules also exhibited a non-toxic effect on human osteoblasts. The MTT also showed that both an extract of KER-c granules and an extract of KER-RAL granules significantly decreased hFOB 1.19 viability to approximately 60%, suggesting that a reduction in cell viability was not caused by released RAL. It is worth noting that bioceramic-based biomaterials may alter the ionic composition of the culture media via their ion reactivity causing reduction in cell viability [45–47]. In turn, CS-c, KER-c, and CS/KER-c granules were non-toxic against Saos-2 cells (cell viability above 83%), whereas extracts of CS-RAL, KER-RAL, and CS/KER-RAL granules caused a significant reduction in Saos-2 viability to approximately 24%, 70%, and 50%, respectively. Thus, in vitro cytotoxicity assessment clearly showed that RAL released from granules was non-toxic against human osteoblasts, but caused a significant decrease in the viability of tumor cells. Considering the results from the cytotoxicity test, both CS-RAL and CS/KER-RAL granules are promising candidates for biomedical applications.

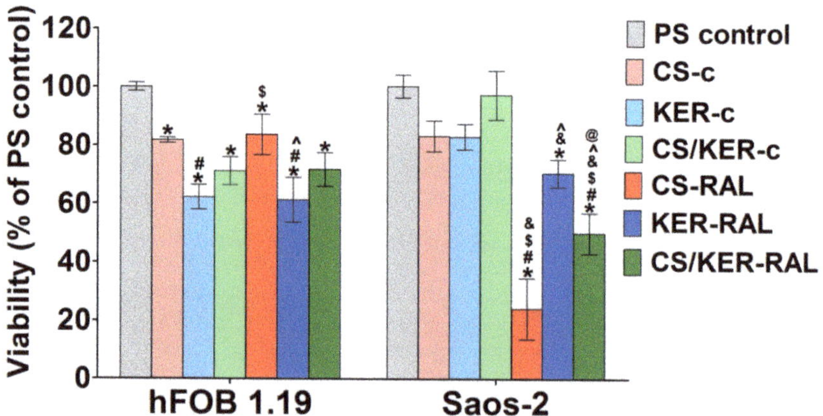

Figure 6. Evaluation of cell viability after 48-h exposure to granule extracts by MTT assay (hFOB 1.19—human normal osteoblasts; Saos-2—human tumor cells derived from osteosarcoma; PS control—polystyrene extract as a negative control of cytotoxicity; * statistically significant results compared to PS control, # statistically significant results compared to CS; $ statistically significant results compared to KER; & statistically significant results compared to CS/KER; ^ statistically significant results compared to CS-RAL; @ statistically significant results compared to KER-RAL; $p < 0.05$, one-way ANOVA followed by Tukey's test).

3. Materials and Methods

3.1. Preparation of Mg,Si-HA

Nanocrystalline, magnesium, and silicon co-substituted hydroxyapatite (Mg,Si-HA) with the nominal composition of $Ca_{9.87}Mg_{0.13}(PO_4)_{5.5}(SiO_4)_{0.5}(OH)_{1.5}$ was synthesized using the standard precipitation method as previously described [39]. Briefly, $Si(CH_3COO)_4$ and $(NH_4)_2HPO_4$ aqueous solutions were added dropwise into $Ca(NO_3)_2$ and $Mg(NO_3)_2$ aqueous solution under gentle stirring at pH 10, and the resultant precipitate was left for 24 h for aging. All reagents were purchased from Sigma Aldrich (St. Louis, MO, USA). The resultant powder was subjected to careful physicochemical analysis, which confirmed the preparation of nanocrystalline hydroxyapatite containing 0.26 wt% of Mg and 0.59 wt% of Si.

3.2. Preparation of the Composite Beads

Three different composite beads were prepared using previously synthesized Mg,Si-HA powder, sodium alginate (SA) (Sigma Aldrich, St. Louis, MO, USA) and two additives: chondroitin sulphate sodium salt (CS) (TCI, Tokyo Chemical Industry Co., Tokyo, Japan) and keratin from wool (KER) (TCI, Tokyo Chemical Industry Co., Tokyo, Japan). The drug used was raloxifene hydrochloride (RAL) (TCI, Tokyo Chemical Industry Co., Tokyo, Japan). The detailed composition of the granules is shown in Table 2.

Table 2. Bead composition.

Sample	Mg,Si-HA (g)	SA (g)	CS (g)	KER (g)	RAL (mg)
CS-RAL	2.8	2.0	1.2	-	100.0
KER-RAL	2.8	2.0	-	1.2	100.0
CS/KER-RAL	2.8	2.0	0.6	0.6	100.0
CS-c	2.8	2.0	1.2	-	-
KER-c	2.8	2.0	-	1.2	-
CS/KER-c	2.8	2.0	0.6	0.6	-

CS, chondroitin sulphate; KER, keratin; RAL, raloxifene hydrochloride; c, control.

Our aim was to produce materials that resemble natural biocomposite bone tissue [48–51]. This is why the mineral phase (Mg,Si-HA powder) was added to an organic matrix, composed of naturally occurring biopolymers. The preparation of the above beads involved the following steps: Mg,Si-HA and one of the selected additives (i) CS, (ii) KER, or (iii) CS/KER (a mixture of both in 1:1 weight ratio) were added to the aqueous sodium alginate solution. Raloxifene hydrochloride (RAL) powder was then added to the resulting suspension and mixed thoroughly. The suspension was added dropwise through a syringe into 6.4 wt% solution of $Mg(NO_3)_2 \cdot 6H_2O$ (Sigma Aldrich, St. Louis, MO, USA) dissolved in a mixture of distilled water and 96% ethanol (EMPROVE, Merck, Germany) in a 60:40 volume ratio. Beads were left for 15 min in the solution for cross-linking and to incorporate additional Mg^{2+} ions on the surface of the beads. The beads were named as follows: CS-RAL, KER-RAL, and CS/KER-RAL.

The obtained granules were then washed several times with distilled water and air dried at room temperature. After the cross-linking, the residual solution was extracted to examine the drug concentration in order to assess incorporation efficiency and raloxifene loss at the granule production stage. The raloxifene content in the cross-linking solution did not exceed 1 mg, so the raloxifene loss was below 1%. The dried beads were further subjected to physicochemical, mechanical, and biological analyses. In the same manner, control granules without raloxifene were obtained and named as follows: CS-c, KER-c, and CS/KER-c.

3.3. Scanning Electron Microscopy (SEM)

In order to analyze the morphology of the obtained samples, a SEM microscope JSM 6390 LV JEOL (JEOL, Tokyo, Japan) at 20 or 30 kV accelerating voltage was used. For SEM analysis, the surface of the beads was sputtered with gold in a vacuum chamber. The SEM images were taken from both the outer and the inner surface after the cross-section of the granule.

3.4. Porosity and Mechanical Strength

The mercury intrusion porosimetry method was used for analysis of the specific surface area and the degree of porosity of the obtained beads. The experiments were carried out using an Autopore IV 9510 instrument (Micromeritics, Norcross, GA, USA), which enables measurements of mercury intrusion pressure in the range 0.0035 to 400 MPa. This method is based on applying controlled pressure to a sample which is immersed in mercury. Briefly, mercury, as a non-wetting liquid is forced into the pores by external pressure. The pressure required to intrude into the pores is inversely proportional to the size of the pores. The dried pieces of tested sample were placed in a measuring vessel (penetrometer) and degassed to a pressure of 50 mmHg. Volumes and pore size distributions were calculated using the Washburn equation.

The beads were also tested to determine the mechanical crushing strength by determining the destructive force of the granules using the Tinius Olsen H10K-S instrument (Tinius Olsen, Horsham PA, USA). Each bead was placed between the stationary plate and the moving measuring head and subjected to a pressure test whilst moving the head at a speed of 5 mm/s. The mechanical strength is a ratio of the pressure at which the bead is destroyed (N) and the diameter of the bead (mm).

3.5. In Vitro Release of Mg and Si Ions from Mg,Si-HA Powder and Beads

In vitro release of Mg^{2+} and SiO_4^{4-} ions from Mg,Si-HA powder and from the obtained beads was evaluated as follows: 1 g of the sample was filled with 12 mL of phosphate buffer saline (PBS) of pH 7.4 and put into a water bath at 37 °C under a gentle stirring. The measurements of the ions released from powder and beads were carried out for four and three weeks, respectively. Sample aliquots of 5 mL were taken at appropriate time intervals, and each sample was replaced with a portion of fresh buffer. For the quantitative measurement of magnesium and silicon released from the samples, the inductively coupled plasma optical emission spectrometer ICP-OES iCAP 7400 Duo Spectrometer (Thermo Scientific, Waltham, MA, USA) was used.

3.6. In Vitro Release of Raloxifene Hydrochloride from Beads

In vitro release of RAL from beads was evaluated in Falcon 50 mL tubes. The studies were performed in PBS of pH 7.4 and ethanol solution (1:1 weight ratio) to create the sink conditions since RAL is poorly soluble in water [52–54]. For the analysis, 0.3 g of each sample was immersed in 50 mL of release medium in the bath shaker and stirred at 100 rpm at 37 °C. Sample aliquots of 5 mL were withdrawn at regular time intervals. Each sample was kept in a separate tube to avoid multiple repeated medium changes. Since the RAL solutions are not stable, the experiment was planned to finish and take samples of all time intervals at the same time. Release was started with the longest measurement (12 weeks) and successive samples were inserted over time. At the end of the experiment, all the samples were filtered through a membrane syringe filter with pore size of 0.8 μm. The samples were then analyzed by the UV-Vis method using a Shimadzu UV-1800 spectrometer (Shimadzu, Kyoto, Japan). The absorbance was measured at 291 nm and all measurements were performed in triplicate ($n = 3$).

The percentage of drug released at each time point was calculated according to Equation (1):

$$\text{Drug release (\%)} = \text{drug in solution } [\mu g/mL]/\text{initial drug content in the sample } [\mu g/mL] \times 100\% \qquad (1)$$

3.7. In Vitro Cytotoxicity Assessment

The in vitro cytotoxicity assessment was conducted using a normal human fetal osteoblast cell line (hFOB 1.19, ATCC-LGC Standards, Kiełpin, Poland) and a human osteosarcoma cell line (Saos-2, ATCC-LGC Standards, Kiełpin, Poland). The 1:1 mixture of DMEM/Ham's F12 growth medium without phenol red (Sigma-Aldrich Chemicals, St. Louis, MO, USA) containing 100 μg/mL streptomycin, 100 U/mL penicillin, 300 μg/mL G418 (Sigma-Aldrich Chemicals, St. Louis, MO, USA), and 10% fetal bovine serum (PAN-Biotech GmbH, Aidenbach, Germany), was used as the culture medium for hFOB 1.19, whereas McCoy's 5A medium (ATCC-LGC Standards, Kiełpin, Poland) containing 100 μg/mL streptomycin, 100 U/mL penicillin, and 15% fetal bovine serum was used as the culture medium for Saos-2. The hFOB 1.19 and Saos-2 cells were cultured in a humidified atmosphere of 5% CO_2 at 34 °C and 37 °C, respectively.

The cytotoxicity evaluation was conducted according to ISO 10993-5:2009 standard against hFOB 1.19 and Saos-2 cells. The MTT assay (Sigma-Aldrich Chemicals, St. Louis, MO, USA) was conducted using 24-h granule extracts as described earlier [55]. Then, 100 μL/well of cell suspension (2×10^5 cells/mL and 3×10^5 cells/mL of hFOB 1.19 and Saos-2, respectively) was seeded into 96-well polystyrene plates. After 24 h of culture, the culture medium was replaced with 100 μl of granule extracts. Polypropylene extract served as a negative control for cytotoxicity. The MTT test was performed as described earlier after 48-h exposure to the extracts [45]. Afterwards, the viability of hFOB 1.19 and Saos-2 cells was calculated according to Equation (2):

$$\text{Cell viability (\%)} = \text{sample OD}/\text{negative control OD} \times 100\% \qquad (2)$$

4. Conclusions

The results enable us to conclude that the obtained granules are materials with favorable parameters, enabling the gradual, local delivery of medicinal substances to the immediate surroundings of the diseased tissue. Results revealed that the beads CS/KER-RAL, composed of both CS and KER, possessed the most sustained drug release and no burst release effect, compared to single-component samples (CS and KER). The obtained results may constitute a starting point for more extensive research, with particular emphasis on in vivo biological tests. Drug delivery systems releasing raloxifene directly to the bone tissue could be a beneficial alternative to its current route of administration.

Author Contributions: Conceptualization, physicochemical analysis, and interpretation, K.S. and J.K.; biological tests, P.K.; main contribution for the writing of the manuscript, K.S. and J.K. All authors have read and agreed to the published version of the manuscript.

Funding: This research received no external funding. J.K. and K.S., are grateful to the Medical University of Warsaw for the financial support FW23/N/20.

Acknowledgments: The authors would like to thank Ryszard Strzałkowski from Electron Microscopy Research Unit, Mossakowski Medical Research Centre, Polish Academy of Sciences, Warsaw, Poland, for SEM measurements, and Paweł Kowalik and Wiesław Próchniak from the Łukasiewicz Research Network, New Chemical Syntheses Institute, Puławy, Poland, for porosity and mechanical strength analyses. We used equipment installed as part of a project sponsored by EU Structural Funds: Centre of Advanced Technology BIM—equipment purchased for the Laboratory of Biological and Medical Imaging.

Conflicts of Interest: The authors declare no conflict of interest.

References

1. De Grado, G.F.; Keller, L.; Idoux-Gillet, Y.; Wagner, Q.; Musset, A.-M.; Benkirane-Jessel, N.; Bornert, F.; Offner, D. Bone substitutes: A review of their characteristics, clinical use, and perspectives for large bone defects management. *J. Tissue Eng.* **2018**, *9*, 2041731418776819.
2. Campana, V.; Milano, G.; Pagano, E.; Barba, M.; Cicione, C.; Salonna, G.; Lattanzi, W.; Logroscino, G. Bone substitutes in orthopaedic surgery: From basic science to clinical practice. *J. Mater. Sci. Mater. Med.* **2014**, *25*, 2445–2461. [CrossRef] [PubMed]
3. Dahiya, U.R.; Mishra, S.; Bano, S. Application of bone substitutes and its future prospective in regenerative medicine. In *Biomaterial-Supported Tissue Reconstruction or Regeneration*; Barbeck, M., Ed.; IntechOpen: London, UK, 2019.
4. Xu, X.L.; Gou, W.L.; Wang, A.Y.; Wang, Y.; Guo, Q.-Y.; Lu, Q.; Lu, S.-B.; Peng, J. Basic research and clinical applications of bisphosphonates in bone disease: What have we learned over the last 40 years? *J. Transl. Med.* **2013**, *11*, 303. [CrossRef]
5. Szurkowska, K.; Laskus, A.; Kolmas, J. Hydroxyapatite-based materials for potential use in bone tissue infections. In *Hydroxyapatite—Advances in Composite Nanomaterials, Biomedical Applications and Its Technological Facets*; Thirumalai, J., Ed.; IntechOpen: London, UK, 2018; pp. 109–135.
6. Chindamo, G.; Sapino, S.; Peira, E.; Chirio, D.; Gonzalez, M.C.; Gallarate, M. Bone diseases: Current approach and future perspectives in drug delivery systems for bone targeted therapeutics. *Nanomaterials* **2020**, *10*, 875. [CrossRef]
7. Gu, W.; Wu, C.; Chen, J.; Xiao, Y. Nanotechnology in the targeted drug delivery for bone diseases and bone regeneration. *Int. J. Nanomed.* **2013**, *8*, 2305–2317. [CrossRef]
8. Newman, M.R.; Benoit, D.S.W. Local and targeted drug delivery for bone regeneration. *Curr. Opin. Biotechnol.* **2016**, *40*, 125–132. [CrossRef] [PubMed]
9. Skjødt, M.K.; Frost, M.; Abrahamsen, B. Side effects of drugs for osteoporosis and metastatic bone disease. *Br. J. Clin. Pharmacol.* **2019**, *85*, 1063–1071. [CrossRef]
10. Vallet-Regi, V.; Gonzalez-Calbet, J.M. Calcium phosphates as substitution of bone tissues. *Prog. Solid State Chem.* **2004**, *32*, 1–31. [CrossRef]
11. Barrère, F.; van Blitterswijk, C.A.; de Groot, K. Bone regeneration: Molecular and cellular interactions with calcium phosphate ceramics. *Int. J. Nanomed.* **2006**, *1*, 317–332.
12. Lin, K.; Chang, J. Structure and properties of hydroxyapatite for biomedical applications. In *Hydroxyapatite (Hap) for Biomedical Applications*; Mucalo, M., Ed.; Woodhead Publishing: Cambridge, UK, 2015; pp. 3–19.
13. Ratnayake, J.T.; Mucalo, M.; Dias, G.J. Substituted hydroxyapatites for bone regeneration: A review of current trends. *J. Biomed. Mater. Res. B* **2017**, *105*, 1285–1299. [CrossRef]
14. Supova, M. Substituted hydroxyapatites for biomedical applications: A review. *Ceram. Int.* **2015**, *41*, 9203–9231. [CrossRef]
15. Kolmas, J.; Krukowski, S.; Laskus, A.; Jurkitewicz, M. Synthetic hydroxyapatite in pharmaceutical applications. *Ceram. Int.* **2016**, *42*, 2472–2487. [CrossRef]
16. Mondal, S.; Pal, U. 3D hydroxyapatite scaffold for bone regeneration and local drug delivery applications. *J. Drug Deliv. Sci. Technol.* **2019**, *53*, 101131. [CrossRef]
17. Bose, S.; Tarafder, S. Calcium phosphate ceramic systems in growth factor and drug delivery for bone tissue engineering: A review. *Acta Biomater.* **2012**, *8*, 1401–1421. [CrossRef]
18. Castiglioni, S.; Cazzaniga, A.; Albisetti, W.; Maier, J.A.M. Magnesium and osteoporosis: Current state of knowledge and future research directions. *Nutrients* **2013**, *5*, 3022–3033. [CrossRef] [PubMed]

19. Zhang, Y.; Xu, J.; Ruan, Y.C.; Yu, M.K.; O'Laughlin, M.; Wise, H.; Chen, D.; Tian, L.; Shi, D.; Wang, J.; et al. Implant-derived magnesium induces local neuronal production of CGRP to improve bone-fracture healing in rats. *Nat. Med.* **2016**, *22*, 1160–1169. [CrossRef] [PubMed]

20. Landi, E.; Logroscino, G.; Proietti, L.; Tampieri, A.; Sandri, M.; Sprio, S. Biomimetic Mg-substituted hydroxyapatite: From synthesis to in vivo behaviour. *J. Mater. Sci. Mater. Med.* **2008**, *19*, 239–247. [CrossRef]

21. Reffitt, D.M.; Ogston, N.; Jugdaohsingh, R.; Cheung, H.F.J.; Evans, B.A.J.; Thompson, R.P.H.; Powell, J.J.; Hampson, G.N. Orthosilicic acid stimulates collagen type 1 synthesis and osteoblastic differentiation in human osteoblast-like cells in vitro. *Bone* **2003**, *32*, 127–135. [CrossRef]

22. Patel, N.; Best, S.M.; Bonfield, W.; Gibson, I.R.; Hing, K.A.; Damien, E.; Revell, P.A. A comparative study on the in vivo behavior of hydroxyapatite and silicon substituted hydroxyapatite granules. *J. Mater. Sci. Mater. Med.* **2002**, *13*, 1199–1206. [CrossRef]

23. Szurkowska, K.; Kolmas, J. Hydroxyapatites enriched in silicon–Bioceramic materials for biomedical and pharmaceutical applications. *Prog. Nat. Sci. Mater. Int.* **2017**, *27*, 401–409. [CrossRef]

24. Taranta, A.; Brama, M.; Teti, A.; De Luca, V.; Scandurra, R.; Spera, G.; Agnusdei, D.; Termine, J.; Migliaccio, S. The selective estrogen receptor modulator raloxifene regulates osteoclast and osteoblast activity in vitro. *Bone* **2002**, *30*, 368–376. [CrossRef]

25. Miki, Y.; Suzuki, T.; Nagasaki, S.; ata, S.; Akahira, J.-I.; Sasano, H. Comparative effects of raloxifene, tamoxifen and estradiol on human osteoblasts in vitro: Estrogen receptor dependent or independent pathways of raloxifene. *J. Steroid Biochem. Mol. Biol.* **2009**, *113*, 281–289. [CrossRef] [PubMed]

26. Rey, J.R.; Cervino, E.V.; Rentero, M.L.; Crespo, E.C.; Álvaro, A.O.; Casillas, M. Raloxifene: Mechanism of action, effects on bone tissue, and applicability in clinical traumatology practice. *Open Orthop. J.* **2009**, *3*, 14–21. [CrossRef] [PubMed]

27. Messalli, E.M.; Mainini, G.; Scaffa, C.; Cafiero, A.; Salzillo, P.L.; Ragucci, A.; Cobellis, L. Raloxifene therapy interacts with serum osteoprotegerin in postmenopausal women. *Maturitas* **2007**, *56*, 38–44. [CrossRef] [PubMed]

28. Gurvich, C.; Hudaib, A.; Gavrilidis, E.; Worsley, R.; Thomas, N.; Kulkarni, J. Raloxifene as a treatment for cognition in women with schizophrenia: The influence of menopause status. *Psychoneuroendocrinology* **2019**, *100*, 113–119. [CrossRef] [PubMed]

29. Soni, N.K.; Sonali, L.J.; Singh, A.; Mangla, B.; Neupane, Y.R.; Kohli, K. Nanostructured lipid carrier potentiated oral delivery of raloxifene for breast cancer treatment. *Nanotechnology* **2020**, *31*, 47. [CrossRef]

30. Smetana, K.; Rosel, D.; Brabek, J. Raloxifene and Bazedoxifene Could Be Promising Candidates for Preventing the COVID-19 Related Cytokine Storm, ARDS and Mortality. *In Vivo* **2020**, *34*, 3027–3028. [CrossRef] [PubMed]

31. Mizuma, T. Intestinal glucuronidation metabolism may have a greater impact on oral bioavailability than hepatic glucuronidation metabolism in humans: A study with raloxifene, substrate for UGT1A1, 1A8, 1A9, and 1A10. *Int. J. Pharm.* **2009**, *378*, 140–141. [CrossRef] [PubMed]

32. Garg, A.; Singh, S.; Rao, V.U.; Bindu, K.; Balasubramaniam, J. Solid State Interactions of Raloxifene HCl with Different Hydrophilic Carriers During Co-grinding and its Effect on Dissolution Rate. *Drug Dev. Ind. Pharm.* **2009**, *35*, 455–470. [CrossRef] [PubMed]

33. Jagadish, B.; Yelchuri, R.; Bindu, K.; Tangi, H.; Maroju, S.; Rao, V.U. Enhanced dissolution and bioavailability of raloxifene by co-grinding with different superdisintegrants. *Chem. Pharm. Bull.* **2010**, *58*, 293–300. [CrossRef] [PubMed]

34. Elsheikh, M.A.; Elnaggar, Y.S.R.; Gohas, E.Y.; Abdullah, O.Y. Nanoemulsion liquid preconcentrates for raloxifene hydrochloride: Optimization and in vivo appraisal. *Int. J. Nanomed.* **2012**, *7*, 3787–3802.

35. Tran, T.H.; Poudel, B.K.; Marasini, N.; Chi, S.-C.; Choi, H.-G.; Yong, C.S.; Kim, J.O. Preparation and evaluation of raloxifene-loaded solid dispersion nanoparticle by spray-drying technique without an organic solvent. *Int. J. Pharm.* **2013**, *443*, 50–57. [CrossRef] [PubMed]

36. Mahajan, A.N.; Surti, N.; Patel, P.; Patel, A.; Shah, D.; Patel, V. Melt Dispersion Adsorbed onto Porous Carriers: An Effective Method to Enhance the Dissolution and Flow Properties of Raloxifene Hydrochloride. *Assay Drug Dev. Technol.* **2020**, *18*, 282–294. [CrossRef] [PubMed]

37. Meme, L.; Santarelli, A.; Marzo, G.; Emanuelli, M.; Nocini, P.; Bertossi, D.; Putignano, A.; Dioguardi, M.; Muzio, L.L.; Bambini, F. Novel hydroxyapatite biomaterial covalently linked to raloxifene. *Int. J. Immunopathol. Pharmacol.* **2014**, *27*, 437–444. [CrossRef] [PubMed]

38. Zhang, M.L.; Cheng, J.; Xiao, Y.C.; Yin, R.-F.; Feng, X. Raloxifene microsphere-embedded collagen/chitosan/β-tricalcium phosphate scaffold for effective bone tissue engineering. *Int. J. Pharm.* **2017**, *518*, 80–85. [CrossRef] [PubMed]

39. Szurkowska, K.; Zgadzaj, A.; Kuras, M.; Kolmas, J. Novel hybrid material based on Mg^{2+} and SiO_4^{4-} co-substituted nano-hydroxyapatite, alginate and chondroitin sulphate for potential use in biomaterials engineering. *Ceram. Int.* **2018**, *44*, 18551–18559. [CrossRef]

40. Vallet-Regi, M.; Izquierdo-Barba, I.; Montserrat, C. Structure and functionalization of mesoporous bioceramics for bone tissue regeneration and local drug delivery. *Phil. Trans. R. Soc. A* **2012**, *370*, 1400–1421. [CrossRef] [PubMed]

41. Polo-Corrales, L.; Latorre-Esteves, M.; Ramirez-Vick, J.E. Scaffold design for bone regeneration. *J. Nanosci. Nanotechnol.* **2014**, *14*, 15–56. [CrossRef] [PubMed]

42. Zhang, J.; Wang, G.; Wang, A. In situ generation of sodium alginate/hydroxyapatite nanocomposite beads as drug-controlled release matrices. *Acta Biomater.* **2010**, *6*, 445–454. [CrossRef]

43. Kaygusuz, H.; Erim, F.B. Aglinate/BSA/montmorillonite composites with enhanced protein entrapment and controlled release efficiency. *React. Funct. Polym.* **2013**, *73*, 1420–1425. [CrossRef]

44. Ching, A.L.; Liew, C.V.; Heng, P.W.S.; Chan, L.W. Impact of cross-linker on alginate matrix integrity and drug release. *Int. J. Pharm.* **2008**, *335*, 259–268. [CrossRef] [PubMed]
45. Przekora, A.; Czechowska, J.; Pijocha, D.; Ślósarczyk, A.; Ginalska, G. Do novel cement-type biomaterials reveal ion reactivity that affects cell viability in vitro? *Cent. Eur. J. Biol.* **2014**, *9*, 277–289. [CrossRef]
46. Malafaya, P.B.; Reis, R.L. Bilayered chitosan-based scaffolds for osteochondral tissue engineering: Influence of hydroxyapatite on in vitro cytotoxicity and dynamic bioactivity studies in a specific double-chamber bioreactor. *Acta Biomater.* **2009**, *5*, 644–660. [CrossRef] [PubMed]
47. Gustavsson, J.; Ginebra, M.P.; Engel, E.; Planell, J. Ion reactivity of calcium-deficient hydroxyapatite in standard cell culture media. *Acta Biomater.* **2011**, *7*, 4242–4252. [CrossRef] [PubMed]
48. Swetha, M.; Sahithi, K.; Moorthi, A.; Srinivasan, N.; Ramasamy, K.; Selvamurugan, N. Biocomposites containing natural polymers and hydroxyapatite for bone tissue engineering. *Int. J. Biol. Macromol.* **2010**, *47*, 1–4. [CrossRef] [PubMed]
49. Venkatesan, J.; Bhatnagar, I.; Manivasagan, P.; Kang, K.-H.; Kim, S.-K. Alginate composites for bone tissue engineering: A review. *Int. J. Biol. Macromol.* **2015**, *72*, 269–281. [CrossRef]
50. Yang, J.; Shen, M.; Wen, H.; Luo, Y.; Huang, R.; Rong, L.; Xie, J. Recent advance in delivery system and tissue engineering applications of chondroitin sulfate. *Carbohydr. Polym.* **2020**, *230*, 115650. [CrossRef]
51. Li, J.S.; Li, Y.; Liu, X.; Zhang, J.; Zhang, Y. Strategy to introduce an hydroxyapatite–keratin nanocomposite into a fibrous membrane for bone tissue engineering. *J. Mater. Chem. B* **2013**, *1*, 432–437. [CrossRef]
52. Fontana, M.C.; Beckenkamp, A.; Buffon, A.; Beck, R.C.R. Controlled release of raloxifene by nanoencapsulation: Effect on in vitro antiproliferative activity of human breast cancer cells. *Int. J. Nanomed.* **2014**, *9*, 2979–2991.
53. Mahmood, S.; Taher, M.; Mandal, U.K. Experimental desing and optimization of raloxifene hydrochloride loaded nanotransfersomes for transdermal application. *Int. J. Nanomed.* **2014**, *9*, 4331–4346.
54. Bikiaris, D.; Karavelidis, V.; Karavas, E. Novel biodegradable polyesters. Synthesis and application as drug carriers for the preparation of raloxifene HCl loaded nanoparticles. *Molecules* **2009**, *14*, 2410–2430. [CrossRef] [PubMed]
55. Kolmas, J.; Pajor, K.; Pajchel, L.; Przekora, A.; Ginalska, G.; Oledzka, E.; Sobczak, M. Fabrication and physicochemical characterization of porous composite microgranules with selenium oxyanions and risedronate sodium for potential applications in bone tumors. *Int. J. Nanomed.* **2017**, *12*, 5633–5642. [CrossRef] [PubMed]

International Journal of
Molecular Sciences

Article

Effective Elimination and Biodegradation of Polycyclic Aromatic Hydrocarbons from Seawater through the Formation of Magnetic Microfibres

M. Susana Gutiérrez [1], Alberto J. León [1], Paulino Duel [1], Rafael Bosch [2,3,*], M. Nieves Piña [1,*] and Jeroni Morey [1]

1 Department of Chemistry, University of the Balearic Islands, Crta. de Valldemossa, Km. 7.5, 07122 Palma de Mallorca, Spain; dominika71@hotmail.com (M.S.G.); albertoperezleon@hotmail.com (A.J.L.); paulino.duel@uib.es (P.D.); jeroni.morey@uib.es (J.M.)
2 Department of Biology, University of the Balearic Islands, Crta. de Valldemossa, Km. 7.5, 07122 Palma de Mallorca, Spain
3 Environmental Microbiology, IMEDEA (CSIC-UIB), Miquel Marquès, 21, 07190 Esporles, Spain
* Correspondence: rbosch@uib.es (R.B.); neus.pinya@uib.es (M.N.P.); Tel.: +34-971-172847 (M.N.P.)

Abstract: Supramolecular aggregates formed between polycyclic aromatic hydrocarbons and either naphthalene or perylene-derived diimides have been anchored in magnetite magnetic nanoparticles. The high affinity and stability of these aggregates allow them to capture and confine these extremely carcinogenic contaminants in a reduced space. In some cases, the high cohesion of these aggregates leads to the formation of magnetic microfibres of several microns in length, which can be isolated from the solution by the direct action of a magnet. Here we show a practical application of bioremediation aimed at the environmental decontamination of naphthalene, a very profuse contaminant, based on the uptake, sequestration, and acceleration of the biodegradation of the formed supramolecular aggregate, by the direct action of a bacterium of the lineage Roseobacter (biocompatible with nanostructured receptors and very widespread in marine environments) without providing more toxicity to the environment.

Keywords: magnetite nanoparticles; polycyclic aromatic hydrocarbon; microfibres; elimination; biodegradation

Citation: Gutiérrez, M.S.; León, A.J.; Duel, P.; Bosch, R.; Piña, M.N.; Morey, J. Effective Elimination and Biodegradation of Polycyclic Aromatic Hydrocarbons from Seawater through the Formation of Magnetic Microfibres. *Int. J. Mol. Sci.* **2021**, *22*, 17. https://dx.doi.org/doi:10.3390/ijms22010017

Received: 26 November 2020
Accepted: 18 December 2020
Published: 22 December 2020

Publisher's Note: MDPI stays neutral with regard to jurisdictional claims in published maps and institutional affiliations.

1. Introduction

The combined and synergistic use of nanotechnology, together with the action of microorganisms originating in the environment, can be a successful combination and a great help to accelerate and improve the overall process of natural bioremediation. The most worrying pollutants for the environment and its integral health have in common a high persistence and toxicity. Removing them completely and recycling and/or degrading them properly, safely, and effectively are not easy tasks. This work must be undertaken from the molecular perspective, developing molecular receptors capable of helping in this task without contributing to the toxicity of the environment. As a primary condition, these receptors must act in demanding environments and form robust supramolecular complexes. In this investigation, we describe the preparation of new hybrid nanoparticles formed by magnetite nanoparticles, Fe_3O_4, FeNP, linked the diimide-dopamine ligands prepared from four electron-deficient π-extended diimides, derived from benzene (BDI), naphthalene (NDI), perylene (PDI), and tetrabromodiimideperylene (BrPDI). This new hybrid nanomaterial is sensitive to the action of an external magnetic field, which makes possible its removal from a liquid suspension in a simple and comfortable way. From the point of view of reuse and sustainability with the environment, this magnetic property is very attractive, precisely because of its simplicity and effectiveness. Polycyclic aromatic hydrocarbons (PAHs) form a group of pollutants considered very dangerous for the environment. They are extremely carcinogenic and persistent contaminants [1–5]. In fact, since 1976, the U.S.

Environmental Protection Agency (EPAH) advises a rigorous control of 16 specific PAHs, which can be generally detected in drinking water and also in certain foods that have a high percentage of oils and fats in their composition. The Toxicity Equivalence Factor (TEF, Figure 1) is used to compare the degree of toxicity among PAHs. For benzo[a]-pyrene (4) the TEF is 1, while the remaining PAHs in the list have values less than 1, except for dibenzo[a,h]-anthracene (6) with TEF = 10 [6]. Despite its low TEF (0.001), naphthalene is the most worrying of the PAHs because it is the most profuse and soluble anthropogenic PAH in water. Therefore, its control is a priority.

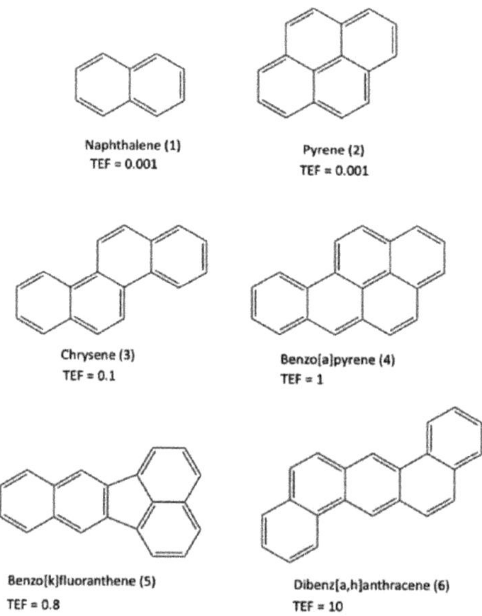

Figure 1. Polycyclic aromatic hydrocarbons used in this study.

The members of the Roseobacter lineage (*Rhodobacteracea* family, *alphaproteobacteria* class) are predominant in the marine ecosystems, where they are ubiquitous, and they represent more than 20% of the coastal planktonic community and 3–5% in surface ocean waters [7,8] being especially abundant in littoral areas contaminated by hydrocarbons (i.e., marinas and harbours) [9,10]. The analyses of their genomes have revealed the presence in Roseobacters of a great biodegradation potential, mainly for monoaromatic hydrocarbons [11,12]. For example, *Salipiger aestuarii* 357, a member of the Roseobacter lineage isolated from the sands of the coast of Galicia (Northern Spain) that were contaminated due to the accidental oil spill of the Prestige oil tanker in 2002 [13], is capable of growing at the expense of salicylate and naphthalene as sole carbon and energy sources.

2. Results

The diimide-dopamine ligands were prepared from four electron-deficient π-extended diimides, derived from benzene (BDI), naphthalene (NDI), perylene (PDI), and tetrabromodiimideperylene (BrPDI). The ligands have a symmetrical substitution, with two dopamine units (DA) at both ends, which allow them to be anchored on the surface of the magnetic nanoparticles of Fe_3O_4 (FeNP), see Figure 2.

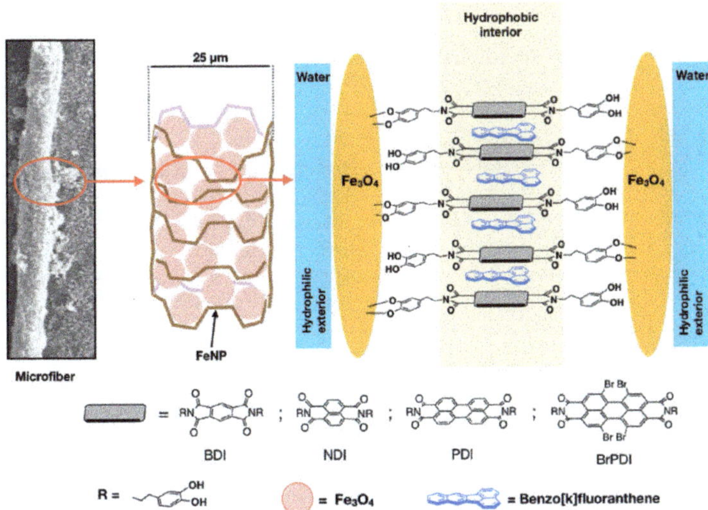

Figure 2. Outline proposed for the formation of microfibres.

The magnetite nanoparticles (FeNP), used as a nanomaterial, were synthesized by the classical co-precipitation method from Fe (II) and Fe (III) chloride salts [14]. The method of obtaining magnetite is quantitative; it takes place in water, without the use of organic solvents, where magnetite is isolated with the help of a boron-neodymium magnet and without the need to filter or centrifuge, obtaining sodium chloride as a reaction byproduct, an environmentally tolerable contaminant. Magnetite is a natural mineral, which has no storage, handling or disposal problems [15].

The binding between the ligand and the magnetite nanoparticles [16] took place in a basic aqueous solution, using a microwave-ready 5 mL tube. Once sealed, it was irradiated for 30 min between 120–130 °C. Under these reaction conditions, a pressure of 3 bar was reached inside the reaction tube. Increasing the pressure within the reaction caused a greater coating of the Fe_3O_4 nanoparticles, substantially increasing the surface coating density and the chemical resistance to extreme aqueous pH conditions. On the contrary, the two hydroxyl groups of the dopamine catechol residues, located in the outermost part of the nanoparticle, formed a hydrophilic layer capable of establishing hydrogen bonds with water and allowed excellent dispersion in these aqueous conditions. These facts were experimentally confirmed by the obtained Dynamic Light Scattering, Zeta Potential Distribution, and Thermogravimetric Analysis (TGA) values (Supplementary Table S1; Supplementary Figures S1–S4).

All nanoreceptors present a polydispersity index (PdI) <0.35 and a low aggregation index. The tests were performed in distilled H_2O at a temperature of 25 °C and pH = 7.0.

The Langmuir constant was measured to quantify the nature of the association between the PAHs and the nanoreceptor surface. Since all PAHs are fluorescent, the Langmuir constant was determined by fluorescence experiments (Table S2).

The following PAHs (together with their excitation wavelengths) were used to determine de Langmuir constant: naphthalene (λ_{exc} = 270 nm), pyrene (λ_{exc} = 338 nm), benzo[a]pyrene, BAP (λ_{exc} = 266 nm), benzo-k-fluoranthene, BKF (λ_{exc} = 308 nm), dibenzo[ah]anthracene, DB[ah]A (λ_{exc} = 290 nm), and chrysene (λ_{exc} = 285 nm). All tests were carried out in an ethanol:water (1:1) v/v mixture.

The results obtained for the Langmuir constant between the functionalised nanomaterial and several PAHs of different geometry were excellent, far superior to those described in the literature [17] and similar to those found when using the FeNP-PDI-DA nanomaterial with the same PAHs [18,19] and the same experimental conditions.

To understand the chemical behaviour of PDI and BrPDI, the electrostatic potential (ESP) surfaces of both diimides were computed at the DFT level (B3LYP using 6-31G*) with Spartan,(Wavefunction, Inc., Irvine, California) (see Figure S9). From the inspection of the Figure S9, more positive ESP values (deeper blue zones) were observed for FeNP-BrPDI-DA, located especially between the six central rings of the tetrabromodiimide perylene, suggesting that it was more electron-deficient than FeNP-PDI-DA.

The formation and growth of FeNP-Diimide-DA microfibres occurred spontaneously—thermodynamic process—by adding an aqueous solution of concentration between 10^{-5} M and 10^{-6} M of FeNP-Diimide-DA, within a 6.5–7.0 pH range, to a specific PAH. Due to the slow growth of the fibres, the dispersion was left in a tightly closed vial at room temperature, at rest, and protected from light. After 7 days, the appearance of elongated microscopic fibres was observed, initially attached to the walls of the vial, which eventually grow and disperse within the bulk of the solution. These microfibres were quickly attracted when a neodymium magnet approached the outer wall of the vial (see the video in the Supplementary Material). If a hydrophilic co-solvent was added to the aqueous solution, such as 50% methanol or ethanol (*v:v*), the formation of the FeNP-Diimide-DA was slower, and the fibres became visible after 2–3 weeks. Conversely, at concentrations greater than 10^{-5} M, fibre formation was not observed. It was thoroughly verified that, under the same experimental conditions, self-aggregation processes of either diimides or non-functionalised FeNP were not observed.

Given the strength of these aggregates, an energetic acid treatment, such as digestion, was necessary to dissolve and break the fibres into their starting components. This was achieved by treating the fibres with concentrated HCl (37%) at 45 °C for 30 min. From this aqueous-acid residue, the organic phase was extracted with n-heptane [20], and the organic phase was concentrated with water using a dry Ar stream. Subsequently, the extract was solubilised in heptane for its HRMS analysis.

Figures S5–S8 show the high-resolution mass spectra together with the isotopic pattern of PAHs isolated from the digestion of chrysene@FeNP-NDI-DA, BAP@FeNP-NDI-DA, BKF@FeNP-PDI-DA, and DB[ah]A@FeNP-PDI-DA fibres, respectively, which experimentally confirmed the presence of PAHs in the supramolecular complex together with the FeNP-diimide-DA hybrid material.

Fibre formation was dependent on the size of PAH—particularly on the number of double bonds of each PAH molecule [18]—and on the geometry and electronic properties of electron-deficient diimides anchored on the magnetite nanoparticle.

In this study, tubular fibres were observed in the solutions prepared with the dopamine-diimide derivative of benzene FeNP-BDI-DA, with chrysene (nine double bonds), benzo[a]pyrene, BAP (ten double bonds), benzo[k]fluoroanthene, BKF (ten double bonds), and dibenzo[ah]anthracene DB[ah]A (11 double bonds).

Figure 3 shows a composition of photomicrographs of DB[ah]A@FeNP-BDI-DA network-growing filamentous fibres, performed with the optical microscope (Figure 3A), and with the confocal fluorescent microscope, with a rhodamine laser at λ_{exc} = 540 nm (Figure 3B–D). The distribution of the host and guest in the network is highlighted by the different fluorescent colouration when using different wavelengths for the excitation of the sample. Figure 3B,C shows, preferably in green and blue, the distribution of the FeNP-BDI-DA nano receptor in the network and the distribution of DB[ah]A, respectively. Finally, Figure 3D is the result of superimposing of all photomicrographs.

Likewise, the formation of tubular fibres was observed with the nanoparticles derived from naphthalene diimide, (FeNP-NDI-DA) with pyrene, chrysene, BAP, BKF, and DB[ah]A.

Figure 4 shows the photomicrographs of the supramolecular assembly formed by the FeNP-NDI-DA with the DB[ah]A, obtained with optical and confocal microscopy. With the hybrid material derived from the tetrabromo perylene diimide, FeNP-BrPDI-DA, flat fibres were observed with chrysene, BKF, BAP, and DB[ah]A.

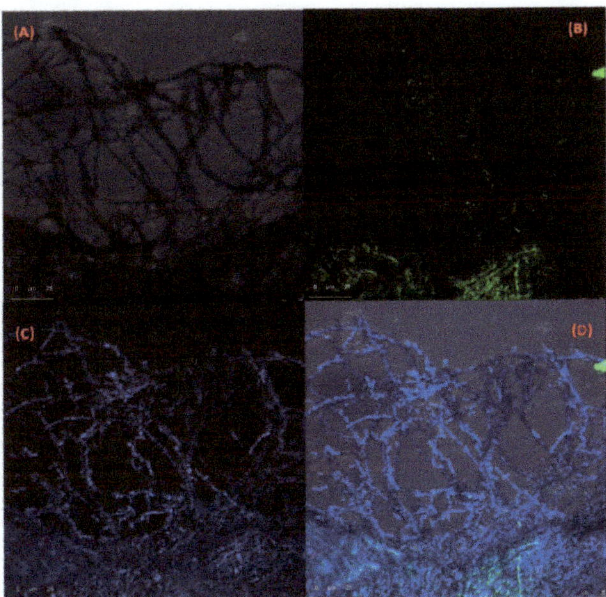

Figure 3. Optical and confocal microphotographs showing microfibres assembled between FeNP-BDI-DA (green) and DB[ah]A (blue) in aqueous media: (**A**) Transmitted light micrograph. (**B**) Fluorescent micrograph λexc = 448 nm, λem = 517 nm (FeNP-BDI-DA). (**C**) Fluorescent micrograph λexc = 405 nm, λem = 454 nm (DB[ah]A). (**D**) Micrograph Composition of A, B, C.

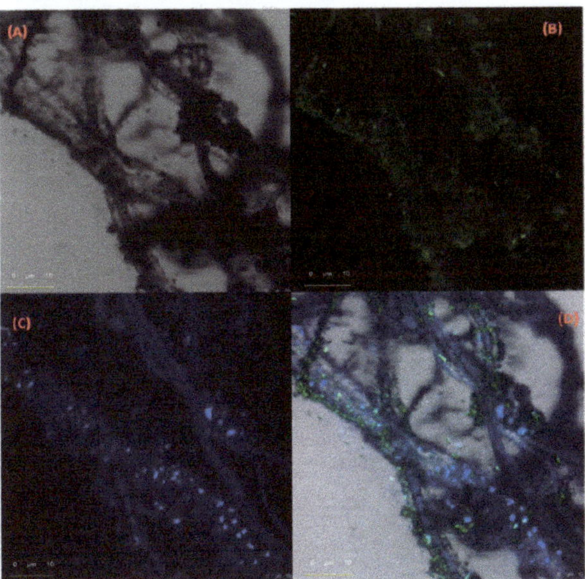

Figure 4. Optical and confocal microphotographs showing microfibres assembled between FeNP-NDI-DA (green) and DB[ah]A (blue) in aqueous media.: (**A**) Transmitted light micrograph. (**B**) Fluorescent micrograph λexc = 448 nm, λem = 517 nm (FeNP-BDI-DA). (**C**) Fluorescent micrograph λexc = 405 nm, λem = 454 nm (DB[ah]A). (**D**) Micrograph Composition of A, B, C.

Figure 5 shows the photomicrographs obtained by SEM for the supramolecular assembly formed by FeNP-BrPDI-DA with BKF and BHA, respectively. From the enlargements of the microphotographs (at 47,000 magnifications), the spherical FeNP nanoparticles of an approximate diameter of 40 nm—light grey—that formed the matrix (skeleton) of the supramolecular complex fibres could be observed.

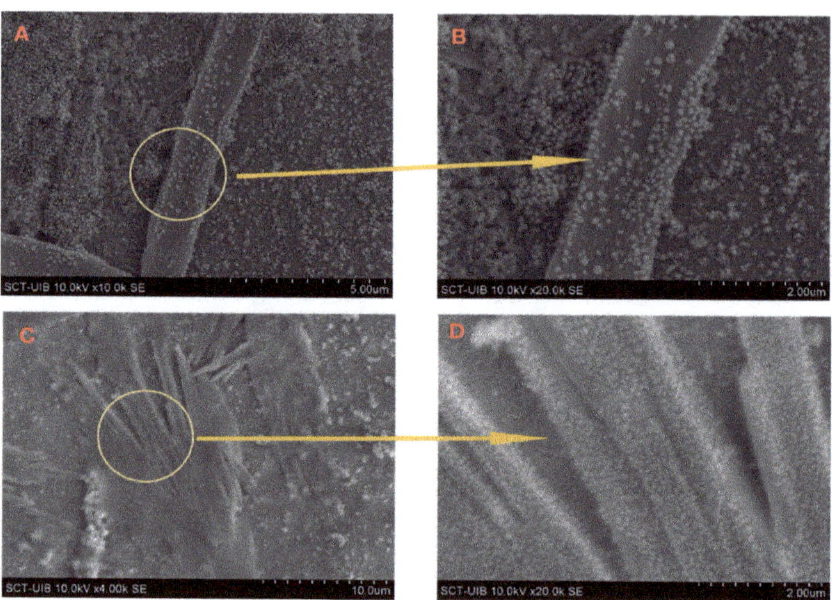

Figure 5. (**A,B**) SEM micrographs showing microfibres assembled between FeNP-BrPDI-DA and BKF. (**C,D**) FeNP-BrPDI-DA and BHA, in aqueous media.

It is worth noting that, in all cases, the fibres showed great chemical stability in aqueous dispersion, since they remain unchanged over time without showing appreciable alterations up to more than 2 years on the laboratory bench, at room temperature, in a hermetically sealed vial closed and protected from light.

The cooperative degradation method consists, first, of the formation of the naphthalene @FeNP-NDI-DA complex. For this purpose, functionalised magnetic nanoparticles, FeNP-NDI-DA, were added to an aqueous medium contaminated with naphthalene. After 30 min with magnetic stirring, the naphthalene@FeNP-NDI-DA complex was formed, which, if desired, could be separated from the solution by magneto filtration. When integrated into the supramolecular aggregate, naphthalene was placed in a molecularly limited space environment. Then, a second stage follows; after seeding with the *S. aestuarii* 357 strain, it was observed that the degradation of the contaminant was accelerated due to greater availability of the substrate. It is worth mentioning that the environmental impact of magnetite nanoparticles was minimal because it is a biocompatible material [21]. The linker dopamine is not a substance considered dangerous, since it is a neurotransmitter widely distributed in nature and easily biodegradable. The remaining organic components, diimide naphthalene, were degraded by the bacteria without altering their integrity, as evidenced by the conducted experiments.

In Figure 6A,B, the growth curves of *S. aestuarii* 357 are shown as a function of time, in three different experimental conditions: in the presence of FeNP-NDI-DA, of naphthalene as the only source of carbon and energy, and of naphthalene and FeNP-NDI-DA (white, grey and black circles, respectively). Figure 6A,B correspond to two different experiments, with different FeNP-NDI-DA nanoreceptor concentrations: 100 mg and 200 mg. The absorbance intensity measurements were recorded at λ_{exc} = 595 nm, and each point represents an average of three experimental replicates.

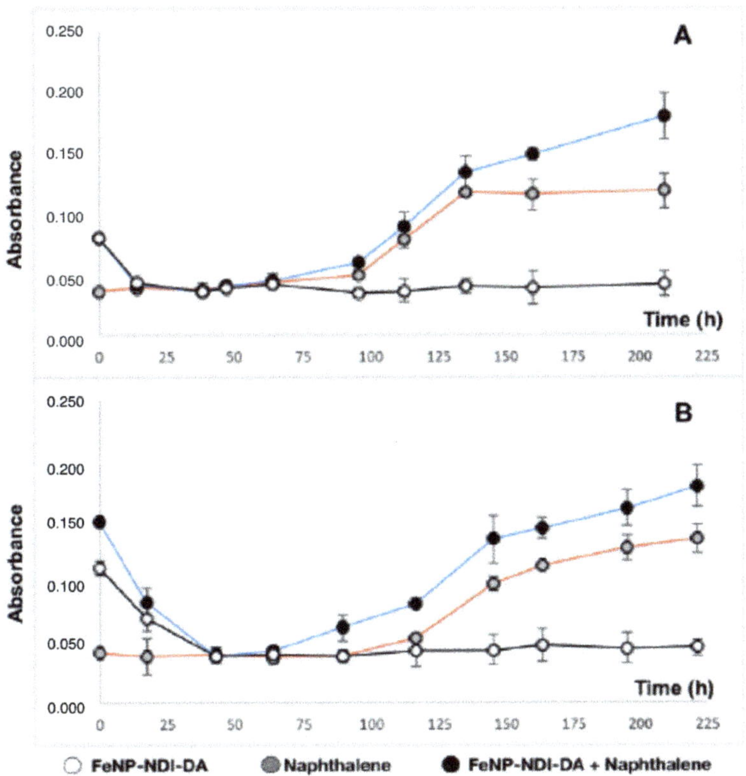

Figure 6. Growth (Absorbance, 595 nm) of *S. aestuarii* 357 with naphthalene (0.1% *w/v*) as unique carbon and energy source supplemented with 100 mg/mL (**A**) and 200 mg/mL (**B**) of FeNP-NDI-DA. Growth values are average and SD of three different experimental replicates.

In these experiments, the bacterium *S. aestuarii* 357 was grown using 100 mL flasks, in 20 mL of a marine mineral medium (see experimental part) [20], fed with naphthalene (0.1%) at 30 °C and under constant agitation (180 rpm). The FeNP-NDI-DA nanoparticles were provided at 1% and 0.5% (*v/v*). The growth was monitored by reading absorbances at λ_{exc} = 595 nm.

3. Discussion

In these systems, the π-π interactions have most likely taken place between the host and the guest molecules. The strength of the π-π interaction will largely depend on the geometry and electronic properties of both the electron-deficient host (FeNP-Diimide-DA) and the electron-rich guest (PAHs). Therefore, if an electron-poorer host is used, as was the case with the substituted tetrabromo perylene diimide, this interaction will be further favoured. The results obtained for the Langmuir constant support this hypothesis; when comparing the values obtained with the FeNP-BrPDI-DA and its synthetic precursor, the perylene diimide FeNP-PDI-DA, equal or higher (by one order of magnitude) Langmuir constants were observed for the former, in the case of association with pyrene, BKF, DB[ah]A, and chrysene. The computed electrostatic potential also supports this statement.

The biocompatibility of the bacterium *S. aestuarii* 357 with the nanomaterial FeNP-NDI-DA was manifested by being able to grow when naphthalene was added as the sole nutrient, the only source of carbon and energy (Figure 6A,B, grey circles). On the contrary, the bacteria cannot grow if FeNP-NDI-DA (white circles) was the only available compound. The addition of FeNP-NDI-DA nanoparticles (black circles), in the presence of naphthalene, accelerated the growth of *S. aestuarii* 357. This phenomenon was more pronounced the higher the concentration of nanoparticles used (Figure 6A vs. Figure 6B). The non-growth of bacteria with only nanoparticles, and the acceleration of growth with the mixture of nanoparticles and naphthalene, via formation of naphthalene@FeNP-NDI-DA complex, suggests not only that the FeNP-NDI-DA nanoparticles are not toxic to *S. aestuarii* 357—otherwise, either there would be no growth, or it would have been delayed—but, also, they increase the bioavailability of naphthalene, the only source of carbon and energy usable by *S. aestuarii* 357 in our experiments.

4. Materials and Methods

4.1. General

Reactions were carried out in oven-dried glassware under an atmosphere of argon unless otherwise indicated. Thin-layer chromatography (TLC) was conducted on aluminum plates coated with silica gel (60 F254, Merck, Merck Life Science S.L.U., Madrid, Spain). Column chromatography was performed using silica gel (Geduran Si 60 from Merck (Merck Life Science S.L.U., Madrid, Spain), particle size 0.040–0.063 mm) as a stationary phase.

4.2. Materials

All commercially available reagents: dopamine hydrochloride, triethylamine, perylene and 3,4,9,10-tetracarboxylic dianhydride, were supplied by Sigma–Aldrich (Merck Life Science S.L.U., Madrid, Spain). All the solvents were purchased from Scharlab (Scharlab, S. L., Barcelona, Spain).

4.3. Instrumentation

^1H and ^{13}C NMR spectra were recorded on a Bruker Advance Spectrometer (Bruker Española S.A., Madrid, Spain) at 300 and 75 MHz at 25 °C. Chemical shifts are reported as a part per million (δ, ppm) referenced to the residual protium signal of deuterated solvents. Spectral features are tabulated in the following order: chemical shift (δ, ppm), multiplicity (s-singlet, d-doublet, t-triplet, and m-multiplet), number of protons. (FTIR) were obtained on Bruker Tensor 27 (Bruker Española, Madrid, Spain) instrument in solid-state. Matrix-assisted laser desorption/ionisation mass spectra (MALDI) were recorded with an Autoflex III MALDI TOF/TOF mass spectrometer provided with a Smartbeam Laser at 200 Hz (Bruker Española, Madrid, Spain). Functionalization of iron nanoparticles was performed on a Biotage Initiator Classic Microwave Synthesizer (Biotage, NASDAQ, Stockholm) at 400 W and 2 bar.

4.4. Synthesis of Magnetite Nanoparticles (FeNP)

For the synthesis of Fe_3O_4 nanoparticles, an adapted procedure of F. Yazdani et al. [14] was used. At room temperature, 50 mL $FeCl_3$ of 0.1 M solution and 25 mL $FeCl_2$ of 0.1 M solution were added to 350 mL distilled deionised water under an argon atmosphere and magnetically stirred. Then 35 mL of the 1 M NaOH solution was added to the reaction vessel and the mixture stirred for 3 min. The black precipitated product was separated with a boron–neodymium magnet. The precipitate was washed 3 times with 50 mL deionised water. After washing, the black product was dried in a vacuum oven for 12 h at 80 °C.

FTIR (KBr): 3406, 2921, 2851, 1591, 1384, and 632 cm^{-1}.

4.5. Preparation of Ligands

Preparation of 1,6,7,12-tetrabromoperylene-3,4,9,10-tetracarboxylic acid bisanhydride [21]: The bromination of perylene-3,4,9,10-tetracarboxylic acid bisanhydride, was carried out with

Br$_2$ in a mixture of sulfuric acid and fuming sulfuric acid. Then, 0.516 g of perylene-3,4,9,10-tetracarboxylic acid bisanhydride, (1.31 mmol) was dissolved in a mixture of 6 mL of sulfuric acid 98 wt% with 1 mL of fuming sulfuric acid, under magnetic stirring for 12 h. After the 12 h, the temperature was raised to 80 °C, and 7.75 mg of iodine (0.03 mmol) and 150 µL of bromine were added dropwise (2.63 mmol). The mixture was heated for 48 h at 80 °C. Then, 7.75 mg of iodine (0.03 mmol) was added, and the temperature was raised to 100 °C, and 150 µL of bromine was added again dropwise. The mixture was heated at 100 °C for 80 h. Finally, the brown-oil was poured onto ice water, and a dark red precipitate was observed. The solid was filtered and washed with 10 mL of sulfuric acid. The red solid was washed with miliQ water until the filtrate became neutral. The product obtained, as a dark red powder, was dried in a vacuum at 120 °C for 8 h, giving 0.409 g (44%, 0.578 mmol). The crude product was used directly in the next step without further purification. ^{1}H NMR (300 MHz, DMSO-d_6). δ: 8.74 (4H, s). FTIR (KBr): ν = 3442, 1716, 1634, 1591, 1557, 1431, 1361, 852, 810, 768 cm^{-1}. HRMS (EI): m/z = 708 (M$^+$).

Preparation of 2,9-bis(3,4-dihydroxyphenethyl)-1,6,7,12-tetrabromoperylene tetracarboxylic bisimide: In a round bottom flask 0.176 g (0.250 mmol) of 1,6,7,12-tetrabromoperylene-3,4,9,10-tetracarboxylic acid bisanhydride was dissolved in 15 mL of a 1:1 (v/v) mixture of H$_2$O and DMF. Then, 2 mL of Et$_3$N was added and stirred for 2 h plus. After that, 99 mg (0.522 mmol) of dopamine hydrochloride was dissolved in 10 mL of a 1:1 (v/v) mixture of H$_2$O and DMF and was added dropwise to the first flask. Once the addition was complete, the mixture was refluxed overnight. The oil obtained was allowed to temper and was transferred to a falcon tube, where concentrated HCl was added until a brown–reddish precipitate was formed. The crude was centrifuged and washed repeatedly with miliQ water, by centrifugation, until neutrality of the supernatant liquid was observed. The product was dried in a vacuum at 120 °C for 8 h and obtained as a reddish–brown powder, 98 mg (0.100 mmol, 40% yield). ^{1}H NMR (300 MHz, DMSO-d_6): δ: 8.7 (s, 4H), 7.9 (s, 4H), 6.90 (m, 6H), 3.65 (s, 4H), 2.67 (t, 4H). ^{13}C NMR (300 MHz, DMSO-d_6). δ: 176.9 (C=O), 137.7 (C=C), 132.4 (C-Br), 128.3 (C=C), 127.9 (C=C), 127.7 (C=C), 120.8 (C=C), 45.3 (CH$_2$), 41.7 (CH$_2$). FTIR (KBr): ν = 3442, 1716, 1634, 1591, 1557, 1431, 1361, 852, 810, 768 cm^{-1}. MALDI-TOF-MS m/z (%): [M]$^+$ calculated. for C$_{40}$H$_{22}$Br$_4$N$_2$O$_8$ 978.2380 found 977.8069.

4.6. Preparation of Functionalised Magnetite Nanoparticles

In a microwave tube 25.26 mg (0.026 mmol) of 2,9-bis(3,4-dihydroxyphenethyl)-1,6,7,12-tetrabromoperylene tetracarboxylic bisimide was introduced with 4 mL of MiliQ water, one drop of 1M NaOH and 1 mL of magnetite nanoparticles suspension (11.4 mg/mL). The mixture was sonicated for 5 min and then introduced in the microwave reactor. The reaction conditions of the microwave reactor were: 120 °C, 3 bar and 30 min of reaction time. Once the reaction was over, the nanoparticles were decanted with the help of a boron-neodymium magnet and washed 3 times with EtOH. Finally, the hybrid nanomaterial was suspended and stored in 10 mL of EtOH in an argon atmosphere.

FTIR (KBr): ν = 3442 (s), 1692 (s), 1643 (m), 1594 (s), 1517 (s), 1403 (s), 1385 (s), 1319 (s), 672 (s) and 588 cm^{-1} (s).

4.7. General Procedure for the Formation of Microfibres

The preparation of magnetic microfibres was carried out as follows: Two hundred microlitres of a 1.14 mg/mL suspension of magnetite nanoparticles functionalised with one of the FeNP-diimide-DA receptors was introduced into a 10 mL vial. Afterward, 4.70 mL of miliQ water was added to the vial. To this solution, 100 µL of a 10^{-4} M solution of chrysene, benzo[a]pyrene, benzo-k-fluoranthene, or dibenzo[a,h]anthracene dissolved in a mixture of water/ethanol, 1:3 (v/v) was added, respectively. The vial was tightly closed and was left at room temperature, in the dark, for 14 days. Enough time for microfibres to develop and observed with the naked eye. After this time, no further growth occurred. The final concentrations were: functionalised nanoparticles: 0.456 mg/mL, and polycyclic aromatic hydrocarbon, PAHs: 10^{-6} M.

4.8. Growth Curves of S. aestuarii 357

The bacterium *S. aestuarii* 357 was grown using 100 mL flasks, in 20 mL of a marine mineral medium (see experimental part) [20], fed with naphthalene (0.1%) at 30 °C, under constant agitation (180 rpm). The FeNP-NDI-DA nanoparticles were provided at 1% and 0.5% (*v*/*v*). The growth was monitored by reading absorbances at λ_{exc} = 595 nm.

5. Conclusions

This study showed that in the global assembly process of magnetic nanoparticles modified with PAHs, resistant supramolecular aggregates are always formed, which in some cases can form fibres of up to several hundreds of micrometres in length, most likely due to the π-π interactions established between PAHs and the electron-deficient surface of the diimides anchored to FeNP, as can be inferred from ESP calculations. On the other hand, it was shown that the formation of the supramolecular aggregates with naphthalene, the most abundant PAH pollutant in the environment, promotes bioavailability, and accelerates the degradation of naphthalene. These results allow us to propose, as a strategy for the removal and biodegradation of PAHs from contaminated aqueous media, the combined use of FeNP-diimides-DA together with a marine microorganism, such as *S. aestuarii* 357, belonging to the bacterial lineage of Roseobacter, as a sustainable measure of environmental decontamination, based on the capture, sequestration, and biodegradation of toxic pollutants.

Supplementary Materials: Supplementary materials can be found at https://www.mdpi.com/1422-0067/22/1/17/s1.

Author Contributions: The experimental work for FeNP-NDI-DA were conducted by M.S.G. The experimental work for FeNP-BrPDI-DA was conducted by A.J.L. and P.D. The experimental work with *S. aestuarii* 357 was conducted by R.B., R.B., J.M. and M.N.P. designed and directed the study, and wrote the article. All authors have read and agreed to the published version of the manuscript.

Funding: This research was funded by the MINECO/AEI of Spain (projects CTQ2017-85821-R and CTM2015-70180-R, both with FEDER funds; and project PID2019-109509RB-I00/AEI/10.13039/501100011033).

Institutional Review Board Statement: Not applicable.

Informed Consent Statement: Not applicable.

Data Availability Statement: Not applicable.

Acknowledgments: The SCT (UIB) is acknowledged for their support in the utilisation of the instrumentation used herein. M. Susana Gutiérrez acknowledges her scholarship from CONACyT (Consejo Nacional de Ciencia y Tecnología de México), Paulino Duel acknowledges his scholarship from GOIB (Govern de les Illes Balears). The authors would like to thank G. Ramis Munar for his help with the confocal microphotography.

Conflicts of Interest: The authors declare no conflict of interest.

References

1. Ghosh, U.; Gillette, J.S.; Luthy, R.G.; Zare, R.N. Microscale Location, Characterization, and Association of Polycyclic Aromatic Hydrocarbons on Harbor Sediment Particles. *Environ. Sci. Technol.* **2000**, *34*, 1729–1736. [CrossRef]
2. Tang, J.; Carroquino, M.J.; Robertson, B.K.; Alexander, M. Combined effect of sequestration and bioremediation in reducing the bioavailability of polycyclic aromatic hydrocarbons in soil. *Environ. Sci. Technol.* **1998**, *32*, 3586–3590. [CrossRef]
3. Werner, D.; Higgins, C.P.; Luthy, R.G. The sequestration of PCBs in Lake Hartwell sediment with activated carbon. *Water Res.* **2005**, *39*, 2105–2113. [CrossRef] [PubMed]
4. Lorimer, W. *Polycyclic Aromatic Hydrocarbons: Binary Non-Aqueous Systems Part I: Solutes A-E*, 1st ed.; Oxford University Press: Oxford, UK, 1995; Volume 58.
5. López, K.A.; Piña, M.N.; Alemany, R.; Vögler, O.; Barceló, F.; Morey, J. Antifolate-modified iron oxide nanoparticles for targeted cancer therapy: Inclusion vs. covalent union. *RSC Adv.* **2014**, *4*, 19196–19204. [CrossRef]
6. Andersson, J.T.; Achten, C. Time to Say Goodbye to the 16 EPA PAHs? Toward an Up-to-Date Use of PACs for Environmental Purposes. *Polycycl. Aromat. Compd.* **2015**, *3*, 330–354. [CrossRef] [PubMed]
7. Buchan, A.; González, J.M.; Moran, M.A. Overview of the Marine Roseobacter Lineage. *Environ. Microbiol.* **2005**, *71*, 5665–5667. [CrossRef] [PubMed]

8. Moran, M.A.; Belas, B.; Schell, M.A.; González, J.M.; Sun, F.; Sun, S.; Binder, B.J.; Edmonds, J.; Ye, W.; Orcutt, B.; et al. Ecological Genomics of Marine Roseobacters. *Appl. Environ. Microbiol.* **2007**, *73*, 4559–4569. [CrossRef] [PubMed]
9. Aguiló, M.M.; Bosch, R.; Martín, C.; Lalucat, J.; Nogales, B. Phylogenetic analysis of the composition of bacterial communities in human-exploited coastal environments from Mallorca Island (Spain). *Syst. Appl. Microbiol.* **2008**, *31*, 231–240. [CrossRef] [PubMed]
10. Nogales, B.; Aguiló-Ferretjans, M.M.; Martín-Cardona, C.; Lalucat, J.; Bosch, R. Bacterial diversity, composition and dynamics in and around recreational coastal areas. *Environ. Microbiol.* **2007**, *9*, 1913–1929. [CrossRef] [PubMed]
11. Buchan, A.; González, J.M.; Moran, M.A. *Handbook of Hydrocarbon and Lipid Microbiology*, 1st ed.; Springer: Berlin/Heidelberg, Germany, 2010; pp. 1335–1343.
12. Newton, R.J.; Griffin, L.E.; Bowles, K.M.; Meile, C.; Gifford, S.; Givens, C.E.; Howard, E.C.; King, E.; Oakley, C.A.; Reisch, C.R.; et al. Genome characteristics of a generalist marine bacterial lineage. *ISME J.* **2010**, *4*, 784–789. [CrossRef] [PubMed]
13. Suarez, L.Y.; Brunet, I.; Piña, J.M.; Christie, J.A.; Peña, A.; Bennasar, A.; Armengaud, J.; Nogales, B.; Bosch, R. Draft Genome Sequence of *Citreicella aestuarii* Strain 357, a Member of the Roseobacter Clade Isolated without Xenobiotic Pressure from a Petroleum-Polluted Beach. *J. Bacteriol.* **2012**, *194*, 5464–5465. [CrossRef] [PubMed]
14. Yazdani, F.; Seddigh, M. Magnetite nanoparticles synthesized by co-precipitation method: The effects of various iron anions on specifications. *Mater. Chem. Phys.* **2016**, *184*, 318–323. [CrossRef]
15. Hongping, H.E.; Zhong, Y.; Liang, X.; Tan, W.; Zhu, J.; Wang, C.Y. Natural Magnetite: An efficient catalyst for the degradation of organic contaminant. *Sci. Rep.* **2015**, *5*, 10139.
16. Gutiérrez, M.S.; Piña, M.N.; Morey, J. Fast microwave-assisted conjugation of magnetic nanoparticles with carboxylates of biological interest. *RSC Adv.* **2017**, *7*, 19385–19390. [CrossRef]
17. Liu, X.; Hu, Y.; Stellaci, F. Mixed-Ligand Nanoparticles as Supramolecular Receptors. *Small* **2011**, *14*, 1961–1966. [CrossRef]
18. Gutiérrez, M.S.; Duel, P.; Hierro, F.; Morey, J.; Piña, M.N. A Very Highly Efficient Magnetic Nanomaterial for the Removal of PAHs from Aqueous Media. *Small* **2018**, *14*, 1702573. [CrossRef] [PubMed]
19. Piña, M.N.; Gutiérrez, M.S.; Penagos, M.; Duel, P.; León, A.; Morey, J.; Quiñonero, D.; Frontera, A. Influence of the aromatic surface on the capacity of adsorption of VOCs by magnetite supported organic–inorganic hybrids. *RSC Adv.* **2019**, *9*, 24184–24191. [CrossRef]
20. Lanfranconi, M.P.; Bosch, R.; Nogales, B. Short-term changes in the composition of active marine bacterial assemblages in response to diesel oil pollution. *Microb. Biotechnol.* **2010**, *3*, 607–621. [CrossRef] [PubMed]
21. Qiu, W.; Chen, S.; Sun, X.; Liu, Y.; Zhu, D. Suzuki Coupling Reaction of 1,6,7,12-Tetrabromoperylene Bisimide. *Org. Lett.* **2006**, *8*, 867–870. [CrossRef] [PubMed]

MDPI
St. Alban-Anlage 66
4052 Basel
Switzerland
Tel. +41 61 683 77 34
Fax +41 61 302 89 18
www.mdpi.com

International Journal of Molecular Sciences Editorial Office
E-mail: ijms@mdpi.com
www.mdpi.com/journal/ijms

9 783036 531397